MEDIATING ORDER AND CHAOS

THE WATER-CYCLE IN THE COMPLEX ADAPTIVE SYSTEMS OF ROMANTIC CULTURE

56

Internationale Forschungen zur Allgemeinen und Vergleichenden Literaturwissenschaft

MEDIATING ORDER AND CHAOS

THE WATER-CYCLE IN THE COMPLEX ADAPTIVE SYSTEMS OF ROMANTIC CULTURE

Rodney Farnsworth

Rodopi

Amsterdam – New York, NY 2001

Cover illustration:
John Constable: Cloud Study (c. 1822)
Oil on paper, 4 1/2 x 7 (11.4 x 17.8)
Inscribed on verso 'Cirrus'
Victoria and Albert Museum

Le papier sur lequel le présent ouvrage est imprimé remplit les prescriptions de "ISO 9706:1994, Information et documentation - Papier pour documents - Prescriptions pour la permanence".

The paper on which this book is printed meets the requirements of "ISO 9706:1994, Information and documentation - Paper for documents - Requirements for permanence".

ISBN: 90-420-1427-X

Printed in The Netherlands

TABLE OF CONTENTS

PREFACE

My study—essentially literary—deals with the hydrological-cycle as viewed by the Romantic culture: the recurring movement of water from glaciers and springs through streams and falls to lakes and oceans and finally to vapor and clouds where the cycle begins anew. I have incorporated not only the arts, but also the social, philosophical, and scientific understanding related to the water-cycle—both as fact and as metaphor. To hold all this together I needed a methodology. It is toward anthropology I had to turn for a complete model. J. Stephen Lansing and his special adaptation of complexity theory has offered me the impetus I needed to break free of binary methodologies (e.g., literature *and* fine arts, arts *and* politics) as well as to frame my larger project: that of centering an extensive, if certainly not comprehensive, view of Romantic culture around the water-cycle. I hasten to add that the water-cycle offers only a center; I am not naively—after the New Critics—proposing it as some sort of higher unity for Romanticism. *Center* is the operative concept; the water-cycle only offers a center of gravity around which to field a study. Lansing holds the complexity view that all systems function in an operative field surrounded by one or more types of chaos; the slightest change of just the wrong/right sort can throw a system into disorder.

The Theory

I can complete, with relative ease, the cross-disciplinary 'import' work because Lansing has already 'humanized' complexity theory out of natural science into the social science, anthropology; moreover, it is fortuitous for my purposes that Lansing happens to apply the theory to hydrology and hydraulics, specifically of Bali. In his account, the Balinese rice farmers use their water-temple system to steer a middle path of optimal order, running, as it does, between the chaos of rodent, insect, bacteria, and viral destruction, on the one hand, and the chaos of drought, on the other. By letting the largest possible tracks of rice fields lie fallow, they keep pest damage at an acceptable level. Actually, letting all the fields in a certain area lie fallow would work best against pests; however, the farmers cannot leave too much area fallow, for if everyone started new plantings in the fields simultaneously, they would deplete the available water supply and most of the new plantings would die. For centuries, even during the Dutch occupation, this optimally balanced landscape continued. No one understood—other than, of course, the farmers and priests—how it worked or how well it worked until government intervention began to unravel it. Members of the Green Revolution got the Indonesian government agencies on their side; they passed laws and went into the fields to show the rice farmers how irrigation, planting, and pest control should be carried out. The Green policies in Bali had disastrous results. Water shortages and insect damages increased, and rice yields went sharply downward. It was in the midst of this chaos that Lansing began a study that not only revealed the nature of the lost optimal balance between insect damage and water allocation of the farmers, but also was to unveil a system of great complexity and power—operating outside of either a traditional or a Marxian conception of power. In addition to offering an approach to complexity theory that I found better adapted to literature and the other arts, Lansing's study gave me innumerable insights into the ways that hydrology and hydraulics shape and infuse any culture.

The Critical Results

Before becoming aware of Lansing's work, my data seemed to push out into other disciplines, such as rhetoric, music, philosophy, technology, politics, and especially, ecology. This expansion seemed dangerously broad. Even the wide implications of ecology, implicit in hydrology and meteorology, did not prove to have sufficient organizing force. I discovered this organizing force in Lansing's application of complexity theory to anthropology and to a study of hydrology/hydraulics in a non-Western culture, which offered precisely the freshness of vision and the intellectual distancing-effect of a new culture that my data needed. Just as Lansing's theory predicted, I found emergent properties began to arise from the data, the most important of these being that the water-cycle offered a metaphor, which in its turn gave rise to several complex adaptive systems. These systems offered to a significant number of individuals within Romantic culture a way of mediating the implications of an age of unprecedented social, artistic, and technological changes—unprecedented in both variety and degree.

The Bali rice farmers and their irrigation canals as viewed by Lansing further enlarged my view of the science of hydrology to include its technological component—hydraulics. Consequently, the profound shaping effect hydraulics had in creating the landscape of Romantic culture arose as another emergent property. England's canals, of course, are for drainage or transportation, not irrigation. Hydraulics seemed essential in the workings of these, too. Although like the water wheel, canal technology was old, it was only in 1759 that extensive canal building began with James Brindley's engineering breakthrough in the construction of the Worsley Canal. The canals and the canalization of streams, by the end of the eighteenth century, had contributed to the human transformation of the English landscape. Actually, this transformation had long been underway because of the need to build towns around mills and the rivers that drove them, determining in great part the landscape of England and of Europe; therefore, it shaped the landscape of the Romantics. To put it another way, the technological needs of placing water wheels near streams and the economic necessity to build the towns there had shaped their landscape. For all its long history before 1760, engineers did not understand the theoretical basis of water wheels until the year before. The cities and towns grew and prospered along the rivers because of both watermills and water transport. Steam, another form of hydraulics, was about ready to have an even greater effect.

As with Lansing, I will treat the human occupations as they fall along the way of the hydrological and hydraulic network—in his case the rainfall, mountain lakes, streams, and human-built canals. My presentation will follow the entire water-cycle. The presentation of my model revisioning the essential way that the disciplines within Romantic culture presented the water-cycle will parallel the model that Lansing and James N. Kremer, a systems ecologist, constructed. I divide my work according to the Romantic conception and division of the water-cycle. Chapter one treats the source, which may be a natural spring, an artificial fountain, or a glacier.

The Romantics chose the most dynamic source set in mountains. This chapter isolates several fountain symbols that demonstrate the paradoxical tension between the mutable and the eternal. In their works, Goethe and several of his contemporaries force the specific image of water to bear symbolic signification that runs counter to its physical qualities. The poet, in these instances, does not take the easy way of representing form; his signifier must be perched at the edges of chaos of implied contradictions; the signifier is fraught with the implication of flux and reflux and, yet, is made to symbolize permanence. The signifier, at once, represents and deconstructs. In spite of or because of its contradictions, this symbol is simultaneously more philosophically powerful and aesthetically focused. This chapter introduces this particularly expressive use of water symbols, along with the cosmic works that contain phenomena from the full gamut of the water-cycle.

Chapter two treats the flow of streams and waterfalls. In this treatment of rivers, Lansing's complexity theory will be of even greater importance. Just as he treats the whole social, religious, economic, and technological complexity of life along the streams of Bali, so too will I discuss how the rivers shape the diversity of lives pursued along these rivers—streams that because of water milling and navigation have shaped the human geography of Europe and America. Today we find it hard to conceive of the fact that rivers were such a physical and mental shaping force. Energy and power could only find expression in the stream's carrying its water from springs, glaciers, or lakes set at a great height, which gives the water its force of potential energy. And natural science was only one of many aspects of Romantic culture shaped by the flow of streams within their cycle context: this chapter examines the rich artifacts treating streams and falls.

Chapter three concerns the sea or ocean as receptacle. Like the flow phenomena, the lakes, seas, and oceans shaped the complexity of life in much of America and Europe. Unlike flow, the receptacle is not so apparently bound up in the hydrological-cycle. The *flux et reflux* of lakes, seas, and oceans were for Romantic culture, however, such a compelling metaphor for change and, even paradoxically, for permanence that it is as extensively treated as were streams—where the water phenomena flows from one place to another. The vast receptacle of water is often used in Romantic culture to represent order perched on the edge of chaos. From the glacier or the ground, water has gushed forth; now, in the form of streams, we will see it flow down to lakes and seas—the two types of receptacles, which comprise this last quarter. Here the water reaches a state of equilibrium—or so it seems. But the flow, or what we might better call *flux*, continues even when the water seems to be held in its receptacle. Sometimes chaos, moreover, comes as a storm on lake, sea, and ocean. Actually, this water is not received, let alone trapped for very long, because evaporation continuously moves the water of this quarter on to the next—the sky.

Chapter four offers the link—the processes of clouds and rain that are the particular Romantic aspect, which in turn make the water-cycle a completed circuit as well as a scientific endeavor. The sky or air quarter adds a new dimension to the hydrological-cycle: the new mode of change. The first three of the four quarters have brought to the model only translational changes—changes of place. This quarter brings the more fundamental and dramatic mode: the substantial changes of evaporation, vaporous condensation, and precipitation—changes of composition. These processes provided the link in the cycle with certain knowledge rather than conjecture. What is more essential is that the key works on hydrology and meteorology were being produced after 1760; there were yet more changes—now, informational changes—that had to be confronted. There were even a few in the realm of the humanities who knew of the change; moreover, these few—including Shelley and Goethe—understood these discoveries to a surprising degree. For the rest, it was—or was fast becoming part of—a climate of opinion or knowledge, to use Whitehead's concept. Yet, there were a few poets and artists who possessed the knowledge closing the link, even at its simplest level: they kept up with developments in science.

The heyday of Romantic culture, then, was the heyday of meteorology. This is no accident. It may be that the emphasis the Romantics placed upon *dynamic becoming* provided the needed philosophical ground from which to study the complex system of weather, with its changeful oscillations between order and chaos. James Hutton, who contributed most to this new meteorology, viewed weather as a great machine. The empowering engine that drove the great cosmic engine of the water-cycle was the heat of the sun; a key fact that relates other sciences and technologies to the water-cycle studied by hydrologists and meteorologists. James Hutton's macro-cycle finds its correspondence in the micro-cycle of his contemporary and fellow Scot—James Watt. The whole process of the steam engine lends itself to being viewed as a water-cycle in miniature. Sections of Watt's single-acting steam engine of 1788 very roughly approximate the quarters of the water-cycle: the fire under the boiler might be the sun; the boiler, the sea; the pipe that carries the steam to the cylinder, the sky/link; the condenser, the rain; the pipe carrying water back to the boiler, the spring or stream. The study concludes on a work where the micro-cycle and macro-cycle are united: Turner's *Rain, Steam and Speed*.

Acknowledgments

This work is a totally transformed and otherwise much enlarged version of my dissertation, Indiana University, Bloomington (1980). Those who contributed significantly to the dissertation were Newton P. Stalknecht, Frederick Beaty, Diana Guiragossian-Carr, Brion Mitchell, Louis Hawes, Henry Remak, and—above all—Ulrich Weisstein. The work was originally titled, 'Permanence and Change: Water Images and Symbols in Western European Poetry and Painting, 1770–1840', and it placed emphasis upon the scientific context—particularly hydrology and meteorology—of the two arts mentioned in the title. It included neither music nor the social-political matrix.

The completely transformed version was written during my sabbatical from Indiana-Purdue University Fort Wayne during the 1993–1994 school year. Significant parts of the new version were read at the American Conference on Romanticism, on two occasions. The opening pages were presented as 'Romantic Society as Complex Adaptive System: A Case Where Complexity Theory Might Apply', in Park City, Utah, on 5 October 1996. Significant parts of the introduction and chapter two were read in a paper entitled 'Humanized Landscape in the Works of Jane Austen and her Contemporaries', at Athens, Georgia, on 23 January 1998. My thanks to Diana Guiragossian-Carr for reading and commenting on the new version. Others who read and commented on significant parts of the work in its early stages were Shirley

Champion and Elaine Stuart. My thanks to Victoria Hoelzer-Maddox. And, of course, there was always the never-failing readership of John Oliver Anderson.

In treating the foreign language quotations from secondary works, I have kept my English translation in the text and placed the original in the footnotes; I have represented primary works in the text itself both by quotations from the most authoritative versions of the original and by my own translations unless, in rare instances, otherwise specified. I have opted in the case of a handful of particularly important prose and poetic works to quote extensively; such a study involving so many writers would demand much more extensive quoting than—say—a study of Keats, which could easily be supplemented by one or more volumes of works. In order to make this study accessible to a wider audience, I have done my own translations, which I consider key adjuncts to my readings of these texts—after all, a sort of interpretation results in the course of a careful, analytic translation. Most of the German and all of the French works from which I have quoted lack English translations that are accurate both in letter and spirit; this fact offers yet another reason for extensive quoting. Although I devoted a tremendous amount of my own effort to these translations, I could not have felt as confident about them as I do, without the advice of expert-and/or native-readers. I deeply appreciate the help of Diana Guiragossian-Carr, Ingrid Widenhorn-Smith, Victoria Hoelzer-Maddox, Laurie L. Corbin, and David Oberstar.

The three German translations that I found of sufficient value to consult seriously in making my own are listed in a Primary Bibliography in the back. There, too, is a complete list of primary sources used and cited; pagination and in some cases volume numbering are offered parenthetically within the texts. In the cases of lines, stanzas, and cantos, I have indicated these; indeed, I have relied on these rather than on page numbers, so that various editions can be consulted as well as the source I employed. Secondary works and primary works that are not artifacts of Romantic culture have been relegated to endnotes.

As for the works of music, I have resisted musical quotations because (rightly or wrongly) I have been told they seriously intimidate the literary and art-historical reader; for the same reason, except where absolutely necessary for my argument, I have kept harmonic analysis to a minimum. For a musicologically trained scholar, this is a restriction that does not please me.

INTRODUCTION: METHOD, MODEL, AND SUB-MODELS

The Method: Complexity Theory Applied

I begin with a Cartesian discourse on method. True to the original pattern set by Descartes, I started in a quandary; mine arose within a line of research that I have been sporadically pursuing since around 1970; then, I made my first attempts at tracing scientific theories on the hydrological (or water-cycle) through Western-European Romantic poetry. My study went through various metamorphoses including becoming a dissertation seven years later, which then drew in Romantic painting.[1] I never was satisfied with the study in its earlier incarnations: it was mere old-fashioned *Stoffgeschichte*. The data were suggesting to me many connections among science, literature, the other arts, and their socio-political context. I was faced with a whole range of facts that my theoretical, critical methods failed to frame and present in any significant order. I, therefore, set aside the work for other research pursuits.

During a sabbatical, in the early 1990s, I again returned to my data in search of a method. For several years, I kept running up against the realization that we are not as *Post*-Structuralist as we might want to think: antinomies or binary opposition still rules. And this remains true in spite of Michel Foucault's attempts to make us see things beyond polarities and toward spatialized thinking—and of course, his ideas remain useful in continuing the attack.[2] Students of culture still seem to need these binaries: literature and painting; literature and science; literature and, of course, politics. New Historians, who seem to advocate a more expansive view, have recently produced interesting and useful binary studies in, say, Jane Austen and economics.[3] Most exciting of all have been those Marxists who have gone in search of what I term the 'missing binary' in, for example, Keats' 'To Autumn'; some critics participating in this great debate have found politics as a great absence, or displacement, within these poems; but displaced or not, it is still the old literature-and-whatever structure.[4]

---

1 Rodney Farnsworth, *Permanence and Change: Water Images and Symbols in Western European Poetry and Painting, 1770–1840*, diss. Indiana University, 1980 (Ann Arbor: UMI, 1993).

2 The important attack on dichotomizations in Michel Foucault's Chapter Five, 'To Classify' / 'Classer', of *Words and Things* [*The Order of Things* in the standard English translation]: *An Archaeology of Human Sciences* will be a central aspect of my study; *Les Mots and les choses: Une Archéologie des sciences humaines* (Paris: Gallimard, 1966) 137–76.

3 The binary title says it all: women writers and economics is the central theme of Edward Copeland, 'Shopping for Signs: Jane Austen and the Pseudo-Gentry', *Women Writing About Money: Women's Fiction in England, 1790–1820* (Cambridge: Cambridge UP, 1995) 89–115; this is part of the series Cambridge Studies in Romanticism, and the binary titles of other (still useful) works in that series will be seen throughout my study.

4 Geoffrey H. Hartman, 'Poem and Ideology: A Study of Keats's "To Autumn"', *Romantic Poetry: Recent Revisionary Criticism*, ed. Karl Kroeber and Gene W. Ruoff (New Brunswick: Rutgers UP, 1993) 421–38; Jerome McGann, 'Keats and the Historical Method in Literary Criticism', Kroeber and Ruoff 439–64; Richard Bourke, 'Custom and Dominion: *Tintern Abbey*', *Romantic Discourse and Political Modernity: Wordsworth, the Intellectual and Cultural*

Even classics of comprehensive envisionings of Romanticism seem to fragment their larger whole into separate binary considerations. H.G. Schenk's *The Mind of the European Romantics: An Essay in Cultural History* is still a landmark comparative study, and of course, M.H. Abrams' *Natural Supernaturalism*—which I unashamedly take as one of several models for my own—wonderfully endures all attacks and changes in the field. Schenk's model is the binary opposition of Christian orthodoxy and nihilism.[5] In the case of Abrams' masterwork, the criticisms against it, which have a certain justice, have overlooked its great virtues; I am happy to note that the distinguished Danish scholar, Claus Schatz-Jakobsen, recently termed *Natural Supernaturalism* 'a milestone in modern Romantic scholarship'.[6] Abrams' flexible eclecticism—which he so admirably defends in his apologia before the critique of Wayne Booth—is nevertheless constructed in discrete and component arguments around these binaries: literature and metaphysics; the cycle (or spiral) and the arrow; English poets and German philosophers (as well as a few poets); Biblical Apocalypse and the French Revolution.[7] It, of course, must be admitted that Abrams' mode of presentation certainly offers an improved alternative to monographically treating one element at a time in discrete chapter units; this common book format, which proposes to give broad cultural visions, manifestly does not offer a desirable alternative to the binary trap.

I have discovered for myself a work that at least moves toward a freedom from binary thinking: a work described by Karl Kroeber and Gene W. Ruoff as a 'bold challenge to the discourse of displacement' that 'has probably defined the major battle zone of Wordsworthian criticism for the remainder of this century'.[8] The work referred to is *Romantic Ecology* by Jonathan Bate. This scholar finds in Wordsworth a concerted attempt to ground his later political views in an environmental and cultural appreciation for the land and people of his Lake Country, and the result is not inconsistent with his earlier revolutionary stance of the early 1790s. Bate tries to forge links 'between the holistic values of his native vales and the 'social meliorism' that underlay the French Revolution'.[9] All too much of the author's energy is taken up with needed polemics against the politics-and-literature bias of critics like Jerome

---

Critique (New York: St. Martin's, 1993) 40–57; Paul H. Fry, 'History, Existence, and "To Autumn"', Kroeber and Ruoff 465–72; Nicholas Roe, 'Keats's Commonwealth', *Keats and History*, ed. Nicholas Roe (Cambridge UP, 1995) 194–211.

5 H.G. Schenk, *The Mind of the European Romantics: An Essay in Cultural History* (Garden City: Doubleday-Anchor, 1969).

6 Claus Schatz-Jakobsen, 'The Romanticism(s) of Contemporary Criticism with Hartman and Wordsworth', *Prism(s): Essays in Romanticism* 5 (1997) 17–39, qtd. from 20.

7 M.H. Abrams, *Natural Supernaturalism: Tradition and Revolution in Romantic Literature* (New York: Norton, 1971); 'Rationality and Imagination in Cultural History', *Doing Things With Texts: Essays in Criticism and Critical Theory*, ed. Michael Fischer (New York: Norton, 1989) 113–34.

8 Kroeber and Ruoff 140.

9 Jonathan Bate, *Romantic Ecology: Wordsworth and the Environmental Tradition* (London: Routledge, 1991) 19. See two other pioneering works: Paul H. Fry, 'Green to the Very Door? The Natural Wordsworth', *Studies in Romanticism* 35 (1996): 535–51; Ralph Pite, 'How Green were the Romantics?' *Studies in Romanticism* 35 (1996): 356–73.

McGann and Alan Liu, and, yet, Bate does suggest the possibility of a trinary relation among literature, economics, and ecology. Since, unfortunately, he views economics as Red politics and ecology as Green politics, the politics/literature binary remains; Bate merely offers a more expansive conception of politics encompassing the environment. It became clear that from neither current literary theory nor criticism would I be able to gain the theoretical framework for which I searched.

This binary structuring is, as I see it, only one of two possible pitfalls in cultural studies; the other is the tendency to compartmentalize disciplines each within its own part, chapter, or section of a study. Twentieth-century history in its strictly academic form has tended to follow this model of separate chapters for economics, for governmental institutions, the ideas, and the arts.[10] While I admit that this expository form has its advantages, it should not be the only way to treat history and is certainly not the best way to offer an integrated, holistic vision of culture. Actually, the Romantic culture offers one of the greatest models of an integrated, indeed organic treatment of culture: the first part of Germaine Necker de Staël's *On Literature as Regards its Interrelation with Social Structures / De la Littérature considérée dans ses rapports avec les institutions sociales*. Staël holds to her stated goal: 'I have set for myself the goal of examining what influence religion, customs, and the law each has upon literature, as well as what influence literature has upon religion, ethics, and the law'.

> Je me suis proposé d'examiner quelle est l'influence de la religion, des moeurs et des lois sur la littérature, et quelle est l'influence de la littérature sur la religion, les moeurs et les lois (1:99).[11]

As examples of her practical application, one might consider her discussion of the way Protestantism has affected Northern European literature (255).

Since Staël was the first great theorist of nineteenth-century nationalism,[12] it is no accident that she considered ethnic diversity a central issue: 'In observing the characteristic differences that occur among the writings of Italians, English, Germans, and French, I have conceived the need to demonstrate how political and religious institutions play the greatest possible role in these recurrent diversities'.

> En observant les différences caractéristiques qui se trouvent entre les écrits des Italiens, des Anglais, des Allemands, et des Français, j'ai cru pouvoir démontrer que les institutions politiques et religieuses avaient la plus grande part à ces diversités constantes (200).

Staël's efforts and achievements are pioneering; therefore, it may seem ungenerous to note that much of the argument is shaped around the binary of Southern literature versus Northern—that is Classic or Homeric versus Romantic or Ossianic poetry.

To do Staël justice, it certainly must be argued that in any comparative/contrastive act, one must initially and along the way consider the binary between what is

---

10 Let me just take as one example the highly-regarded and still-in-print series The Nelson/Norton Library History of English under the editorship of Christopher Brooke and Denis Mack Smith.

11 *Oeuvres complètes et posthumes de Madame la Baronne de Staël-Holstein*. 3 vols. (1861; Geneva: Slatkine, 1967). Throughout the text of this entire book, all quotations from the works of Staël will come from this edition.

12 See Umberto Marcelli, 'Il Gruppo di Coppet e il concetto di nazionalità', *Le Groupe de Coppet: Actes et documents du deuxième Colloque de Coppet, 10–13 juillet 1974*, eds. Simone Balayé and Jean-Daniel Candaux (Geneva: Slatkine; Paris: Champion, 1977) 401–15.

compared and to what it is compared. This great pioneer of comparative methodology, however, never remains content with antinomies. Staël goes further to avoid the simplificationism implied in this comparative binary by finding differences within each larger group—for example, she contrasts English and German literature as much as she finds similarities. Staël, however, best transcends the limitations of the binary by the diversity of variables that she brings to bear on her comparison: in her practical application of her theory concerning the effect on poetry of climate (see, for example, 255), which is added to the above list of 'religion, custom, and laws'. In my own study, I will add the natural sciences, which she omits from consideration, as well as the non-literary arts, which she neglects. Each of the variables is tested around the focus—literature. Their application is holistic rather that fragmented into separate chapters. Unfortunately, Staël falls into the component approach in the second part of *On Literature* and in another literary study *On German Culture / De l'Allemagne*. In the realm of literary criticism, however, none have yet matched the grandeur and wholeness of Staël's vision.

Art history offers a contemporary example close to what I want to achieve. And it is partly contained in the corpus of Sir Kenneth Clark. His *Civilization* television series and accompanying book come to mind first; however, it offers a popular work that—for its methodological armature—relies upon, as its subtitle says, *A Personal View*. Rather, I have found a very useful standard in his monographic study on Leonardo da Vinci. Since, of course, this analysis limits itself to one artist, it has limited application. A recent art-historical study by William H. Galperin, *The Return of the Visible in British Romanticism,* uses the Romantic Panorama and Diorama experiences as a center from which to move across the disciplinary lines of painting and poetry—and in one instance, across national boundaries (Caspar David Friedrich).[13] The important difference between what Galperin has achieved and what I sought is that he uses the Diorama conventions of treating the visible as a paradigm with which to test the ability of artist and poets to render the visible. Such an imposition of something like the old controlling norm of the New Critics runs counter to my hoped-for method, which should allow properties that may or may not prove normative to emerge from the data.

It is toward anthropology I have had to turn for a complete method, together with model. Lansing and his special adaptation of complexity theory have offered me the impetus I needed to break free of binary methodologies and frame my data. Having the hydrological-cycle at the center of his study, Lansing was able to extend out from this focus into the areas of agriculture, religious customs, social structures, and other aspects of Balinese life. Similarly, I have centered an extensive, if certainly not holistic, view of Romantic culture on the water-cycle. I hasten to add that the water-cycle offers only a center; I am not naively—after the old fashion of the New Critics—proposing it as some sort of higher unity for Romanticism. *Center* is the operative concept; the water-cycle only offers a center of gravity around which to

---

13 The three holistically framed works are: Kenneth Clark, writer and presenter, *Civilization: A Personal View,* BBC 1969; script (New York: Harper, 1969); Clark, *Leonardo da Vinci: An Account of His Development As an Artist,* rev. ed. (1939; Baltimore: Penguin, 1967); William H. Galperin, *The Return of the Visible in British Romanticism* (Baltimore: Johns Hopkins UP, 1993).

field a study. Lansing holds the complexity view that all systems function in an operative field surrounded by two or more types of chaos; the slightest change of just the wrong/right sort can throw an entire system into disorder.[14] Complexity, then, attributes the optimal functioning of a system to several crucial aspects hidden away—at least without complexity analysis—among many seemingly insignificant components. Just precisely what the crucial aspects are is only determined by considering every aspect possible; any methodological predisposition to automatically rule out any aspect as irrelevant before testing is carefully monitored and prevented.

This concept seeks to embrace a multitude of causes, of which even the minutest one can throw the rest into a complete realignment—often, into chaos. Let me just insert here that the essence of this theory is not as far from the heart of the humanities as might be thought: just such a view of the multiplicity of causes behind each historical event is the guiding thesis of Tolstoy's *War and Peace*. Tolstoy embodied his thesis in the form of a novel—filled with persons on all levels of society. Concerning *War and Peace*, R.F. Christian writes: 'Very seldom is a person seen or described in isolation, just as in real life human beings cannot be divorced from the infinite number of animate and inanimate phenomena which make them what they are and determine what they do'. This scholar continues by stating, 'few authors have had Tolstoy's power to show the multiplicity of interacting phenomena in the lives of their imaginary men and women'. Christian expresses Tolstoy's views on the writing of history this way: 'The task of the historian—and what a task it is!—is to try to integrate all the infinitesimally small differentials of history'.[15] In the eyes of his characters, even Napoleon, historical events are seen as an intricate network of

---

14 What follows in the text is my attempt to digest, or more precisely to extract, those parts useful to the construction of my methods from these writings by Lansing: J. Stephen Lansing, *Priests and Programmers: Technologies of Power in the Engineered Landscape of Bali* (Princeton: Princeton UP, 1991). There are many other models of complexity from which I chose Lansing because his theory best fits my data. Complexity models for computers, genetics, and fluid dynamics (this will prove useful later in the study) are discussed in Peter Coveney and Roger Highfield, *Frontiers of Complexity: The Search for Order in a Chaotic World* (New York: Fawcett, 1995). Complexity in the theory and praxis of literature has been aptly handled by Stephen H. Kellert, 'Science and Literature and Philosophy: The Case of Chaos Theory and Deconstruction', *Configurations* 4.2 (1996): 215–32; Carl Matheson and Evan Kirchhoff, 'Chaos and Literature', *Philosophy and Literature* 21.1 (1997): 28–45. While putting my manuscript for this book through its final stages of preparation, I managed to acquire Theodore E.D. Braun and John A. McCarthy, eds. *Disrupted Patterns: On Chaos and Order in the Enlightenment*. Internationale Forschungen zur Allgemeinen und Vergleichenden Literaturwissenschaft 43 (Amsterdam: Rodopi, 2000). The particular literary adaptation of complexity theory that these scholars use works beautifully on the Enlightenment authors, even Goethe and the aspects of his works treated. Reading the study of these scholars, however, convinces me even more that the Lansing model of complexity theory works best in treating Romantic Culture and, more specifically, treating it in light of hydrology and meteorology. Most applicable of all to the post-1789 Romantic culture is Lansing's approach to Balinese culture as a complex adaptive system, which negotiates among the various man-created and nature-created manifestations of chaos. I hope these scholars will view my methodology as supplemental to theirs and not oppositional.

15 R.F. Christian, *Tolstoy: A Critical Introduction* (Cambridge: Cambridge UP, 1969) qtd. from 127, 175, respectively.

unknowns beyond human direction—it seems a chaos; implicitly. Tolstoy believed that what was chaos to mortal eyes was complexity to the eye of God. In place of accepting human ignorance, the complexity theory encourages an optimistic heuristic in each discipline.

Since I announced my own application of Lansing's complexity theory to the Romantic data in a 1996 conference paper, other applications of the theory to aspects of Romanticism have appeared. In a very recent study of Shelley, Hugh Roberts employs chaos theory—another term for complexity theory, suggesting, I think, a subtly different approach. Roberts, rather like Tolstoy, concentrates on 'factals' or the concepts of an infinite complexity of factual data.[16] Roberts sums his approach to chaos theory somewhat more succinctly in an article in the *Keats-Shelley Journal*. In this article, Roberts writes that

> Chaos theory, the study of 'determinate indeterminacy,' provides us with a model with which we can conceptualize [...] flux and indeterminacy. [...] From the standpoint of chaos theory, the limits to our ability to know the world have less to do with epistemology than with the hitherto unsuspected complexity—the 'factal' complexity—of the world itself. To trace the path back from effects to causes through the feedback cycles of nonlinear dynamic systems (which is to say virtually all natural dynamic systems [...]) becomes infinitely complex, not because of Hume's discrediting of the principle of causality itself, but because the determining [conditions] of the system will escape any arbitrarily accurate attempt at definition. The chaotic world is an entopic world; information is constantly lost from it because nature itself is unable to conserve infinite accuracy in its own interactions.[17]

'Chaos theory', as Roberts seems to employ the term, suggests that there is no escape from ignorance in a chaotic world of baffling multiplicity and of disappearing, illusive data. 'Complexity theory', as Lansing employs it, suggests that an underlying order may be found—although the 'emergent properties' most often surprisingly will not resemble what was expected. If a model of Romanticism emerges, it will be one of diversity rather than of unity. The term 'complexity' allows for moments of order in nature immediately knowable to human science; these moments of order mediate between two (among more) sorts of chaos, which are more remote, as yet from human science.

Lansing's use of the complexity heuristic, then, allows the balancing or stabilizing aspects of a system to become apparent and to take the researcher by surprise as 'emergent properties'. This is important because it seeks to allow the object of the study to speak for itself rather than imposing on it theoretical expectations; for example, he uses aspects of Marxist social theory, but he uses them in the spirit of the Critique rather than the Dogma of Marxism, and as I shall later observe, his central emergent property would not make a dogmatic Marxist happy—or a dogmatic Capitalist, either, for that matter. His particular slant on complexity theory is that he has taken it out of the natural sciences into the social sciences, and mine is that I propose to bring it from his field into the humanities.

---

16 Hugh Roberts, *Shelley and the Chaos of History: A New Politics of Poetry* (University Park: Pennsylvania State UP, 1997).

17 Roberts. 'Chaos and Evolution: A Quantum Leap in Shelley's Process', *Keats-Shelley Journal* 45 (1996): 156–94, qtd. from 163.

My cross-disciplinary import work, as it were, is easily accomplished because Lansing has already 'humanized' it out of natural science, into an anthropological study of hydrology and hydraulic agriculture on the Indonesian island of Bali. In Lansing's account, the Balinese rice farmers use their water-temple system and its managerial system of priests to steer a middle path of optimal order, running, as it does, between the chaos of rodent, insect, bacteria, and viral destruction, on the one hand, and the chaos of drought, on the other. By allowing the largest possible tracks of rice fields to lie fallow, they keep pest damage at an acceptable level. Actually, letting all the fields in a certain area lie fallow would work best against pests; however, the farmers cannot leave too much area fallow at one time, for if everyone started new plantings in the fields simultaneously, they would deplete the available water supply of the irrigation system and most of the new plantings would die. For centuries, even during the Dutch occupation, this optimally balanced landscape continued. No one understood—other than, of course, the farmers and priests—how it worked or how well it worked until government intervention began to unravel it. Members of the Green Revolution got the Indonesian government agencies on their side; they passed laws and went into the fields to show the rice farmers how irrigation, planting, and pest control should be carried out. These Green policies in Bali had disastrous results. Water shortages and insect damages increased, and rice yields went sharply downwards—as much as 50% in some areas.

It was in the midst of this chaos that Lansing began a study that not only revealed the nature of the lost optimal balance between insect damage and water allocation of the farmers, but also was to unveil a system of great complexity and power—operating outside of either a traditional or a Marxist conception of power. Lansing achieved this by doing what many on the Right might see as unprofitable neoprimitivism and many on the Left as *laissez-faire* populism: he got the government planners off the backs of the farmers and let them go back to the old ways—and subjected these old ways to rigorous testing and cross-checking with the modern ways of the Green Revolution. He found that the rice farmers—not the modish ecologists—had it right all along. And he discovered that the system of these farmers was controlled by a network of water-temples that spread across the Balinese landscape and that find their spiritual nexus at the temple of the 'Goddess of the Lake', *Dewi Danu*, and her priest, the *Jero Gde*. Lansing found an interconnecting system of rituals emanating like rays outward from this center. This system, in addition, was found by him to possess a sophisticated network functioning with both committee meetings and operative memoranda—indeed, he found it hard to distinguish where ritual ended and management began. The goddess and her divine underlings (and there are many lesser gods and goddesses because Bali is Hindu, which in turn draws on older strata of chthonic deities) found a human correspondence in a perfected managerial system of priests and committees of rice farmers. Not understanding the whole, the modernizing Green Revolutionaries from the West and Westernizing bureaucrats in Java attacked the elements of this ancient Balinese tradition, seeing them as useless and superstitious nonsense.

Looking at Lansing's complexity praxis, I see it as a holistic presentation of that part of Balinese society centered on the hydrological-cycle and—this is absolutely crucial for my purposes—as one which coordinates into sub-models different series of data or 'factals': rainfall; rainfall-runoff relationship; rice growth-stages and harvest-time; water-stress or usage; pest damage; human management; then, above it all stands the Goddess of the Lake and her priest at the core of a large divine and human network. The goddess may be a myth, but not the agricultural optimum and its supremely effective management that she presides over as it were; in so far as the system has a head, this was and is her 'powerless' priest and not the ancient Balinese kings, nor the Dutch colonizers, nor the Jakarta bureaucrats. Imagine how shocking this goddess-centered management notion is to the manufacturers of fertilizer and herbicides or, let alone, to those who would like to see the goddess and her water-temple system as a mere 'superstructure' irrelevant to the power and value of labor. Lansing's complexity methodology prevented him from making the mistake of the Green Revolutionaries and the Westernizing Javanese bureaucrats—namely, that of imposing his concepts on the Balinese culture; rather, Lansing's approach allowed unexpected properties to emerge freely from the culture: these are the *emergent properties*.

To me the complexity theory, as employed by Lansing, offers a counterweight to binary thinking—even a way of taking the widest possible view of the hydrological-cycle as the core of a complex adaptive network, which draws into itself literature, rhetoric, painting, music, philosophy, economics, and politics—in short all the arts and sciences. This adaptive network offered Romantic culture a great collective coping mechanism to face the greatest degree of change then yet known to human-kind: the Industrial Revolution; the era of the American and French revolutions; Napoleon. Abrams summed up the situation this way:

> The Romantic era was one of technical, political, and social revolutions and counter-revolutions—of industrialization, urbanization, and increasingly massive industrial slums; of the first total war and postwar economic collapse; of progressive specialization in work, alterations in economic and political power, and consequent dislocations of the class structure; of competing ideologies and ever-imminent social chaos. To such a world of swift and drastic change, division, conflict, and disorder, the inherited pieties and integrative myths seemed no longer adequate to hold civilization together.[18]

Abrams goes on to show how the Romantics sought consoling alternatives in the Christian Apocalypse, in Neoplatonism, and in other metaphysical structures; so much so that he might better have titled his book *Supernatural Naturalism* than the other way about, as he did. His work envisions Romanticism as a reaffirmation, if restatement, of the Judeo-Christian tradition, which for some of its adherents it certainly was. Lansing's method avoids such a one-sided view.

There is a key difference between Lansing's praxis and mine that cannot be over-emphasized: the adaptive systems that I treat met *psychological needs*, and the ones studied by Lansing met essentially *physical ones*; yet there the difference ends. Both center on the hydrological-cycle. It has emerged from my data that this cycle offered to many Romantics a primarily natural structure with which to represent a world view that reconciles the user to mutability; this cycle, in short, embodies a faith in continuity or permanence-in-change. It will be part of my purpose to show to what degree they did use the water-cycle to signify this faith, and I can only do this up to a point simply because they used it so frequently. Goethe, Wordsworth, Shelley, and

18 Abrams, *Natural Supernaturalism* 292–93.

Hugo—to take particular instances—found in it a way of facing revolution and change. It was for them what I have already termed a coping mechanism, or to put it another way, the water-cycle metaphor offered an expressive way for artists to negotiate their way through particular circumstances arising in a revolutionary age. I feel that I have discovered an important transformation of political hopes and fears (ideological, but not crystallized ideologies) into symbolic forms. This transformation can take place at a conscious level, or—far less probably—at an unconscious level, but I suspect it takes place on a barely super-liminal level of consciousness. Such use of any visual or verbal image or metaphor is a way of responding to a particular state of being where generalization is just not enough; the feeling must find its embodiment in as tangible a form as visual or verbal signs can offer. Gods and saints no longer mean enough for many. Science offered a source for the physically tangible. And to embody the seemingly cosmic changes going on around them, only a large scientific construct like the hydrological-cycle would do as an alternative to the Judeo-Christian God. Another emergent property is: this water-cycle is connected with related cycle rhetoric, like that of Vico, Hegel, or the poets—treated so sensitively by Abrams in *Natural Supernaturalism*. Because the water-cycle partakes of the holistic vision of these other cycles and, like them, contains as well as embodies a whole complexity of thought, this insight gives me the certainty that the subject transcends *Stoffgeschichte*; examining the hydrological-cycle is different—let us say—from just cataloging that the volcanoes Etna and Vesuvius appear in the works of Turner, Goethe, Lyell, Hölderlin, Shelley, or Staël.

To try to keep this study from falling, at its very outset, into the currently very fashionable lure of theoretical abstractions, let me take one of the many possible examples—one that lies on the very cusp that Romanticism shares with the Enlightenment: J. Hector St. John de Crèvecoeur's *Letters from an American Farmer*. Both Crèvecoeur and his semi-fictional, semi-autobiographical farmer partook of a conservatism perched on the edge of the potential chaos of change; taken as a whole, this can represent Romantic responses to change. An impoverished French aristocrat from Normandy, he offers an example of that great Norman type that Charles Homer Haskins generalizes as the sort that combines conservative thoughts with adventuresome actions.[19] He came to Canada, when it was still under French control, to work as a surveyor and explorer on the far side of the Great Lakes. He fought against the English in the 'French-and-Indian' theatre of the Twelve-Years War; then, he moved south to the colony of New York to become a competent farmer on the frontier, traveling on occasion up and down the seaboard. His earlier life shows courage and willingness for change; indeed, his choice to settle right on the edge of Western civilization, on the frontier, if you will, demonstrates a fascination for existing poised on the edge of chaos, but also full of hope to create order there. When the American-Revolutionary debate heated up and especially when the conflict started, he showed himself to be more conservative, more Tory, than many of the small-holding farmers around him; however, he failed to be as Loyalist as the British wanted and was imprisoned by them, then shipped back across the Atlantic. If there is anyone who

19 Charles Homer Haskins, *The Normans in European History* (New York: Norton, 1915) 11–17.

knew the vicissitudes of change in all its manifestations—from natural to human, from personal to political—then, it was Crèvecoeur.[20]

Crèvecoeur was a very adventurous pioneering spirit, who had set out to seek a life of change in the New World—indeed on the very frontier of that world: He was clearly not afraid of change. His preface to the first edition of his *Letters* (1782), therefore, is very revealing as to what a shock the American Revolution was and to what degree it was a transformational experience personally, socially, and politically. The carefully chosen adjectives he uses to characterize the changes of the revolution and his reactions to them speak volumes:

> Our farmer had long been an eye witness of the *transactions* which have *deformed* the face of America; he is one of those who *dreaded and has severely felt the desolating consequences of a rupture* between the parent state and her colonies, for he *has been driven* from a situation the enjoyment of which the reader will find pathetically described in the early letters of this volume (27; my emphasis).[21]

While these 'early letters'—which are said by our partly fictitious narrator to have been drafted earlier than the later war pieces—give a sense of a lost paradise, they are not without their ambivalence concerning permanence and change.

What is even more significant for my study is the metaphor with which Crèvecoeur embodies this ambivalence: he uses the ocean, which is one of the central symbols of Romantic culture and will receive a sizable proportion of the attention of this study. Several of the letters treat the geography and culture of Nantucket—a low-lying island perched on the edge of the continental shelf before the wild Atlantic. It seems, in part, a place of escape and even quietude from social changes of the 'great world' and its 'perturbations'. Escape and certainly quietude are the last feelings to arise when reading the rest of Crèvecoeur's description. This connection of the human 'perturbations' and the natural ones that follow are significant, and they are not simple oppositions either: the sea causes 'perturbations' in the mind of the 'American Farmer', the narrator:

> The *ever-raging ocean* was all that presented itself to the view of this family; *it irresistibly attracted my whole attention*: my eyes were involuntarily directed *to the horizontal line of that watery surface,* which is *ever in motion* and *ever threatening destruction to these shores* (157; my emphasis).

The ocean, then, is reduced geometrically and physically to its basic essentials. Space becomes a threat in the horizontal line. Flux is an even greater threat. To Crèvecoeur's narrator, this family living at the edge of the Atlantic Ocean is poised on the edge of chaos.

Such consciousness of motion and space as Crèvecoeur evidences may not be peculiar to the frontier-shaped consciousness of the American mind; however, it

---

20 Crèvecoeur's personal, social, and political vicissitudes are traced from his immigration to the American Colonies, through the Revolutionary War, then, through the French Revolution and Napoleonic Hegemony, by Wilson Allen and Roger Asselineau, *St. John de Crèvecoeur: The Life of an American Farmer* (New York: Viking, 1987).

21 J. Hector St. John de Crèvecoeur, *Letters From an American Farmer and Sketches of Eighteenth-Century America* (New York: Signet, 1963) 27. Throughout this entire book, all quotations from Crèvecoeur's works will be taken from this edition.

certainly exists in him with considerable force. It is striking that something similar appears in *Moby Dick* (1851). The mental impact of the ocean upon the mind of Ismael offers a parallel significantly close to that made upon Crèvecoeur's American farmer. Ismael, when he goes on board the Pequod to sign-on for his first voyage, is sent forward to view the ocean—as a psychological test as to his seaworthiness:

> Going forward and glancing over the weather bow, I perceived that the ship swinging to her anchor with the flood-tide, was now obliquely pointing toward *the open ocean. The prospect was unlimited, but exceedingly monotonous and forbidding; not the slightest variety that I could see* (6: 72; my emphasis).[22]

These two reactions are more than American; they are part of a new (proto)ecological awareness of how the ocean, like all water phenomena, shapes—not just the material lives—but the human consciousness of the people who live alongside them. We are as much what we hear and see, as what we eat.

Crèvecoeur's narrator—presumably raised on a inland farm—goes even further in attempting to dissect the effect of the sea upon human consciousness, almost to the point of making it into a catalogue of sense-data sources:

> My ears were *stunned with the roar of its waves rolling one over the other,* as if impelled by a superior force *to overwhelm the spot on which I stood.* My nostrils involuntarily inhaled *the saline vapours* which arose from the dispersed particles of *the foaming billows* or from the weeds scattered on the shores (157; my emphasis).

The tremendous effect that all this has on the American Farmer's mind is conveyed by the catalogue of its effect on the senses, in a series of parallelisms: 'my eyes'; 'My ears'; 'My nostrils'. And yet, the resolution of his profound dis-*ease* is sounded by the phrase 'ever in motion', and it is achieved by the narrations couching the change in a series of destruction-creative cycles:

> My mind suggested a thousand vague reflections, pleasing in the hour of their spontaneous birth, but now half forgotten, and all indistinct; and who is the landsman that *can behold without affright so singular an element, which by its impetuosity seems to be the destroyer of this poor planet, yet at particular times accumulates the scattered fragments and produces islands and continents* fit for men to dwell on! Who can observe *the regular vicissitudes of its waters* without astonishment, now *swelling themselves in order to penetrate through every river and opening and thereby facilitate navigation,* at other times *retiring from the shores to permit man to collect that variety of shell-fish* which is the support of the poor? Who can see the storms of wind, blowing sometimes with impetuosity *sufficiently strong even to move the earth,* without feeling himself *affected beyond the sphere of common ideas*? Can this wind which but a few days ago refreshed our American fields and cooled us in the shade be the same elements which now and then *so powerfully convulses the waters of the sea, dismasts vessels, causes so many shipwrecks and such extensive desolations*? How diminutive does a man appear to himself when filled with these thoughts, and standing as I did *on the verge of the ocean!* (157–58; my emphasis).

Lansing could have hardly portrayed a more complex vision of how the ocean shapes the daily activity and thoughts of humans that eke their living at its chaotic margins. In light of this ocean vision, how diminutive seem the cycle of wars and revolutions,

---

22 Herman Melville, *The Writings of Herman Melville*, eds. Harrison Hayford et al. 15 vols. (Evanston: Northwestern UP; Chicago: Newberry Library, 1968–89) 6: 72. Throughout this entire book, all quotations from Melville's works will be taken from this edition.

and I think one comes close to understanding the consoling nature of Crèvecoeur's vision. It approaches in spirit the central tenet of the ethics of Baruch Spinoza: in order to reconcile oneself to the hard ways of the world, one must look at it under the aspect of the eternal (*sub specie eternitatis*) rather than dwelling on the specific pains of temporality. Most importantly, humans must realize that their destiny lies in the great whole, alternatively termed God or Nature (*deus sive natura*).

It is no accident that it was the Romantics who first gave Spinoza his proper place as one of the greatest of all Western philosophers. His presence has long been felt amid the 'factals' of my study—profoundly in the works of Goethe and distortedly in those of Wordsworth. In Karl Kroeber's *Ecological Literary Criticism,* he writes: 'Philosophically, of course, it was Spinoza who served as the chief intellectual support for the development of such proto-ecological views as Wordsworth espoused'. Spinoza directly and indirectly influenced Romantic culture. Kroeber, therefore, bemoans the fact that 'Spinoza's role in all of romantic thought [...] has been unduly slighted in most recent romantic criticisms'.[23] The importance of Spinoza is one of the many 'emergent properties' of that complex aggregate called all-to-simply Romanticism. This seventeenth-century Jewish philosopher's concept of *sub species eternitatis,* his presumed unity of *deus* with *natura,* his monism that joined mind and matter, and his resulting brand of pantheism were welcome building material for Romantics seeking to create an all encompassing environmental vision advocating the unity of human and natural history. As Henry E. Allison writes, Spinoza took on the 'project of replacing the traditional conception of God as a superhuman person with the scientifically inspired conception of nature as an infinite, necessary, self-contained, and, above all, thoroughly intelligible system'.[24] Any cosmic idea or inclusive metaphor that fortified them to better meet their unprecedented world of change and negotiate along the edges of threatening chaos—both physical and spiritual—was welcome. This metaphor is—to employ the terminology of Lansing—an 'emergent property' that I have found from among the Romantics. What they were reacting to might be reduced to a binary opposition; however, the way they reacted to it was far more complex.

Often, paradoxically, the most prevalent feelings about movement and change during the Romantic age were expressed by conservatives. During the Napoleonic era, an aide of Metternich, Friedrich von Gentz, wrote that there was 'too much movement' in his world.[25] It is no accident, too, that one of the most important of these conservative thinkers, Edmund Burke, held as the tenets of his political belief a prevision of complexity theory, which he had expressed as early as 1770:

> Our constitution stands on a nice equipoise, with steep precipices and deep waters upon all sides of it. In removing it from a dangerous leaning towards one side, there may be a risk of oversetting it on the other. Every project of a material change in government so complicated as ours, combined at the same time with external circumstances still more complicated, is a matter

23 Karl Kroeber, *Ecological Literary Criticism: Romantic Imagining and the Biology of Mind* (New York: Columbia UP, 1994).

24 Henry E. Allison, *Benedict de Spinoza: An Introduction*, rev. ed. (New Haven: Yale UP, 1987) 82.

25 Qtd. in Frederick B. Artz, *Reaction and Revolution, 1814–1832* (New York: Harper, 1934) 67.

full of difficulties; in which a considerate man will not be too ready to decide; a prudent man too ready to undertake; or an honest man too ready to promise (1: 368).[26]

Clearly, Burke saw the political situation of his time as a dialectic complexity of balancing permanence and change on a 'nice equipoise'. Burke has anticipated the complexity/chaos approach to politics not only of conservatives, but also as I will later demonstrate, of progressives like Byron, Hugo, and Lamartine.

Lansing's research and writings have given me these firm convictions: that I can best proceed by revolving my study around the central axis of the hydrological-cycle and by centering on literature, especially poetry, as the main vector of my training will lead me to do; that I will, therewith, produce a model that approximates the subject of my study—a partial, focused vision of the European and, to a small extent, the North/South American world between 1760 and 1870. And while it could be argued that the polynary vectoring of my theory may be closer to a more holistic picture than the binary or polarities of monographic works on Romanticism, it will still not be a complete picture. 'And a picture of what?' I might further ask. Will I portray, let alone analyze, the noumena of Romanticism itself? No—at the very best, I might conceitedly hope to create a more complete model of aspects from Romantic culture than ever presented before. It, however, will be only that: a model.

This leads me to a theoretical flight of fancy. Complexity research does not seem to demand model creation by means of the computer, but the machine and its programmers do crop up with a certain inevitability. As a later and corollary component designed to test the findings of his research and the resulting theory, Lansing has drawn upon the computer modeling skills of James N. Kremer, a systems ecologist.[27] When run over many cyber-generations and compared to the actuality, the computer model was found to parallel with remarkable accuracy the one derived by field research. Now I, for one, am not going to attempt a computer program that envisions as an end product the emergent property of even mediocre—let alone great—Romantic poetry, painting, music, scientific, or political theory. It is significant—in this regard—that Lansing brought in Kremer and his program later. I imagine that a later scholar might create a similar computer model to 'predict' and thereby more deeply understand the physical and psychological importance of hydrology and hydraulics in Romantic culture.

My own model must remain hardly more than a metaphor. For my methodological needs, for framing data, however, metaphors are enough. After all, science has had its metaphors, from the central fiery hearth world-system of Copernicus, which Rutherford adapted as a model for explaining early sub-atomic theory, up to Schrödinger's cat in a sealed box. Literary criticism, similarly, has had its own guiding metaphors, from the well-wrought urn as the fetish of New Critics, up to the dialogical carnival of Bakhtin. A new literary metaphor is precisely what the rice farmers of Bali—via Lansing and Kremer—have provided for me. These two

---

26 *The Works of Edmund Burke*, 8 vols. (London: Bell, 1906–10).

27 James N. Kremer, 'Technical Report on the Ecological Simulation Model', Appendix B, Lansing, *Priests and Programmers*, 153–159; Lansing and Kremer, 'Emergent Properties of Balinese Water Temple Networks: Coadaptation on a Rugged Fitness Landscape', *American Anthropologist* 95 (1993): 97–114.

complexity researchers' emphasis on model creation has made me even more aware that all scholars expound their own models of a specific culture rather than present the thing itself. Abrams' Romantic England and Germany is his model, based on his readings/his models of texts. The same might well be said of the Marxist historians who assail Abrams' scholarship—except that they have a more precise notion of the archetype giving form to their modeling. Working at the textual or even monographic level allows one to sustain the myth that we analyze; writing a comprehensive view of a culture makes one aware that we model.

The presentation of my model—as well as its revisioning of the essential ways that the disciplines within Romantic culture and the modern disciplines studying it have presented the water-cycle—will parallel the Lansing-Kremer model. The work will be divided according to the Romantic conception and division of the water: chapter one treats the source, which may be a natural spring, an artificial fountain, or a glacier. Chapter two, the flow of stream and waterfalls; chapter three the lake, sea, or ocean as receptacle; chapter four the particular Romantic aspect that makes the water-cycle a completed circuit as well as scientific. As with Lansing, I will treat the human occupations as they fall along the way. I have followed Kremer's division of his model into sub-models to deal with data: rainfall, rainfall-runoff relationship, rice growth-stages and harvest-time, water-stress or usage, pest damage, and human management. I will begin by presenting the sub-models representing the various approaches to the hydrological-cycle, and since it is after all a concept that—as I shall indicate—was defined by science for the other disciplines, I will start there.

I have found that an openness to the inherent complexity within each culture has allowed me to conceive of and I hope deal with the intricate way that the styles of Neoclassicism (a return to the Greco-Roman ideas in the arts, as well as politics) interweaves with the various strands of that complex tapestry that is Romanticism. The tension between the two extreme camps of Romantic criticism has been to use the term—as a bed of Procrustes—cutting the subject down to size or of throwing the term out all together, or hedging the point as I was led to do when this study was in its *Ur*-form. I now see Neoclassicism as a frequently emergent property within water images that might otherwise seem Romantic, or just as often, I see Neoclassical elements as tensions pulling in the opposite directions from the Romantic elements—creating thereby a harmonically taut and very resonate string, as it were. And rather than finding a defined border between Neoclassicism and Romanticism, I find a magnificent borderland of intermingled titans of two tribes. Emergent, too, is a similar cusp between Romanticism and Realism.

I will sum up this section on method by pointing out that it began with a series of data showing a recurring pattern—a fascination within the Romantic painting and poetry, as well as their contemporary natural-scientific endeavor, for the water-cycle—usually in parts, but surprisingly often in its entirety. An immense amount of positivist endeavor (see the bibliographical essay at the end of the volume) helped me add to the data I had already found. Put into a dissertation format in 1980, the data indicated a concern with permanence and change; however, neither the full significance of the data nor their relation to these larger concepts was clear. These data also seemed to push out into other disciplines, such as rhetoric, music, philosophy, technology, politics, and especially, ecology. This expansion seemed dangerously broad. Even the wide-ranging implications of ecology, implicit in hydrology and meteorology, could not have sufficient organizing force. I found this force in Lansing's application of complexity theory to anthropology and to a study of a hydrology/hydraulics in a non-Western culture. The freshness of vision offered by the Balinese rice farmers gave the distancing-effect and viewpoint outside the West that my study needed: it gave me a less limited sense of ecology. As will be seen, Lansing's glosses on Marx helped, too. And further—just as Lansing's inflection of the original theory predicted—I found an emergent property arise from the data: the water-cycle offered a metaphor, which in its turn offered a complex adaptive system that allowed for a significant number of individuals within Romantic culture a way of mediating permanence and change, mediating order and chaos, as well as negotiating the implications of an age of unprecedented changes—unprecedented both in variety and in degree.

It seems, of course, that I have traded one set of binaries for others: order and chaos, permanence and change. Complexity, however, does not employ or construct these as mere binaries. The chaos Lansing saw among his Balinese farmers came in two basis types—the chaos of drought and disease—with order not as antinomy, but as a means between the two sorts of chaos. He showed chaos coming from other sources, such as government intervention and even Western-type environmentalist meddling where traditional 'environmentalists' had been at work for many centuries. The additional sub-models bring other considerations, even disciplines into consideration. Emergent properties always allows for many more surprising ones to arise from the data. My order, also, is one that lies between the chaos of too much change in politics, aesthetics, or technology, and not enough change in one or all of these areas. My sub-models are spread widely: through science, where technology is included with theory-based scientific research; to literature in the context of the other arts; to politics and even poet as politician; ending finally on the vast and complex cycle-rhetoric of Romantic culture, where my focal point, the hydrological-cycle takes its place, as one among many. Even the cycle is not allowed to stand in opposition to a linear model of development; the two become synthesized by Vico and others into a spiral.

The Science Sub-Model, from Hydrology and Meteorology

I hope my readers—having had some exposure to the holistic vision of complexity theory—will be on their guard when I say that the water-cycle, which is a great continuity in nature, is divided between two separate sciences. Since the water-cycle is alternatively referred to as the hydrological-cycle, it would seem logical that the science of hydrology would cover its full circuit. While this is true in theory, practice involves a bias of emphasis. Meteorology views the cycle from a skybound perspective; hydrology, from an essentially earthbound perspective. Rain, for example, is of interest to the latter only as something 'fallen'. Even the historical evolutions of meteorology and hydrology as scientific disciplines fall into two separate periods; for this reason and perhaps because of the fracture between the disciplines in scientific praxis, their treatment in the history of science is not only split, but uneven. Since the hydrological-cycle is heat or sun driven, it will also be necessary to consider the coeval discoveries concerning heat—made because of the development of the steam engine.

Between 1760 and 1870 was a harvest time for meteorological theory; unfortunately, I will not be aided by any really comprehensive histories on the subject of meteorology. Luckily, on the other hand, for understanding the history of hydrology, Asit K. Biswas has written a masterful history of this science, which I used to compose a considerable portion of the following overview.[28] The crucial discoveries of hydrology—which involve the spring, stream, and sea quarters of the cycle, as well as the sky quarter only indirectly, by implication—were made in the hundred-year span before 1760.

Important discoveries in the science of hydrology—that is, the study of the movement of the Earth's waters—were being carried out during the sixteenth and seventeenth century.[29] In viewing sixteenth- and seventeenth-century science, we too often focus on astronomy and its Copernican Revolution with its subsequent moments of high drama; but hydrology has a breadth of scope that suggests, if not actually rivals astronomy, and more clearly demonstrates the impact of the new emphasis on empiricism. It was empirical observation, more than other considerations that caused the revolution in hydrology. In astronomy, the throwing-off of the ancient Ptolemaic and Aristotelian prejudices concerning the universe entailed considerable aesthetic considerations (like the beautiful simplicity of the heliocentric universe) and ignored observational data (the failure to achieve a parallax); but in hydrology, a similar overthrow of the dominant ancient views was based almost entirely on empirical observations and methods. The ancients offered all sorts of erroneous theories, but the most common of these was that water repercolated from the seas to the springs by way of underground conduits. Lucretius, for example, had it both ways: he held that while the water that fell in rain arose from the sea, this water had nothing to do with the springs, which were fed by underground conduits. The ancients thought that rain was insufficient to supply springs and that the ground was too impermeable for water to soak in deeply enough to make a difference in the hydrological table. The one extant ancient work that came close to the truth was by the practical, hands-on Vitruvius; the theorists were partly or altogether wrong. Practice in hydrology is crucial to pure theory. It is no accident that, in a pivotal dialogue published in 1580, a *Monsieur Pratique* broaches the modern concept of the hydrological-cycle and a *Monsieur Théorique,* the Aristotelian, disagrees.

This dialogue was by Bernard Palissy—the founder of modern hydrology. He was the first scientist to advance a correct explanation of how water moved on the largest scale within the hydrological-cycle. The water moves from sea to sky to sea again.

---

28 Asit K. Biswas, *History of Hydrology* (Amsterdam: North-Holland, 1970), see esp. 208 ff.

29 For this survey of the history of hydrology, I have employed, in addition to Biswas' *History of Hydrology,* Oscar E. Meander, Introduction, *Hydrology,* vol. 9 of *Physics of the Earth* (New York: McGraw, 1942) 1–31, see esp. 13–15; Frank Dawson Adams, 'The Origin of Springs and Rivers', *The Birth and Development of the Geological Sciences* (1938; New York: Dover, 1954) 426–60; A. Wolf, *A History of Science: Technology, and Philosophy in the 16th and 17th Centuries*, 2 vols. (New York: Torchbooks-Harper, 1959) 2: 361–62; Lynn Thorndike, *A History of Magic and Experimental Science*, 8 vols. (New York: Columbia UP, 1932–58) 5: 596–99, 7: 586–87.

Part of the cycle involves water that evaporates from sea to sky, falls in rain or snow, that runs off immediately, or that soaks into the ground, bubbles forth from a spring and flows in ever-growing streams down to the sea again. It is the replenishment of springs that Palissy and his seventeenth-century followers were seeking to provide with a scientific explanation, and it is this part, too, that before the early modern era caused the greatest confusion.

It remained for Pierre Perrault to test the hypothesis with empirical observations—including quantitative calculations (albeit often estimations) concerning the rainfall and drainage of the headwaters of the Seine Basin. Even though both Edmé Mariotte and, somewhat later, Edmund Halley perfected the experimental techniques of Perrault, there were still conservatives like Robert Plot and Athanasius Kircher who continued to hold that precipitation was too slight to account for the volume of water that passed through the rivers to the seas. In *Mundus subterraneus* (1664), the Jesuit polymath Kircher follows most of the ancients in describing how springs are supplied by a great network of subterranean hydrophylacia—that is, caverns transporting water to the spring openings. Kircher's views, equivalent to those of the most fanciful of modern science popularizers, were widely published and read. Such views continued to rival the real scientific notions into the eighteenth century, so it was only by the latter part of that century that the workings of the spring quarter of the cycle were able to become widely understood outside a limited scientific community. Almost everyone would then have been aware of the links among the source, flow, and receptacle quarters, as well as the general principles behind them. By the end of the seventeenth century, two Italians put the capstones on the knowledge of springs and their connection to the rain and snow of the mountains; possibly because after Galileo's humiliation by the Roman Church, Italians looked to the surface of the Earth, rather than to the Heavens, for the subjects of their science. This situation was, at least, a boon to hydrology. Bernardino Ramazzini and Antonio Vallisnieri both took advantage of the close proximity of high mountains and spring-rich planes in their native land to finally determine the connection between precipitation, mountains, and springs. So, by the beginning of the eighteenth century, knowledge concerning the movements of waters on or below the earth's surface was relatively complete.

By 1760, the knowledge needed to link springs to precipitation was no longer missing. In the following decades, scientists added the final elements needed to lend full closure to the hydrological-cycle: the precise workings of evaporation, condensation, and precipitation. Historians of science have tended to ignore or brush past this landmark moment. Distracted by the more controversial discoveries in geology and biology between 1760 and 1870, scholars of Romanticism have tended (with a few important exceptions) to ignore the quieter, but more decisive, definitive revolutions taking place in meteorology. It may be that history, like fiction, demands the narrative tension that controversy can provide; scientific history would be inclined to ignore a science that moved from interrogative hypothesis to established theory with few arguments.

The heyday of Romantic culture was the heyday of meteorology. This is no accident. It was only the emphasis Romanticism placed upon dynamic becoming that was able to provide the needed philosophical ground from which to study a complex system such as weather, which is a 'nonlinear' or 'natural dynamic system' as Hugh

Roberts would have it.[30] It was only in the early decades of the nineteenth century at the height of Romantic culture that the sky link—including the processes of evaporation, cloud formation and movements, and precipitation—was fitted into the rest of the hydrological-cycle with a fully developed theory. This theory of the sky quarter's place in the cycle offered the final link in understanding how the entire water-cycle worked. It was, in great part, the creation of two individuals: James Hutton, a Scottish scientist most known today for his geological theories, and Luke Howard, an English scientist. It is true that Howard's work has received some significant attention; however, the meteorological work of Hutton, *Theory of Rain,* on which Howard in part based his work, has been ignored, while the geological work by Hutton—thanks to a popularization by John Playfair—finally received and continues to receive a certain notoriety. I, therefore, will concentrate my discussion on Hutton's *Theory of Rain*. Before I do, however, it is important to point out that others made important contributions to weather before Hutton. In his history of early-modern meteorology, H. Howard Frisinger writes: 'Halley was the first to connect the general circulation of the atmosphere with the distribution of the sun's heat over the earth's surface. The sun is important to the theory of rain in that it is the prime motive force for transporting the moisture'.[31] The sun, then, is the prime mover of water.

As with Hutton's *Theory of the Earth,* the scientific findings and theories of *Theory of Rain* had to find their way out into the popular scientific culture via other scientists—Howard being the most important of these. Both of Hutton's works suffer from a dense style and a lack of rhetorical sense of audience. His *Theory of Rain* proposes the rule that 'the action and effect of heat and cold' lies at the base of all rainfall. As he says, he seeks 'to form a theory of rain upon that investigated rule, concerning the evaporation and condensation of water' (41). Evaporation is the process by which 'water may be rendered an invisible elastic fluid, by means of heat alone'; condensation is the process by which 'this fluid [vapor or steam] in being cooled, will be condensed into water, and appear visible'. The crux of the matter is a proportional one: 'In order to determine the actual ratio of the dissolving power of air, in relation to water in different degrees of heat, or the ratio in which the power of heat converts fluid water into elastic steam, we must consider the several ratios in which this operation may proceed' (42). Backed with a considerable array of facts on humidity, temperature, wind currents, and precipitation, Hutton propounds that the amount of humidity in the atmosphere increases proportionally to an increase in the temperature; precipitation occurs where a colder air mass collides with a hotter air mass, thereby cooling it and forcing it to expel its moisture:

> let us consider the summer hemisphere of the globe, warmed by the influence of the ascending sun. From the laws of hydrostatics, it will appear, that there should be formed, on this occasion, two opposite currents in the atmosphere about this half of the globe, the one moving along the surface of the earth, from the polar region towards the equator, the other flowing above, in a contrary direction.

---

30 Roberts, 'Chaos' 163.

31 H. Howard Frisinger, *The History of Meteorology: to 1800*. Historical Monograph Series of the American Meteorological Society (New York: Science History, 1977) 123.

He continues to trace the weather process of this cycling humidity-bearing air through the cycle of seasons:

> On the one hand, the evaporation of the winter's moisture from the surface of the continent, warmed by the summer sun, must tend to saturate with humidity the polar atmosphere, as it acquires an evaporating power from its increasing heat; on the other hand, the progress of the upper current, from the tropic towards the pole, in having its degree of heat diminished by the general cooling cause, will naturally bring the mass to the point of saturation with the aqueous vapour which it has received. In this state of things, the two opposite currents in the atmosphere, while separate, might pass on without condensing humidity sufficient to produce rain; but the moment that sufficient portions of those saturated streams shall mix, not only cloud, but showers will be produced [...] (50).

All this, of course, does not fully explain the manifold ways in which the movements and interactions of cold and warm air masses create rain. There are many others—some of which he proposes.

Mountain ranges, he proposed, can have the same raising and cooling effect on warmer, moisture-carrying air—thereby, causing precipitation; mountains aid 'cold portions of the atmosphere to be mixed with the warmer portions, saturated with humidity, and ready to produce rain' (49). While this is only a partial view of the actual process, Hutton claimed that rain can be predicted scientifically by a careful study of geography and of wind currents, which combine in set patterns. Although this was revolutionary science when Hutton first propounded it, his proofs and supportive data on the main points were so conclusive when he read the paper to the Royal Society of Edinburgh in 1784 that the revolution was a very quiet one. Disseminated throughout the following decades, this theory of weather was one of the great landmarks of Romantic science and had a great effect on Romantic literature and the other arts. Yet the definitiveness of Hutton's victory has caused it to be overlooked.

Many details within Hutton's theory were quietly altered by later scientists and many needed points were added by them. On the question of water vapour, John Dalton, more famous for his contributions to the theory of atomic weight, converted Hutton's qualitative science to a more quantitative one during the 1820s and early 1830s. A decade earlier, Sir Humphry Davy proved the important concept that temperatures decrease proportionally with height and the atmosphere becomes thinner—and, further, both effects cause the air to release water vapour. Sir John Leslie proved that there is a constant movement of high cold currents downward and warm ones upward, that the cooler the air the more readily it will part with its moisture, and these concepts led to the important awareness that air moving over mountains will be forced upwards and will part with its moisture as precipitation.[32]

It must be emphasized that a crucial contribution to the knowledge of precipitation made by Hutton's theory, suggested by Halley, was that the engine that drives the great cosmic engine of the water-cycle is the heat of the sun; a key fact that relates other sciences and technologies to the water-cycle studied by hydrologists and meteorologists. Indeed, the whole process of the steam engine might be seen as a

---

32 This historical summary is based on W.E. Knowles Middleton, *A History of the Theories of Rain and Other Forms of Precipitation* (New York: Watts, 1966).

water-cycle in miniature. James Watt's single-acting steam engine of 1788 might be taken as a paradigm.[33] While sections of Watt's engine offer only a very rough approximation to the quarters of the water-cycle, there is, however, a certain striking correlation: the fire under the boiler might be the sun; the boiler, the sea; the pipe that carries the steam to the cylinder the sky/link; the condenser, cloud-making and rain as moisture returning to a receptacle; the pipe carrying water back to the boiler, the spring or river. The steam engine can thus be viewed as a hydrological-cycle in miniature. Bate penned a phrase that hints at this process: 'Science has packed the clouds into iron cylinders and made the steam engine that enables man to travel at a cloud pace [...]'.[34] In what happened next, it was really technology of the ironmongers and engineers like Richard Trevithick that did the packing.

It was also an engineer, Sadi Carnot, who first gave to science two laws of thermodynamics in two works: the 1824 paper *Reflections on the Motive Power of Heat and on the Best Machines for Developing this Power / Réflexions sur la puissance motrice du feu et sur les machines propres à developper cette puissance*; in addition, a manuscript note of 1830.[35] The *Reflections* open with a celebration of the steam engine, after which Cadi gets down to introducing the scientific reasoning on the principles of the motive power of heat. His note of 1830 sets forth the law that 'Heat is nothing but motive power or rather it is motion that has changed its form' / 'La chaleur n'est autre chose que la puissance motrice ou plutôt que le mouvement qui a changé de forme'.[36] This cycling between heat and motion corresponds to the cycling from water to steam. Heat can raise and expand water by converting it into steam; steam, then, is only lifted and expanded water, which when forced into a cylinder can become a motive force. The perfected steam engine and its theoretical explanation was one of the great creative culminations of Romantic culture. Engineers and scientists all over Europe worked out the details. The steam engine, Frederick Engels was to observe later during this century of steam, was the first truly international invention, and it is one that testifies to a tremendous historical advance.[37] It is significant that this invention and its concomitant scientific theories were being achieved while meteorologists were working on the details of the hydrological-cycle theories.

33 See H.W. Dickinson, 'The Steam-Engine to 1830', *The Industrial Revolution, c. 1750 to c. 1850*. Vol. 4 of *A History of Technology*, ed. by Charles Singer et al. (Oxford: Clarendon, 1958) 168–98, diagram p. 184); see also: Paul Mantoux, *The Industrial Revolution in the Eighteenth Century: An Outline of the Beginnings of the Modern Factory System in England* (New York: Torchbook-Harper, 1961) 311–38.

34 Jonathan Bate, *Romantic Ecology: Wordsworth and the Environment Tradition* (London: Routlege, 1991) 82.

35 J.D. Bernal, *Science and Industry in the Nineteenth Century* (1953; Bloomington: Indiana UP, 1970) 40–49; Charles Coulton Gillispie, *The Edge of Objectivity: An Essay in the History of Scientific Ideas* (Princeton: Princeton UP, 1960) 357–70.

36 Text of note is published in the French original and in translation in Bernal 41; I quote from both.

37 Frederick Engels, *Dialectics of Nature*, trans. Clemens Dutt (New York: International, 1940) 80.

The clouds of the water-cycle are also only water raised by the heat of the sun (evaporation). The clouds driven by winds—driven in their turn by the heat of the sun—are the motive links in the movement of the water-cycle, carrying the water up over land, where the lowering of the heat causes final condensation into precipitation. These clouds and their various metamorphoses became the focus of another great work of Romantic culture: Howard's classification of clouds.

As will become clear later, it is of the greatest significance that Howard has divided cloud formations into three larger classes (5–6).[38] The first are the 'simple modifications', which seem at first to form discrete categories:

1. The cloud, commonly termed 'racks', that he called *cirrus* is described by him as: 'Parallel, flexuous, or diverging fibres, extensible in any and in all directions'.
2. The *cumulus* is: 'Convex or conical heaps, increasing upward from a horizontal base'.
3. The *stratus* is: 'A widely extended, continuous, horizontal sheet, increasing from below'.

The definitions, with their sense of continuous flux, convey the sense that Howard realizes he is dealing less with entities and more with processes.

The second category makes clear the profound understanding Howard had of what he was trying to categorize: he fully comprehended the great complexity of a process where every transformation can only be plotted on a gradual continuum. Howard described the dynamic nature of his phenomena not as cloud *formations,* which we do in everyday English, but as cloud *modifications*. In arriving at his sense of categorizing, Howard has traveled a tremendous distance from the rigid, discrete 'class'-system of Linneus's early Enlightenment botany. This class is used to describe the 'intermediate modifications':

4. The cloud Howard terms *cirro-cumulus,* he defines as: 'Small, well defined roundish masses, in close horizontal arrangement'.
5. The *cirro-stratus* is: 'Horizontal or slightly inclined masses, attenuated towards a part or the whole of their circumference, bent downward, or undulated, separate, or in groups consisting of small clouds having these characters'.

It is very striking just how active and how verbal the adjectives are that Howard uses to describe the *cirro-stratus*. Once again, it is clear that this scientist sees clouds more as a process caught, in an otherwise passing moment, by the act of description, than as a formation or a class.

Howard's third category is called compound modifications and again demonstrates the attempt to organize fluid phenomena that intertwine and oscillate among different states:

6. *Cumulo-stratus* is, as the name implies, a mixture of mid-level type, cumulus, and of the lower, stratus clouds.

---

38 Luke Howard, *On the Modifications of Clouds, and on The Principles of Their Production, Suspension, and Destruction,* facsimile printed in *Neudrucke von Schriften und Karten über Meterologie und Erdmagnetismus,* ed. G. Hellmann (1893; Nendeln: Klaus-Thomson, 1969).

7. *Cumulo-cirro-stratus,* or *nimbus,* is a mixture of all main types of clouds found at the three levels; added to the cumulo-stratus of number 6, there are the higher level cirrus clouds (5–6).

This cloud, which draws all three shapes, levels, and processes into itself, makes a very impressive pile of varied forms, changing at a tremendous rapidity. Because it is the bringer of thunderstorms, this cloud will prove especially dear to Romantic poets, painters, and composers.

Implicit in the categories is the desire to make meteorology into a useful method of predicting weather. As Howard says, when discussing the cirrus cloud type: their

> modification, although in appearance almost motionless, is intimately connected with the variable motions of the atmosphere. Considering that clouds of this kind have long been deemed a prognostic of wind, it is extraordinary that the nature of their connection would not have been more studied, as the knowledge of it might have been productive of useful results (6).

In speaking of this Romantic hope of predicting weather 'with the same accuracy that we predict the movement of heavenly bodies', Hugh Roberts writes,

> The insight chaos science gives us [...] is that our inability to predict the weather is not a reflection of our limited ability to gather relevant data, but a fundamental fact of nonlinear dynamic systems. No matter how precise our information about the current state of the system, our ability to predict its future development will be limited; significant information will always be below the scale of our measurement.[39]

I would only qualify his over-emphasis of the chaos side; meteorologists are able to predict weather in a limited way, but even then, with a degree of accuracy that would have surprised even the sanguine Howard.

More successful than this practical application of meteorology was hydraulics—the technological applications of hydrology. Although the canal will be the focus of a consideration of engineering relating to water, we must not forget the watermill, the great bridges to carry road traffic, even canals, and later the steam-driven trains over rivers and bays. It was only near the end of Romantic culture that steam had driven this age-old form of power out of its pride of place—particularly in clothing manufacture. The watermill is at least as old as the Roman Empire. It was crucial to the Medieval and Early Modern worlds. The need to build towns around mills and the rivers that drove them determined in great part the landscape of Europe; therefore, it shaped the landscape of the Romantics. To put it another way, their landscape was shaped by the technological need of placing water-wheels near streams and the economic need to build the towns there. Actually, for all its long history before 1760, the water-wheel did not receive its theoretical basis until the year before. On 2 and 24 May of 1759, John Smeaton read two papers before the English Royal Society that set forth his experimental data based on his two models for testing and comparing the two types of water-wheels: the undershot wheel (with the water flowing under); the overshot wheel (flowing over). Smeaton determined that it was the weight of the water rather than its forward momentum that could do the most work; he also determined the optimal size for wheels. Add to these developments the shaping effect of the great iron bridges. These were great Romantic iron spans

---

39 Roberts, *Shelley and the Chaos of History* 252–53.

thrown like acts of imaginative idealism across breath-taking spaces—from Abraham Darby's Ironbridge at Coalbrookdale in 1779 to the early nineteenth-century masterpieces of Thomas Telford and of Isambard Kingdom Brunel.[40]

The Bali rice farmers and their irrigation canals as viewed by Lansing were what enlarged my view of the science of hydrology to include its technological adjunct: the hydraulics relating to canals. As a consequence, the profound shaping effect that hydraulics had in creating the landscape of Romantic culture arose as another emergent property. I viewed more clearly than before the positive and humanizing effect that hydraulics had had upon the landscape. England's canals, of course, are for drainage or transportation, not for irrigation. The techniques of hydraulics are essential in the workings of all three types. Although, like the water wheel, the canal technology was old, it was only in 1759 that extensive canal building began with James Brindley's engineering breakthrough in the construction of the Worsley Canal. This same Brindley went on to conceive and initiate the construction, between 1766 and 1777, of the Grand Trunk Canal; actually, it was to become the core of a system of canals linking the Irish Sea and the North Sea trade, as well as the Thames trade. Thus, the canals and—most important for this study—the canalization of streams had, by the end of the eighteenth century, further contributed to the human transformation of the English landscape.[41] The cities and towns grew and prospered along the rivers because of both watermills and water transport. Steam, really another form of hydraulics, was about ready to have a different effect; when placed on rails and wheels, steam engines were able to transform the cross-country landscapes between the great streams and canals. Seas and oceans would be less barring and more freeing to human transport and trade.

Even in the New World of the 1780s, where canals were unknown and watermills scarce, Thomas Jefferson makes it clear how cultural geography was shaped by the physical geographical considerations of where the water is. Out of a total of twenty-two scientific queries contained in Jefferson's *Notes on the State of Virginia,* four directly treat water:

II. Rivers
III. Sea-ports
IV. Mountains
V. Cascades (27).[42]

The human landscape of American North America, like that of Europe, was shaped by water. This became less true in the later half of the nineteenth and during the twentieth century: the cities that have grown up since the latter part of the nineteenth century had other considerations in their placement and growth than water-based transportation or power; mineral deposits, train routes, or airline flight patterns create, for example, a modern Denver. For the Romantics even in North America, however,

40 A. Stowers, 'Watermills, c. 1500–c.1850', *History of Technology* 199–213.

41 Mantoux 120–35.

42 *The Portable Thomas Jefferson,* ed. Merrill D. Peterson (New York: Viking, 1975) 27. Throughout the text of this entire study, all references to the works of Jefferson will be from this collection.

human landscape—in ways that we find hard to understand—depended upon a water landscape.

Jonathan Bate, in his 1991 *Romantic Ecology,*[43] oversimplifies Karl Marx's sophisticated views on nature or, at least, takes over the misreadings of other scholars. Nature merely as a source of raw materials for human industry does not do justice to Marx's ecological perspective. A solid reading of what the inventor of the Economic Critique really said about nature is not to be found among the literary theorists but rather among the anthropologists. As a sidelight to his great field application of complexity theory to the irrigated rice fields of Bali and its hydrologic system, Lansing finds in Karl Marx's 1844 papers a division of natural landscape into two types—or, implicitly, of modern landscapes into three: the dehumanized and denaturalized urban landscape; the wild and untouched 'eternal nature'; and between these two, Marx placed a 'humanized landscape'—but not denaturalized landscape. What Marx thought of the urban landscape, *Capital* makes clear; what he seems to have meant by 'eternal nature', to my mind, is found in the representations of the Alps by Wordsworth or by Turner; what he means by 'humanized nature' Lansing finds in the irrigated fields of Bali and, I feel, is represented by Constable's and Austen's English farm country.

The science and technology sub-model is of course crucial to my endeavor because hydrology and meteorology are the basis modes of understanding the workings of the water-cycle and its parts. Much of the scientific work on the source, flow, and receptacle quarters of the cycle had been completed before 1760. The knowledge of these quarters, moreover, had time to become part of general knowledge. The fourth quarter, linking the other three into a cyclical unity, involves the interconnection of three complicated processes of evaporation, condensation, and precipitation. Edmund Halley and others had made contributions before 1760; enough was available for a rudimental working-model. The process was very rapid in the following decades. A general theory and categories were worked out by Hutton and Howard, as well as corrected by many others. Only then could the water-cycle become a model, not just for science, but for Romantic thought and expression, as well. The technologies of hydraulics and hydrodynamics spawned by the hydrological-cycle shaped cultural geography and the lifestyle of the age in town and country quite as much as it has shaped the life of the Balinese. These technologies helped to add to the great changes and the stress they were causing; these did so, in part, by adding yet another revolution to an unprecedented variety and plethora of revolutions: the Industrial Revolution. This, further, was an age when science and technology, the mind and the hand, worked together to understand and exploit the motions and interactions of water and heat; after the 1760s, scientists were 'in contact with things' through its artisans and their technology to a degree not known since the days of Ionian science, before Plato.[44]

---

43 Bate, *Romantic Ecology* passim.

44 I have taken this concept from Benjamin Farrington, whose views on science and technology have shaped my own; in addition to his various studies on Bacon, *Greek Science: Its Meaning For Us* (Baltimore: Penguin, 1953) 18–19, 134–48, 301–13.

The steam engine and its concomitant science of thermodynamics—both the work of engineers—were achieving at the micro level what the water-cycle performed on the macro. All this work had a direct, but more often an indirect, effect upon the conception of this great cosmic model—the hydrological-cycle—and upon how scientific actualities were adapted and used by the Romantics in the humanities.

The Humanities Sub-Model, from Literature, Painting, and Music

Before the water-cycle was fully worked out, the visions could only be partial. To take one early example from before the two stages of the scientific revolution, Robert Herrick employs the hydrophylacia as a symbol in 'Proof to No Purpose'[45] in his *Hesperides* (1648). The water-cycle, which is used to represent the cycle of human mortality, is traced from the stream to the sea and then back up through the hydrophylacia—subterranean caverns filtering out salt and carrying water from the sea up to the sources, by the pressure of tides it was believed:

> Next, from that world of waters, then
> By poares and cavernes back agen
> Induc't that inadultrate same
> Streame to the Spring from whence it came (ll. 7–10).

Herrick, thereby, employs the traditional scientific theory of hydrology; further, he chooses the most incorrect aspect of the older theory—the hydrophylacia. This older theory offers only a partial cycle with the rain connected to run-off water, but not to spring water. Literature, as well as the other humanities, must, as it were, wait for science to create a full vision of the water-cycle: a symbol of archetypal import.

In his discussion of types of cyclic movements as themes in literature, Northrop Frye cites the water-cycle as the seventh and the last:

> Water-symbolism has also its own cycle, from rains to springs, from springs and fountains to brooks and rivers, from rivers to the sea or the winter snow, and back again.
>
> These cyclical symbols are usually divided into four main phases, the four seasons of the year being the type for four periods of the day (morning, noon, evening, night), four aspects of the water-cycle (rain, fountains, rivers, sea or snow), four periods of life (youth, maturity, age, death), and the like.[46]

The water-cycle as outlined by Frye offers itself as a logical and systematic way to organize the study of water images and symbols in Romantic culture. This is true in spite of the fact that the scientifically shrewd among these poets and artists employed a much more complex vision of the cycle with all sorts of little epicycles, disruptions, and other discontinuities—in short, one which more approximated the scientific actuality.

Frye's outline is a useful compromise between scientific theory and poetic usage, and it will therefore prove useful in moderating and grouping the creative tension between poetic significations and scientific qualifications. Frye places snow with the

---

45 Robert Herrick, 'Proof to No Purpose', *The Poetical Works of Robert Herrick*, ed. L. C. Martin, rev. ed. (Oxford: Clarendon, 1963) 244–45.

46 Northrop Frye, *Anatomy of Criticism: Four Essays* (Princeton: Princeton UP, 1959) 160.

sea—that is, representing both as receptacles where water is stored and the processes of the cycle are suspended; but in the scientific schema of the hydrological-cycle, snow and ice should properly be placed with rain, springs, and *run-off* (i.e., water that runs into seas and streams before soaking into the ground). This type of run-off happens in temperate zones during spring thaws and in mountains from glaciers or ice-fields during the summer. Frye, of course, is validating ice or snow poetically in the sense that it offers a symbol for the freezing-up of life in death. For the rest, hydrology qualifies Frye's generalized water-cycle: only a small portion of the earth's water moves in this way; most of it 'short circuits' this generalized spring/stream/sea/rain cycle. Actually, less than ten percent of the water moving within the cycle flows in streams to the ocean. Far less than that really passes through springs; rather, much of it passes directly into the streams through run-off or ground-water seepage from the rain/snow and does not pass into the earth and out through these springs. Moreover, approximately three-fourths of the water in the cycle never reaches the land at all but falls directly into the sea from which it rose. Similarly, approximately one fifth of the water, which is precipitated on the land, evaporates directly into the sky.[47]

Science aside, in its main outline, Frye's water-cycle schema fits *mutatis mutandis* the poetic cycle used by the writers who fall between 1760 and 1870. Except for modifications demanded by the poems themselves, the following sequence is based on that cycle:

1) The earthly sources are most often natural springs that are left untrammeled by human construction, or they are modified by retaining walls or the like, for the purpose of drawing water. Suffice it merely to observe that, although both are natural sources, the former are loosely and generally termed by the poets as *springs* and the latter, *fountains*. Frye lists the rain as the first of his water-cycle 'phases'. For our poets and poetry, springs represented the first stage of the life cycle—birth.

Poets and painters, as well as composers, closely connect the small brook, produced by the spring, combined with the spring itself. Other than the Romantic trait that allows for a sense of movement with a rivulet's bubbling forth and flowing off from the source, the spring with brook was often employed to signify birth and infancy. For artists, run-off is only important to the cycle when it derives from glaciers or mountain ice-fields. Generally, even here, the water from this melted ice is shown as running from the glacier itself or, having seeped through rocks below, bursting forth out of clefts; often, it is at once a spring and a waterfall. Poetic springs, which form the headwaters of streams, are usually found in mountains that have high enough elevations to retain snow or at least catch rain.

2) The spring-born rivulet joins with other rivulets and issues into streams, which tend to flow downward to sea-level. The poets I will be discussing were attracted to the variety of phenomena that go beyond process in the water-cycle. Periodically, during their seaward course, sudden drops in elevation cause a second and more

---

47 In this summary, the following standard references concerning hydrology and the hydrological cycle were employed: Oscar E. Meander, ed., *Hydrology,* vol. 11 of *Physics of the Earth* (New York: McGraw, 1942); Richard J. Chorley, ed., *Water, Earth, and Man: A Synthesis of Hydrology, Geomorphology, and Socio-economic Geography* (London: Methuen, 1969); Alan D. Tweedie, *Water and the World* (Sidney: Nelson, 1966).

common type of waterfall. There are two important symbols that are corollary to the waterfalls: sunlight shining through the spray creates a miniature 'rainbow'; the rising and falling spray creates a miniature 'water-cycle'. This latter symbol is one of several ones whereby any sense of a linear progress from quarter to quarter is short-circuited.

3) The lake offers another interruption in the seaward flow of the stream and, therefore, in the cyclical process broadly outlined by Frye. Importantly, the poets between 1760 and 1870 tended to emphasize the natural dams and like obstructions in the stream's path that created the lakes. These poets, moreover, often conceive of a lake as a receptacle of water having the important attributes of a sea, but on a smaller scale—particularly, the way both lake and sea when becalmed can reflect the sky. A stream flowing into a lake is used symbolically like a stream flowing into the ocean. The latter, as Frye points out, represents old age and death; so does the former. This cycle of life, combined with the image of a boat moving downstream and carrying a maturing human, is employed, during the height of Romantic culture, in order to produce artifacts striking enough to be considered, even though in the strictest scientific sense a boat has little to do with the water-cycle per se. In keeping with the method of Lansing, however, I will not seek to avoid the complex dimension offered by the presence of human figures in the landscape along shores of streams, lakes, and oceans; this includes corollary motifs like shipwrecks, human dwellings, agriculture, water mills, and other signs of human industry alongside or otherwise affected by the quarters of the cycle. Following Lansing, I will mention religious or mythological adjuncts to these quarters: sirens, nymphs of springs, or river gods.

4) Frye slightly oversimplifies his cycle by failing to mention the processes of evaporation, cloud formation, or precipitation. He merely alludes to the first two processes with the phrase 'and back again'. Generally, the cloud is too often forgotten as an important part of the water-cycle and, indeed, of water imagery in general; for it still remains water, albeit in a vaporous state, just as ice and snow remain water, in a frozen state. A few Romantic poets significantly treat the cloud as a comprehensive symbol that includes all three related processes.

Such, then, is the division of the water-cycle into four quarters, which I have used as the basis of the four chapters of this study, as well as their sub-divisions within the chapters; however, I have adapted Frye's schema or that of science to fit the complexity of the artistic works under analysis. Of course, to be fair to Frye, he only desired to sketch broadly the poetic uses of the water-cycle—for poets of all periods or cultures of literature; I am writing about three arts of one very special culture, with a unprecedented fascination for and full understanding of the hydrological-cycle. It is appropriate that he has limited himself merely to drawing the water-cycle/life cycle connection for an individual. Among our poets between 1760 and 1870, there is a significant type who uses the water-cycle to represent the life cycle of groups, peoples, and nations.

To offer a background to these poets, a short discussion of each water-quarter in painting and poetry before 1760 will open each of the four chapters. Such a survey will facilitate the isolation of peculiar aspects of water images and of water symbols between 1760 and 1870. Of equal importance, such an understanding will indicate the similarities between the latter usages and the former. This survey is perforce based heavily on important secondary works, whose authors use various terms which might tentatively be defined in their relations to water images and symbols: (1) *Baroque* poets as a rule, in using water symbolically, show a marked predilection for water in a state of turbulence; (2) *eighteenth-century neoclassic* poets tend to use water as mere images and show a predilection for placidly flowing water. The latter poets place water either in an ideal, composite landscape—for this use, the general term *descriptive* is employed—or they use specific loci that include significant water phenomena—the more specific term *topographical* is applied here.

Like the four-part division of the water-cycle, the chosen time-frame (1760 to 1870) offers a convenient structuring device; and like the former, it is by no means an arbitrary one; not only has my data led me to include certain works bearing striking resemblance to one another, there are also certain historical turning points that fall around these dates. I have chosen to begin at 1760 because this date is roughly equivalent to the ending of three decades of world war, which when it was not a 'hot' war in Europe, it was a 'cold' war in the colonies of India and the Americas. These wars mark the end of the high point of Mercantilism, Dynasticism, and Feudalism in the Ancient Regime. The Treaty of Paris of 10 February 1763 brought an end to the War of the Austrian Succession and the Seven Years War, not to mention all the minor ones like the War of Jenkins' Ear—this sequence of wars constituting one long and unusually exhaustive outlay of national energies even in a century of several other such dynastic struggles. The forces of change were wearing away what was—especially outside England—an essentially feudal world of social hierarchy, guild restrictions, and patrimonial control. I am following the lead of Marilyn Butler who sees 'social and economic pressures [...] building up from the 1760s which would lead to revolutions from America to Poland.[48] And actually beyond Poland to the far reaches of Catherine the Great's Russian Empire: Franco Venturi makes it clear the Pugachev Rebellion (1773) was part of this European ferment—a threat of change that by fear arrested a real social change in Russia.[49] The initial setbacks for reform with the fall of Johann Friedrich Struensee, however, were a new beginning for significant social change in Denmark, as well as the rest of northern Europe.[50] With Struensee, a movement was begun that would end in 1787 and the year following with servile serfs getting (in theory anyway) equal legal rights with all other Danes and, then, getting their full freedom.[51] There were also the well-known changes wrought on central Europe by Leopold II in Tuscany and by Joseph II in the rest of the vast Hapsburg direct domain. While it may be argued that this despotic centralization of power went against developing individualism and capitalism, it was a necessary stage in the breaking down of feudal guild and trade barriers. In short, it is wrong to see change only in terms of the American Revolution

---

48 Marilyn Butler, *Romantics, Rebels, and Reactionaries: English Literature and its Background 1760–1830* (Oxford: Oxford UP, 1981) 11.

49 Franco Venturi, *The End of the Old Regime in Europe, 1768–1776: The First Crisis*. Trans. R. Burr Litchfield (Princeton: Princeton UP, 1989) 154–65.

50 Venturi 247–63.

51 Leo Gershoy, *From Despotism to Revolution, 1763–1789* (1944; New York: Torchbook-Harper, 1963) 141.

and the French Revolution, with its Napoleonic aftermath; change was rearing its desirable/terrible head all over Europe and some of its colonies.

I realize that the date of 1870 is a bit more problematic than 1760. Yet, the date of 1840 held to by Butler has its problems too; not only does it have a limited fit to English Romanticism; even then, while it encloses the Reforms of 1832 and many of the second generation of Romantic poets, it omits the greatest canvases of Turner and the 1850 publication of *The Prelude*. 1840 works, perhaps, better in Germany—except for the great music dramas of Wagner. And I must add that even 1870 fails to fully encompass the creation-years of the great *Ring* cycle—but it does encompass those of the first drama of the cycle, which is more important for this study. As, however, for the late blooming Romantics of France and the even later flourishing Becquer of Spain, 1840 is hopeless. 1870 includes the third French Revolution of 1848 and allows a closure just before the 1871 Paris Commune. Karl Marx best expressed it by saying that the Commune and its aftermath was the first manifestation of the new era, different from what I have termed Romantic culture: he saw an unprecedented brutality by which the bourgeois put down the secular martyrs of the Commune, which was the forerunner of a new social order.[52] The enemy of progressive social change was no longer monarchs and aristocrats, but was now the bourgeois; as Frederick Engels would have it—the state remained 'a machine for the oppression of one class by another'; the oppressing class had, however, evolved, changing the shape of class struggle. For him, too, the Paris Commune had been the first 'Dictatorship of the Proletariat'.[53] The two great barometers of their age have marked off 1870 from 1871 and after. The writings of Marx and Engels remind us that Romantic Culture fits, for the most part, within a transitional period, in which bourgeois capitalism had not yet cinched for once and for all its victory over the forces of feudalism. This is a point often overlooked by Marxist critics of Romantic culture.

Complexity theory has motivated the above discussion of political matters in the sub-model for literature and the other humanities, rather than in its own; this is true for several reasons. Lansing's methodology views things holistically even within each sub-model. I wish, moreover, to make it clear that the properties emergent from my analysis of the humanities have guided my chronology. Emergent, too, are uncanny correlations with the economic and political events in Europe. A systematic consideration of these changes, along with those in the science-technology model, is part of the complexity approach. Yet, just as Lansing placed an emphasis on Balinese agriculture, I focus on Romantic literature and the other arts.

No one has ever firmly set the period boundaries of the Romantic humanities by replying to Jacques Barzun's *Classic, Romantic, and Modern*. This work is one of the most neglected contributions to the great debate 'What is Romanticism?' Barzun's polemic offers, among other things, a suggestive overview of Romanticism. In answer to those modern critics who seek to say 'Romanticism lasted from such

---

52 Karl Marx, 'Address of the General Council of the International Working Men's Association on the Civil War in France, 1871' [the 30 May 1871 address], *The Civil War in France* (New York: International, 1940) 36–84, concepts and quotations from 76, 78–79, 81; see Isaiah Berlin, *Karl Marx: His Life and Environment*, 3rd. ed. (New York: Galaxy-Oxford, 1963) 258.

53 Frederick Engels, Introduction, Marx, *Civil War* 9–22, qtd. from 22.

and such a date—or consisted of this or that', he views it as an on-going comprehensive and seminal movement that gave rise to most that came after it:

> True to its inclusive purpose, romantic art was simultaneously idealistic, realistic, and symbolic; impressionist, expressionist, and surrealist. It produced forms and amassed contents only now nearing exhaustion, after furnishing the models for the movements which we enumerate through the past century as Realism, Symbolism, Impressionism, Naturalism, and Post-Impressionism.[54]

If these 'Isms' are not Romantic, but something else, then, 1870—although a bit late for excluding Realism as a self-sufficient movement, which produces some masterworks in the 1850s and 1860s—works for the rest. The 1860s, then, can best be viewed as an overlapping of the Romantic with the Realistic—one style or the other dominant in this country or in that art.

When there are so many Neoclassical works appearing between 1760 and 1870, one might ask: why do I use the reductionist term offered by the term 'Romantic Culture' of my title? It is true that within the study itself I am careful and selective in employing the term 'Romanticism' in part because this term is reductive and often meaningless when applied in an international context.[55] It is also true that the chosen time-frame includes anticipatory 'movements' sometimes called 'pre-Romantic'—I prefer to call these Early Romantic. Well within the time-frame is Weimar Classicism, which combines Romantic elements with those derived from Greco-Roman civilization (or at least as it was interpreted by Winckelmann). Neoclassicism, as I have noted earlier, is included as a dialectic style within Romantic culture. I admit that the term 'Neoclassical-Romantic culture' would be more accurate; however, it is so awkward and cumbersome, and 'Romantic culture' is a simple means for describing a fair amount of what happened between 1760 and 1870—more on this later. '(Neo)Classical-Romantic culture' is a further improvement. The parentheses are not just added out of some Postmodern affectation but represent a necessary qualification for any general cultural study of Romantic culture, which seeks to incorporate music with the other arts. In literature and visual-arts, the term Neoclassical delineates an art that has Greco-Roman coordinates, placing an emphasis on the mimetic qualities of art; music has few such Classical coordinates or referents; and its examples of mimesis employ signs of such stylized natures that their ambiguity makes them marginally mimetic at best.

The Classic style of music shares the same dialectic relation with the Romantic style that the Neoclassic of the other arts has. Yet, the two styles must not be seen as rigid binaries: as one period following in reaction to the other. Rather they must be seen as styles that mix freely. It, indeed, is from a musical historian—Friedrich Blume—that I have taken and adapted the term (Neo)Classical-Romantic; however, he uses it as a period term, whereas I employ it to call attention to the stylistic diversity and complexity of Romantic culture. Moreover, the *(Neo)*-prefix must be

---

54 Jacques Barzun, *Classic, Romantic, and Modern* (Chicago: U of Chicago P, 1961) 137; see also 98–99.

55 Arthur O. Lovejoy, 'On the Discrimination of Romanticisms', in *Essays in the History of Ideas* (Baltimore: Johns Hopkins UP, 1948) 228–53; Ulrich Weisstein, 'Epoch, Period, Generation, and Movement', in *Comparative Literature and Literary Theory: Survey and Introduction*, tr. William Riggan (Bloomington: Indiana UP, 1973) 66–98, esp. 80–82, 92–96.

added when dealing with art; yet, there is no Neoclassicism in the music of Romantic culture. (The term 'Neoclassicism' is, of course, reserved by musicologists for a twentieth-century style, and it otherwise could only fit either the era in Florence around 1600 when composers felt they were resurrecting Ancient Greek Drama, only to create the very different form of Baroque opera, or the Gluckean reforms of opera during the 1760s and 1770s.) There is a similar absence of a real relationship to the Classicism of Greco-Roman culture in landscape painting; it is as hard to find the source of Poussin in the landscapes of Pompeii, as it is to find that of Mozart—even considering the more complex issue of his operas—in the extant shreds of Greek music. While Blume gives separate articulation to the terms Romantic and Classic to delineate with each a different nexus of stylistic traits, he—throughout his highly regarded survey of the time—calls the whole period 'the Classic-Romantic age' and seeks to show the essential dialectic unity of the two antinomies contained in this phrase:

> The unity of Classicism with Romanticism is so pronounced that basically there is neither a purely Classic nor a purely Romantic period. What appears in Haydn and Mozart as the fullest maturity of Classicism is permeated with the provocative charms of Romanticism. These [...] lend the 'Classic style' that indescribable atmosphere of characterful sensuousness, of the unique and the 'infinite', which constitutes the essence of High Classicism, while what appears in Schumann or Brahms as most sublimely Romantic rests firmly on the underpinnings of this 'Classic style'.

I would only qualify what Blume has said by saying that this 'unity' is an interactive process between the (Neo)Classicism with the Romanticism of all the arts. Blume, whose survey draws from literature and visual-art history, next employs an image from the early German Romantic, Novalis: 'Not until Romanticism's "blue flower" blossomed in the pure mountain air of Classicism did its music come to perfection in the highest classicity'. Musicologists have been somewhat more successful in imposing their periodization, first, of Classicism in the later eighteenth century, followed by Romanticism in the second or third decade of the nineteenth century. Blume shows how the best of the so-called Viennese Classical group are also prized for their Romantic elements: 'Not until Romanticism's charms have worked their way into the clear severity of the Classic style's forms do we get that which in Haydn and Mozart perforce won over their contemporaries as well as their followers, a style breathed upon by the Romantic spirit and that we, looking back from the twentieth century, call the "High Classic" style'. For Blume, the Romantics never developed their own fully independent style, but they 'simply continued to further shape this High Classic style'.[56] I think the term 'High Romantic' style better fits so-called 'Early German Romanticism' around the opening decades of the eighteenth century. Much of the works of the Second Generation of English Romanticism fit, as do most of the lyrics of Victor Hugo. The dialectical relationship between Neoclassic waters

---

56 Friedrich Blume, *Classic and Romantic Music: A Comprehensive Survey*, trans. by M.D. Herter Norton (New York: Norton, 1970) 124–25. Fritz Novotny presents fine arts of essentially the same time frame as (Neo)Classical styles or schools existing simultaneously with Romantic ones: *Painting and Sculpture in Europe 1780–1880*, trans. R.H. Boothroyd, 2nd ed. (Harmondsworth: Penguin, 1971).

and Romantic ones in works like *Werther* is one of the emergent properties of my analysis.

Lansing's complexity theory and praxis with the Bali rice farmers will help reveal yet another interesting aspect of Eurocentric culture between 1760 and 1870. This has to do with the tendency to see these works in the context of dialectically opposed periods—Neo(Classicism) and Romanticism—rather than seeing where the two or more styles work in dialectical tension, within a single period. While period terms have a certain merit showing a particular consensus of interest at a particular time, they have their reductive dangers; this is especially true when one is dealing with period terms that can also be employed as style terms—such as (Neo)Classical and Romantic. (I am aware that there are those who deny that Romantic elements constitute a style; I think it is because these scholars were too close in time to it to see it as such. I think that the view from the twenty-first century will make Romanticism more resemble Baroque or Rococo in this regard.)

Crucial here, I think is the important attack on dichtomizations in Michel Foucault's *The Order of Things [Words and Things]: An Archaeology of Human Sciences / Les Mots and les choses: Une Archéologie des sciences humaines,* where he shows that eighteenth-century biology should not be divided in any strict sense between the believers in rigid classification by stasis, like Linnaeus, and—on the other hand—in the looser identification by dynamism, like Buffon:

> One can see to what degree it is superficial to set in opposition—as if they were two alien and hostile opinions—a 'fixism' that is content to classify the beings of nature in a permanent picture and a sort of 'evolutionism' which would believe in a immemorial history of nature and in an underlying urge of beings to carry forward the continuity. The solidity, without any lacunae, of a network of species and of genera and the series of events that have obscured the network, both belong, at the same level, to the epistemological platform that made a body of knowledge like natural history possible in the Classical age. They are not two modes of perceiving nature, radically opposed because deeply rooted in philosophical choices older and more fundamental than any science; they are two simultaneous requirements in the archaeological network that defines the knowledge of nature in the Classical age.[57]

Why has this valuable contribution to a breakdown of binary thinking not had an effect? Partly because—even though it calls into question our ability to reduce the ideas of Linneaus and Buffon to reductive either/or slots in the history of science—he still deals here with a binary opposition. Above all, he is seeking to define 'the Classical age'. Yet, he does see this age as being woven out of two very different types of threads. I will view Romanticism as being woven from more than just the set of

---

57 My translation of the following original: 'On voit combien il est superficiel d'opposer, comme deux opinions différentes et affrontées dans leurs options fondamentales, un 'fixisme' qui se contente de classer les êtres de la nature en un tableau permanent et une sorte d'évolutionnisme' qui croirait à une histoire immémoriale de la nature et à une profond poussée des êtres à travers sa continuité. La solidité sans lacunes d'un réseau des espèces et des genres et la série des événements qui l'ont brouillé font partie, et à un même niveau, du socle épistémologique à partir duquel un savoir comme l'histoire naturelle a été possible à l'âge classique. Ce sont pas deux manières de percevoir la nature radicalement opposées parce qu'engagées dans des choix philosophiques plus vieux et plus fondamentaux que toute science; ce sont deux exigences simultanées dans le réseau archéologique qui définit à l'âge classique le savoir de la nature'. Foucault 163.

threads I term (Neo)Classical and Romantic. This, of course, is only one step beyond binary thinking and only one step toward comprehending the diversity and complexity of Romanticism.

The acknowledgment of the existence of these (Neo)Classical-Romantic strands of Romantic culture, therefore, celebrates the aesthetic complexity of an aspect of the culture and offers a useful label that preserves us from seeking to blanket contradictions with a single style term—indeed, it helps us isolate, analyze, and emphasize these seeming contradictions within the cultural fabric. Above all, it contains the essential terms that represent the dialectics of the age. Such contradictions found between 1760 and 1870 in science—as well as in its rich, useful, but despised (because applied) family relation, technology—are interesting in themselves; however, most fascinating of all is the way science was employed by poets, painters, and musicians to cope with all the other mass of social, aesthetic, and intellectual contradictions. And yet, while I posit (Neo)Classical-Romantic as a more precise term, I still reserve the right to use for shorthand the terms *Romanticism* and *Romantic culture*; after all, Romanticism contains its dialectic opposite. Drawing from the work of recent art historians, Marilyn Butler conceives of 'a redefined Neoclassicism [...] co-existing with Romanticism' in such a way that it 'allows from a dialogue within the arts, for conflict and even contradiction'. [58] I see it as an ever-flexible relationship between the two—where it often becomes a matter of a style choice to reflect a particular world view. While it is true that every period contains its pervasive antinomies in a certain conflict, I can conceive of few periods before the Romantic that celebrated those contrasts and sought so systematically to resolve—but not dissolve—them with irony and paradox.

Complexity theory, moreover, causes me to question the application of the binary (Neo)Classicism and Romanticism to the case of the subtle cultural fabric between 1760 and 1870. After all, Realism was a very important part of the art. With the exceptions of a few later canvases, Constable's oeuvre has little to do with the Romanticism of Turner, let alone a Friedrich (his oily, smooth facture only misleads one into thinking his works realistic). Realism seems to fit *Hay Wain* or *Flatford Mill.* Is there a correlation with writers? If so, Realism would then seem to describe a third set of threads found in Romantic culture, as well as a way beyond binary thinking. (Neo)Classicism-Realism-Romanticism, then, is an even more precise way to delineate the complexity of Romantic culture—with the Romantic elements in predominance. By raising the matter from a binary to a trinary one, we have moved more significantly in the direction pointed out by Foucault. Lansing has further made us aware that we must not stop here but allow more properties to emerge from the complexity method. Since what emerges will create some tension with present theories on Romanticism, some pains needs to be taken in dealing with these tensions.

It cannot be said often enough that there is a certain arbitrary taint even in the most judicial application of terms like 'Romanticism', 'Romantic culture', and 'Romantic period'; however, it is only partly arbitrary; for, there is a certain chronological logic to the term. All three terms, then, have their use. Of the three, however, I

58 Butler 11.

prefer Romantic culture because to me it is much less arbitrary. An *Ism* implies a dogmatic consensus among its adherents—there is a sort of danger in that; the danger is allowable to a certain extent, and Romanticism is more useful than the plural form use advocated by that maker of Gordian knots: Arthur O. Lovejoy.[59] Although wherever possible (as already stated), I will refer to the many school and style terms like 'Storm and Stress', 'Weimar Classicism', '(Neo)Classicism', or 'Lake School'. These style terms, which connote a certain series of elements, offer an important affirmation of the complexity of Romantic culture. And yet 'Romanticism' is a charm against chaos. The umbrella-term 'Romanticism', then, is a luxury I would like to forego, but in rare instances cannot. 'Romantic period' might seem at first the most desirable of the three, as a term somewhat more conformable than 'Romanticism' to the complex overlappings of human cultures; but this term—being temporary determined by a beginning date and an ending date—bears more than a taint of inflexibility. 'Culture' can be made to conform to a spatial as well as a temporal significance, or as I use the term, to both. As I define the term, Romantic culture is one in a whole diachronic progression of European cultures. A culture can chronologically overlap with other cultures, and best of all, it may demand aesthetic and ideological conformity, but it most likely will fail to get it. Culture is a state of mind or climate of opinion, and as such it suggests something shared by a generality, not by an entirety. Culture, I posit, is held together by a sort of collective psychology reflected in its arts.

I will continue to hold that the real social, as well as aesthetic, questions of Romantic culture are summed up in the concept of 'permanence and change'—politically expressed, this is a question: 'Permanence or change or a synthesis of both?' This concept is not a New Critical norm or value, but rather a unified quest by a group of seekers. The questioning and the seeking after permanence and/or/in change unite many in Romantic culture. Each, however, breaks off from the group to hunt after one or more among a whole series of possible value systems. And if one decides for change, say, in politics, then the question is: 'How far, how fast?' Reaction depends on social placement, but especially on the human intangibles of talent, genius, and personal disposition. Actually, what I will seek to show is that, to requote Foucault, permanence and change 'are not two modes of perceiving nature, radically opposed because deeply rooted in philosophical choices older and more fundamental than any science; they are two simultaneous requirements in the archaeological network that defines the knowledge of nature' and, I might add, of the human position within that nature as it was found in Romantic culture. These simultaneous requirements should not only be perceived as analytic tools but also be seen in all their various combinations—which constitute a plurality of adaptive systems, as well as idiomatic transformations for personal use by many writers and artists of the Romantic culture.

---

59 Lovejoy 228–53. I must say that I greatly enjoyed the way that the editor and contributors to a recent guide for students deconstructed the term 'Romanticism' while, at the same time, they continued to use it with a certain careless abandon; I refer to Stuart Curran, ed. *The Cambridge Companion to British Romanticism* (Cambridge: Cambridge UP, 1993). Far more honest (and useful for students) is Marilyn Gaull, *English Romanticism: The Human Context* (New York: Norton, 1988) 171.

No one between 1760 and 1870 wrestled so hard to resolve the tension between permanence and change—indeed, to break down the antinomy altogether with a synthesis—than did Goethe. These dates contain the creative career of Goethe, whose poetry contains various types of water symbols—all used with great complexity and subtlety. Not only does he make significant use of every quarter and its various corollary phenomena, but also he generally uses them in such a way as to suggest the larger context of the water-cycle. In Goethe's series of poems honoring Howard the British meteorologist, the German poet and scientist places the workings of the cycle at the center of his symbolic expression of the main concepts of the poems. In addition, he makes considerable use of water, from descriptive landscape images to full-blown symbols—a distinction that, as will be seen, is a focal point of this study. A significant evolution from the descriptive usage of water to its symbolic employment can be found in the various styles and stages of his career: Storm and Stress / *Sturm und Drang* and the closely related poetry of the early Weimar style prior to his Italian journeys; Weimar classicism; the amalgamation of Greco-Roman elements with those of Persian poetry found in the *West-östlicher Divan*. Running through all these stages is the combination of science and poetry, which is especially prevalent in his later poetry. So central are the works of Goethe to my study that I was sorely tempted to entitle it 'Water Images and Water Symbols in the Works of Goethe and His Contemporaries' or '—of the Age of Goethe'.[60] As it is, Goethe's wide range of poetic usages of water forms a *tertium comparationis* around which many of my analyses are constructed.

Although Romanticism implies a breaking up of old rigid forms, a broadening of vocabulary and techniques within the various art forms, essential facts must be remembered: a Classicist like Mozart and a Neoclassicist like David could contribute elements to the Romantic vocabulary; Beethoven and Goethe created works upon which the labels fail except in a careful combination; Romanticism has never stuck very efficiently to Ingres, Rossini, or Brahms, in spite of their traditionally high place in their respective arts. One must conclude that Romanticism as a style does pervade and even distinguish, but does not define the full range of the culture. From an international as well as an interactive perspective, Romantic culture cannot be said to begin at one date and end at another. But the appearance of important works in the style of Romanticism can be plotted to a mean curve as having greater frequency at certain dates, in particular countries and arts: in England the first two decades of the nineteenth century show a richness in Romantic poetry; there, between the 1810s and the 1840s, Romantic painting flourished; as a movement German Romanticism flourished around the same time, but in literature and music, the stature of essentially (Neo)Classical Goethe and Beethoven qualified any question of Romantic predominance in the strictest sense; while allowing for earlier and later masterpieces, the French see the 1830s and 1840s as the only decades of Romantic prevalence, but never dominance; Spanish and Italian masterpieces come even later.

---

60 A rather old-fahioned but still solid study has been done: Ernst Loeb, *Die Symbolik des Wasserzyklus bei Goethe* (Paderborn: Schöningh, 1967); my study of the water-cycle in Goethe complements Loeb's because each has emphasized, for the most part, different works.

The time-frame includes what will later be shown to be an important phase of innovative use of water images and symbols—the Storm and Stress / *Sturm und Drang* movement of the 1770s and allied groups (like the *Göttinger Hainbund*). In addition to Goethe's works in this style there are the works of Friedrich Leopold von Stolberg. In my subsequent discussion of comparative studies, it will become clear that Stolberg's works are important in the German poetic usage of water. Moreover, several of the poems by Stolberg to be examined show a significant, direct influence on Coleridge, thus indicating a possible channel of influence on other English poets. From a close study of water in Storm and Stress lyrics, the notion has emerged for me that the Storm and Stress are the proper 'early' Romantics of German poetry. In addition to Goethe's poems of Weimar classicism, I refer to two works by Schiller merely to round out an argument and shed light on other poems; he seems to have felt only a rare need to employ water images and symbols. Again, for the same purposes, I have included passages by Hölderlin, a maverick poet who wrote in his own style, which parallels Weimar Classicism only in its strong Hellenic concerns.

In what I find to be 'High' Romantic poets of Germany, like Brentano, the water images and symbols characteristically employed by these poets are water-droplets, fountains/springs, water-sprites, and voyages on ships or boats.[61] From this group, I have chosen portions of Novalis and Eichendorff. The later German Romantics are represented above all by Lenau, whose works are rich in water symbols. Primarily on the basis of its similarities to more important works by Shelley, Hugo, and Blake, I have included a short lyric by Rückert. The early, proto-political period of Heine will be represented by the poetic seascapes in *Die Nordsee*.

Although it is uncharacteristic of his output, a fragment by Burns is included as an interesting example of the use of water images in the topographical tradition. Blake is included for his complex usage of cloud and water-cycle symbols in *The Book of Thel*, in spite of the fact that he has a highly individual symbolic, indeed mythic, 'vocabulary'. The creative activity of the early, loosely knit group of English Romantics, often collectively known as the Lake School, falls conveniently within the time-frame of my study. Wordsworth's two versions of *The Prelude* is important to my study because his poetic usage of water shows a development and variety equal only to that of Goethe. It varies from including water as images in verbal landscapes found in *Descriptive Sketches* to the complex, multifaceted range of uses in *The Duddon River* sonnet series. Coleridge's *The Rime of the Ancient Mariner* is only touched on because its water images/symbols are structured too elaborately around the ocean-voyage motif to offer an exhaustive treatment within a comparative study.[62] The same might be said of Melville's extremely complex *Moby Dick*. Southey might have been included on the basis of his poem 'The Cataract of Lodore'; but, in great part because this lengthy lyric is hardly more than a virtuoso piece of pure onomatopoeia composed for his children, I have decided to omit it;

61 See the following: Mildred Tanner Andrews, 'The Water Symbol in German Romanticism Culminating in Fouqué's *Undine*'. Diss. Washington, 1969; Harry Tucker, 'Water as Symbol and Motif in the Poetry of Clemens Brentano', *Monatshefte*, 45 (1953): 320–23.

62 See Stewart C. Wilcox, 'The Water Imagery of the *Ancient Mariner*', *The Personalist* 25 (1954): 285–92.

while efforts by poets to capture the water sounds and a sense of flux are germane to this study, these efforts are only important when they embody something larger.

From among the later English Romantics, I have chosen works by Byron, Mary Shelley, and Percy Shelley for particular attention, as well as several passages from *Endymion* by Keats for points of comparison. Byron, like Goethe, defies periodization. Although he set the tone and created many of the themes of English and Continental Romantic poetry, he always considered himself an imitator of Neoclassic poets, which he clearly is in *Don Juan*; thus his water images and symbols resemble those of both Neoclassic and Romantic poets. For the most part, I have found *Childe Harold* to be a compendium of Byron's poetic usages of water phenomena and, with two exceptions, I have focused attention on passages from the third and fourth cantos of this work. Shelley's complex symbols in 'The Cloud' and 'Mont Blanc' are central to my argument.

The Lac de Bienne passage, composed by Rousseau in the 1770s, holds an important position for an innovative prose-poetic usage of water phenomena in French literature, which in time and in importance rivals that held by Stolberg's Storm and Stress lyrics in Germany. For the representative of the first wave of French romantic poetry, I have chosen De Staël, as well as Lamartine, partly for his considerable qualities as a poet and partly for his corresponding political career. In Chateaubriand's *Atala* (1801), the Mississippi and the roar of the Niagara are more than what they at first seem: exotic decorations in the novel. The bulk of Chateaubriand's water images and symbols are connected with voyages and shipwrecks, and will not be treated.[63] Two remarkable Latin American lyrics—by Heredia and Pombo, respectively—are included because of the influence they each received from Chateaubriand's Niagara Falls passage. Hugo stands out as one of the leading exponents of French Romanticism in its full-fledged manifestation during the 1830s; yet it is not for this reason alone that I have chosen to represent this movement by his poetry: two lyrics from an 1830 collection, along with works of Goethe, Shelley, and Byron, form the core of my chapters on the sea and sky quarters of the water-cycle. A treatment of the waterfalls in a late poem of Hugo is an intrinsically interesting work, which also bears remarkable resemblance to elements in the works of Goethe, Byron, and Lenau, which form the heart of my comparison of waterfalls.

The time-frame between 1760 and 1870 is not one of the great ages of poetry in Spain; however, there are a few poets of considerable merit and interest for this study. The first of these is Meléndez Valdés. He is, at once, the most considerable poet of Spanish neoclassicism and one of Spain's first Rousseauists.[64] Brook passages from the series of lyrics by Zorilla are taken as representative of Spanish Romanticism. The late dates of the poems of 'the greatest and the purest Romantic

---

63 See Auguste Dupouy, *La Poésie de la mer dans la littérature française* (Paris: Ariane, 1947) 98–102. Rosa Vallese, *Le Thème de la mer dans l'oeuvre de Chateaubriand* (Naples: Albright, 1934).

64 William E. Cornford, *Juan Meléndez Valdés: A Study in the Transition from Neo-Classicism to Romanticism in Spanish Poetry* (New York: Hispanic Institute, 1942). For a critical consensus on Meléndez' status in Spanish literature, see 11–12, for Rousseau's influence on Meléndez, see 226–27; also see J. R. Spell, *Rousseau in the Spanish World before 1933: A Study in Franco-Spanish Literary Relations* (Austin: U of Texas P, 1938) 157–59, 164–65.

of Spain's nineteenth century',[65] Bécquer, are even stronger reasons (like those of Wagner's *Ring*) than politics for extending the date of this study to 1870; a sea poem of his contains strong parallels to the ocean passage in Byron's *Childe Harold* and other works influenced by the latter.

Finally, the time-frame draws into it the lives of the three leading landscape painters—Friedrich, Constable, and, except for the last decade of his life, Turner. In sepia series, *The Cycle of Life,* Friedrich employs the water-cycle along with that of the seasons as correlatives of the human-life cycle. The American painter Thomas Cole more directly employed the four quarters of the water-cycle in his *Voyage of Life*. Beethoven includes parts of the cycle in the program for his Sixth Symphony (*The Pastoral*) and, after a fashion, in the music. Under the inspiration of a water-cycle poem by Victor Hugo, Franz Liszt may well have tried to approximate the cycling effect in his instrumental effects and use of the harmonic cycle of fifths.

In addition, the time-frame encompasses the works of two of the important topographical painters of the late eighteenth and early nineteenth century, who made frequent and, for the purposes of this study, significant use of water-phenomena—J.A. Koch and J.-P. de Loutherbourg. The time-frame will include selected paintings of de Loutherbourg from the late 1760s and relevant passages from Diderot, especially his critique of de Loutherbourg in his *Salons* of those years. This great art criticism is another non-political reason for the 1760 dating of the study: in it Diderot announces the entire program of Romantic land-, sea-, and skyscapes of Turner and other Romantic painters, as well as those of Goethe and other poets (and implicitly those of Liszt and Wagner). The widespread use of this symbol demonstrates a desire on the part of painters as well as poets of integrating the changing phenomena around them and their own life-to-death change into a larger pattern of permanence. Water and its cycle become a symbol that would give expression to their paradox of permanence within change.

My humanities sub-model—which is frankly Eurocentric—employs a schema of the water-cycle that has been outlined by Northrop Frye; however, he is viewing it synchronically. Actually, its configurations are very time-dependent; indeed, the developments of that sub-model within Romantic Culture reach a culminating transformation; the Romantic scientists and other thinkers provided the linking knowledge that completed it. I have also taken considerable pains in this section to demonstrate that the relationship between the hydrological-cycle and Romanticism works both ways: the cycle offers a key, I think, to the periodicity of Romantic culture. All too much of the Romanticism(s) debate has taken place at the levels of generalities or upon the all too specific level of—is an artist Romantic or not? The hydrological-cycle is sufficiently comprehensive enough to provide a scaffolding broad enough to contain many of the recent questions of the temporality of Romanticism. It allows for examining Romantic Culture without losing sight of the (Neo)Classical and Realistic strands that are interwoven with the pervasively dominant Romantic strands. It further allows for our avoiding seeing Romantic culture solely in terms of the binary of literature and politics.

---

65 Edgar Allison Peers, *A History of the Romantic Movement in Spain* (Cambridge: Cambridge UP, 1940) 2: 358.

The Political Sub-Model, from Revolutions and Counterrevolutions

It has been often said that every age is an age of transition—true; but in some ages, transition seems to hold a greater sway: the metaphysical Axis Age in sixth-century BCE Eurasia; Julian and Augustan Rome echoing the beginnings of the Han Empire in China; the twelfth-century renaissance of medieval Europe found its proximate counterpart in the Chinese Sung Dynasty. The European Romantic culture takes its place among these transitional epochs.

Transition and change result from and cause contradictions—so certainly these ages of change were also ages of contradiction. It was, however, between 1760[66] and 1870 that contradictions not only became legion but also appeared in an intellectual and aesthetic climate that was developing self-awareness to such a pitch that these contradictions became the subject of worry or, indeed, of relishing delight. The second half of the seventeenth century, commonly called the Age of Reason, had ushered in the spirit of a *critique,* which in the Enlightenment of the eighteenth century reached full spate. In the decades around 1760, this critique began to attack Reason itself. Reason, or at least rationalism, was roundly beaten down by first Hume, then Rousseau, and finally Kant—each of course using different weapons; religion was battered; monarchy and aristocracy were nearing the end of a long and exhausting battle, while the middle class was waiting in the wings to deliver the *coup de grace* to both combatants. This carnival of transition and contradiction includes the shock of the American Revolution of 1776, the French Revolution of 1788/89, and the Napoleonic World War ending in 1815, as well as a series of other French revolutions. Even the Far East was beginning to suffer from European Imperialism, which employed the technology originally invented in China against it and its neighbors.

Between the 1760s and the 1870s, contradictions and changes pervaded even in that most sacredly absolutist realm: aesthetics. Paradoxically, the Neoclassicists of the Enlightenment had been busy dismantling the concept of 'classicism' itself by the application of cultural relativism and historicism to the study of art. Nothing better sums up the aesthetic spirit of the age than the fact that traditional historians of literature and the other arts parcel up the years between 1760 and 1870 into that period marked out as Neoclassical and that marked out as Romantic—with Rococo and Realism tugging at each edge. As I have discussed earlier, my study will help to develop the theme that (Neo)Classical-Romanticism, Realist-Romanticism, and Romanticism are not separate periods in the old traditional sense of the work, but rather strands contained in the same period: the Romantic period, or better yet the same culture.

Of course, the most obvious aspects of socio-political change dominating the Romantic imagination was the French Revolution—or better to say, the series of

66 In the original form of this study the date here was 1770; however, I have now changed it to 1760, partly out of a desire to include within the dates the crucial selections from Diderot's *Salons,* which will be treated at the outset of Chapter Four; partly too, in keeping with Marilyn Butler's rereading of the 1760s not only in England, but in France and other parts of the Continent as well (11–38).

French Revolutions stretching from 1789, through 1830 and 1848, to at least 1871; further, these revolutions found their roots in the American Revolution of 1776. The middle-class revolutions were part of a large context best isolated and described by Raymond Williams, at least in terms of the two generations of English Romantics:

> these two generations of poets lived through the crucial period in which the rise both of democracy and of industry was effecting qualitative changes in society: changes which by their nature were felt in a personal as well as in a general way. In the year of the French Revolution, Blake was 32, Wordsworth 19, Coleridge 17 and Southey 15.[67]

The French Revolution has been the most treated of all the moments of change—because of its great dramatic impact; also, because it marked in Europe the great birth-pangs of the bourgeois democracy that guides our political and literary histories.

Williams not only reminds us of the English socio-political aftermath of the French Revolution, but also of Napoleon's final defeat at Waterloo. During the 1960s and early 1970s there was a tendency (shared by Abrams and only partly offset by Williams) to emphasis the French Revolution's effect on Romanticism over that of Napoleon, or at the very least, to see the latter only as one part of the working-out of the earlier event. Napoleon was only partly a continuity of this revolution, but also in many ways something very different; if in no other way than that he symbolized the power of an individual, whereas the revolution apparently projected the power of classes. The impact of Napoleon—which certainly was as strong as or perhaps even stronger, in the long-run, than the French Revolution—has recently been the subject of several fine, if binary studies.[68] Binary is quite often appropriate for treating one work or one writer.

After the war with Napoleon, Britain was subject to the inevitable post-war slump, in which the urban poor were the sufferers and suffered their own defeat in 1819, called Peterloo by the triumphant ruling class—their second victory against change; this was a bloody attack upon massed urban workers in Manchester by a militia composed mainly of small landholders: 'In the year of Peterloo', writes Williams, 'Byron was 31, Shelley 27, Keats 24. The dates are sufficient reminders of a period of poetical turmoil and controversy fierce enough to make it very difficult for even the least sensitive to be indifferent'. Peterloo pointed up problems only partially addressed and solved by the Reforms of 1832. Concerning the slow but more surely and more drastic series of changes we term the Industrial Revolution and that underlie Peterloo, Williams is eloquent:

> Of the slower, wider, less observable changes that we call the Industrial Revolution, the landmarks are less obvious; but the lifetime of Blake, 1757 to 1827, is, in general, the decisive period. The changes that we receive as record were experienced, in these years, on the senses: hunger, suffering, conflict, dislocation; hope, energy, vision, dedication. The pattern of change

---

67 Raymond Williams, *Culture & Society—1780–1950* (1958; New York: Torchbook-Harper, 1966) 31.

68 Warren Roberts, *Jane Austen and the French Revolution* (1979; London: Athlone, 1995); Simon Bainbridge, *Napoleon and English Romanticism* (Cambridge: Cambridge UP, 1995); John Claiborne Isbell demonstrates how Germaine de Staël's propaganda against Napoleon contributed to the development of a Romantic ideology, in his *The Birth of European Romanticism: Truth and Propaganda in Staël's 'De l'Allemagne', 1810–1813* (Cambridge: Cambridge UP, 1994).

was not background, as we may now be inclined to study it; it was, rather, the mould in which general experience was cast.[69]

Williams correctly implies that this slow and quiet revolution may be more important than either the French Revolution or the Napoleonic Wars.

M.H. Abrams offers his debt to this writer and his *Culture and Society*. Abrams, of course, goes far beyond the suggestions of Williams to construct a model of Romantic culture—revisioning and reconstructing it as a reaction to the French Revolution; however, he (and this is a serious omission) brackets out, from significant consideration, the Industrial Revolution and its concomitant technological advances. And yet what Abrams does offer is remarkable. He shows us how High Romantics defined themselves in terms of the events following 1788. He offers the cultural impact of the French Revolution in the broadest terms by observing that for the Romantics it 'was an intellectual, moral, and aesthetic as well as political crisis'.[70]

Neither Williams nor Abrams acknowledge their great forerunner in correlating the French Revolution to literature—Germaine Necker de Staël.[71] She writes,

> while contemplating both the ruins and the hopes which the French Revolution has—so to speak—jumbled together, I thought how important it would be to understand the power that this revolution exercised over the Enlightenment and what possible effect one might hope to derive from it, if someday it might be possible to wisely and effectively combine liberty with order, ethics with republican independence.
>
> en contemplant, et les ruines et les espérances que la révolution française a, pour ainsi dire, confondues ensemble, j'ai pensé qu'il importait de connaître quelle était la puissance que cette révolution a exercée sur les lumières, et quels efforts il pourrait en résulter un jour, si l'ordre et la liberté, la morale et l'indépendance républicaine étaient sagement et politiquement combinés (1: 200).

If Williams and Abrams are correct in assigning to the French Revolution a paramount position in the genesis of European Romanticism in its fullest form, Staël's views are crucial because of her important contributions to that very genesis, both in France and the rest of Europe.[72] It might even be said that Staël based Romantic literary theory upon the political and cultural premise that was the French Revolution. One, of course, gains a partial view if one merely considers how the French Revolution might affect political progressives; we also should consider its effect on conservatives.

Constable, Lenau, and later Wordsworth can be taken as representative of conservative Romanticism. There has, of course, been over the last decade a veritable mäelström of New Historical debate, pro and contra, centering on Wordsworth's

---

69 Williams 31.

70 Abrams, *Natural Superaturalism*, qtd. from 429, 335, 428, respectively.

71 At least Williams and Abrams are not guilty of the dismissive and unjust treatment given to Staël by Jerome J. McGann, *The Romantic Ideology: A Critical Investigation* (Chicago: Chicago UP, 1983) 59.

72 See Isbell passim.

suppression of politics in 'Tintern Abbey'.[73] I feel no need to dwell on Wordsworth's politics since it has also been the focus of two of the most thought-provoking essays on Romanticism of the 1980s—those by Jerome J. McGann and Alan Liu.[74] McGann shows a class sensitivity in gauging his methodology to distinguish Wordsworth the orphan or Keats the son of a stable manager, from the mavericks of the ruling class, Shelley and Byron. Later in this section, I will devote a considerable amount of time to a group of Romantics who because of the Marxist determinants of class combined with a personal inclination towards social justice were able to contribute to progressive political development. Through their economic and social positions of privilege, Lamartine, Byron, Hugo, and others were able to effect political change. Their contributions come off rather well even by the most rigorous Marxist standards—if properly placed in their Marxist historical context.

In considering the political milieu between 1760 and 1870, it is hard to determine what is progressive politics in light of the historical dialectics: does it fall in the feudal stage or the capitalistic stage? when can we say in any single country, let alone all the West, when the bourgeois has won out over aristocratic privilege? In the new United States during and just after 1776, feudal privileges were only partly and unevenly trimmed in the various colonial/state legislative bodies, and landed interest remained dominant until the 1860s—and even then only losing that dominance in a few states. Anytime between 1698 and 1832 offer possible dates in England; the Reform Acts of 1832 may be the latest point where we might say that anything like a Marx-defined bourgeoisie based on true capital and not on land wins actual hegemony. 1830, 1848, and 1871 are all likely dates in France; later writings of Marx and Engels suggest the latter date. If we are going to fit our analysis to anything like historical objectivity based on context, then we must make sparing and judicious use of synchronic value judgments cutting across a diachronic progression.

There are, however, some in Romantic culture who can be judged by the measure of their age. Constable, for example, fails to come off well by either the Marxist or the Capitalistic standards for judging his historical stage. He, a devoted Tory, affirmed the feudal glory of old English traditionalism by rendering its rural agrarian landscape. Michael Rosenthal has traced Constable's disillusion over the later and more liberal Corn Laws, which threatened his land- and mill-owning family, and the Repeal of the Text and Corporation Acts, which shook the Anglican supremacy from which family members were benefiting: 'Constable understood attempts at political change as revolutionary and inspired by Satanic agency, and would have perceived the events of the later 1820s as irrevocably destructive'.[75] His family's hold on the

---

73 Kenneth R. Johnston, 'The Politics of "Tintern Abbey"', *The Wordsworth Circle* (1983): 6–14; rpt. Karl Kroeber and Gene W. Ruoff, *Romantic Poetry: Recent Revisionary Criticism* (New Brunswick: Rutgers UP, 1993) 123–138. Marjorie Levinson, 'Insight and Oversight: Reading "Tintern Abbey"', *Wordsworth's Great Period Poems: Four Essays* (Cambridge: Cambridge UP, 1986) 14–57. For others in the debate see Johnston's Addendum—1993 (136). M.H. Abrams, 'On Political Readings of *Lyrical Ballads*', in *Doing Things with Texts*, ed. Michael Fisher (New York: Norton, 1989) 364–91.

74 Jerome J. McGann, 'Wordsworth and the Ideology of Romantic Poems', *The Romantic Ideology* 81–92; Alan Liu, *Wordsworth: A Sense of History* (Stanford: Stanford UP, 1989).

75 Michael Rosenthal, *Constable* (London: Thames, 1987) 85–85, 159, qtd. from 162.

local mills of their part of the Stour was only an early nineteenth-century variation on the medieval domination of an area by the aristocrats in charge of the mill. Superficially, the Constables seemed to dominate the market by simple capitalistic monopoly of being the owners of the only available mills; however, in net effect this was not far from what a historian of water-milling describes as the pre-capitalistic or feudal condition: 'The feudal law of milling soke gave the lords the sole right of building and working mills on their estates. Peasants were theoretically compelled, according to the right of multure, to have their corn ground there'.[76] Constable, therefore, grew up in a world where the lines of feudalism and capitalism were not clearly drawn and depended upon the slow replacement of a legal monopoly with an effective one, which is based on centrality to rural life at the time and on the absolute indispensability of the mill. It seems to me that Romantic culture was in great part the superstructure of a world in a transitional state between Feudalism and Capitalism. How else could one explain Lenau, who looked upon democracy and other signs of early nineteenth-century liberalism with all the aristocratic disdain of a medieval nobleman: he was, in short, the perfect Metternichean poet.[77]

This pattern of change, at least in the works of the Germans that so influenced Coleridge and other English Romantics, runs strong before the French Revolution. Early German Romantics were frustrated—the *Werther* frustration—by an inability to integrate themselves, as members of the upper middle class, gentry, and lower nobility, within a society dominated at the top by the court nobility: in a minor poet like Stolberg this can lead to a unqualified glorification of change; however, in the case of a Goethe the attitude to change is mixed with a fear of it and a desire to tie it to some elements of the old stasis. Weimar Classicism, really an often paradoxical combination of various old and new elements that can best be called Romantic-Neoclassicism, came from the engagement of Goethe in the basic administration of a small principality—the Duchy of Weimar, which was relatively speaking a model of progressive reform. Goethe—while very sympathetic toward aspects of the bourgeois Liberalism of his time—remained evolutionary and staunchly anti-revolutionary on the subject of political changes. Revolution in distant America was acceptable; in Europe, it was a horror. It is probably a coincidence, but striking nevertheless, that Goethe began the lengthy process of retiring from direct political engagement as a privy councilor just before and during the outset of the French Revolution. Liberal views of the later years expressed themselves in two widely divergent extremes: local engagement for change on a very practical level in *Hermann and Dorothea*; in the Utopian visions contained in both the second part of *Faust* and that of *Wilhelm Meister*—set, respectively, on newly diked, drained lands or in the New World.

The high tide of Romanticism came after the reign of terror of the French Revolution and flourished, at least in England and Germany, during the Napoleonic period following. In the broadest sense, Romanticism can be viewed with a certain degree of accuracy as a conservative reaction against the philosophical views of the Enlightenment and against the political radicalism of the French Revolution. Liberal

---

76 Stowers 200.

77 Hugo Schmidt, *Nikolaus Lenau* (New York: Twayne, 1971) 18–19.

artists generally gave themselves over to a cultural isolationism from the social norm. Artists no longer worked in, of, and for society, but apart from it in splendid solitude—the 'ivory tower' being a Romantic creation.

This patently explains a Constable, a Lenau, as well as Wordsworth or Keats, but what then of those Romantics who remained engaged in the cause of anticlericalism, antimonarchism, antifeudalism, in behalf of libertarian values? How, for example, can one explain the following: Byron's fight in the House of Lords against the monarchistic reactionaries of Georgian Tories; his staunch, if idealistic and nationalistic progressivism that found embodiment in his martyrdom in the cause of Greek Independence? [78] Finally, we must explain the revolutionary, even Utopian visions of *Don Juan*:

> Raise but an arm! 'twill brush their web away,
>   And without *that*, their poison and their claws
> Are useless. Mind, good People, what I say—
>   (Or rather Peoples), *go on* without pause!
> The web of these Tarantulas each day
>   Increases, till you shall make common cause:
>   None, save the Spanish Fly and Attic Bee,
>   As yet are strongly stinging to be free (emphasis is Byron's; canto 9: stanza 28 [5: 417]).[79]

The 'tarantulas', as other passages in the poem indicate, are the ruling counterrevolutionaries of the age of Castlereagh, Wellington, Czar Alexander, and Metternich. The last references are to two of the great outbreaks of revolutionary feelings in the years around 1820—the Riegoist Uprising in Spain,[80] as well as the Greek nationalists against the Ottomans. Castlereagh, in particular, comes under special attack in the 'Dedication' to *Don Juan*:

> The vulgarest tool that tyranny could want,
>   With just enough of talent, and no more,
> To lengthen fetters by another fix'd
>   And offer poison long already mix'd (Dedication: 12 [5: 7]).

The particular horror that Byron feels for Castlereagh is that the latter is a 'bungler'—an agent of stasis and an enemy of change.[81]

Louis Compton, in his study on the bisexuality of Byron, offers a powerful vision of a deeper jurispolitical irony linking the destiny of Byron and that of his *bête noir* Castlereagh: they both had same-sex drives at a time when any manifestation of this

---

78 Geoffrey Carnall treats the complexity of Byron's politics in a complicated context: 'Byron as Unacknowledged Legislator', *Byron: Wrath and Rhyme*, ed. Alan Bold (London: Vision, 1983) 125–41. While not complete, this is the best monograph on Byron's politics.

79 Lord Byron, *The Complete Poetical Works*, ed. Jerome J. McGann. 7 vols. (Oxford: Clarendon, 1980–93). Throughout the text of this entire book, all quotations from Byron's works will come from this edition.

80 For an overview and European cultural context of the revolts in Spain and elsewhere around 1820, see Frederick B. Artz, *Reaction and Revolution, 1814–1832* (New York: Harper, 1963) 149–83.

81 Jerome Christensen, *Lord Byron's Strength: Romantic Writing and Commercial Society* (Baltimore: Johns Hopkins UP, 1993) 217.

sexuality, particularly by a male, was punished with burning the victim or beating his brains out with a brick while he was bound in a public stock—this, in the mythically just and free England, too. Around the same time that Byron penned the above lines, Castlereagh committed suicide in an act of personal despair over the probability of public exposure of his sexual interest in other men.[82]

I feel that this deeper, more private political concern must surely have given a passion to Byron, the man of action and of words, in his fight for freedom. Was not his death in the then seemingly lost cause of Greek independence a sort of suicide, if of a courageous sort? He certainly makes very clear the political purpose of his poetry in these powerful lines:

> For I will teach, if possible, the stones
> To rise against Earth's tyrants. Never let it
> Be said that we still truckle unto thrones;—
> But ye—our children's children! think how we
> Showed *what things were* before the world was free (emphasis is Byron's; canto 8: stanza 135 [406]).

This is the task of satire; the irony is that if idealistic aspiration is achieved the work will only be a historical document of the way things were before the change:

> That hour is not for us, but 'tis for you:
> And as, in the great joy of your millennium,
> You hardly will believe such things were true
> As now occur, I thought that I would pen you 'em (136).

It will be in light of these views that we must read the code of his water symbolism—too often taken by critics as part of Byron's privately expressive use of symbols, part of the expression of the Byronic personality; however, an important part of the poetic personae he projects, so in love with the dynamic and mutable, is his sincere attachment to revolutionary change. I find it odd that, so far as of this writing, none (to my knowledge) have attempted a complete analysis of Byron's politics, whereas poets without any practical experience in the political arenas have been admirably so treated by studies: Wordsworth, say, or recently, Shelley.[83]

Even more vital and practical engagements with politics than Byron's considerable ones (particular in Greece) were made by Alphonse de Lamartine in the cause of progressive French politics. In the following pages, I will devote considerable space to Lamartine because I think his career, more than that of the others dealt with in this section, embodies the Romantic situation of negotiating between the chaos caused by too much change and that resulting from an excessive adherence to the permanence of tradition.

Lamartine's careers in the Chamber of Deputies (the elective legislative body of France during the first half of the nineteenth century), in the governments of the 1848 Revolution, and in the Second Republic all constitute the praxis side of his politics. Talleyrand said of Lamartine that he had the theoretical acumen to penetrate to the

---

82 Louis Crompton, *Byron and Greek Love: Homophobia in 19th-Century England* (London: Faber, 1985) 12–62, 301–06.

83 Betty T. Bennett and Stuart Curran, *Shelley: Poet and Legislator of the World* (Baltimore: Johns Hopkins UP, 1996).

heart of his country. Learning from France's fate under an earlier Napoleon, he forewarned France of her imminent return to dictatorship under Louis Bonaparte, later Napoleon III. While his compatriots had come to think of revolutions as middle-class directed, like those of 1789 and 1830, Lamartine was alone in reading an important message from the unsuccessful workers' riots in Lyons and in Paris during 1831 and 1832. This statesman-poet foresaw from these riots the necessity for the political education of the working class, whom he believed would initiate all future revolutions; the events of 1848 and 1871 proved him, in great part, correct.[84]

The workers' riots of Paris and Lyons, which provoked a marked turn to the Left in Lamartine's politics, also inspired a poem that gives an eloquent expression to Lamartine's mature views on permanence versus change in politics: *The Revolutions / Les Révolutions* (510–19).[85] This lyric, which was first published in a review, expresses a genuine contempt for the backward-looking conservatism of the majority of his European contemporaries. He contrasts them metaphorically with the nomadic peoples of Arabia, who physically packed up their belongings and passed on to new horizons. By contrast, the European conservatives are, as Lamartine tells them in the first canto, 'men petrified in your timid pride':

> Hommes pétrifiés dans votre orgueil timide [...] (511).

To these hidebound slaves of tradition, the poet has God give them a command to move forward: '"Advance!" His voice says this to the whole of nature'.

> 'Marche!' Sa voix le dit à la nature entière (512).

Lamartine's preoccupation with the necessity of change—previously applied to nature and to human relationships—is here applied to politics and religion; these human institutions are subject to the mutability that is part of a divine plan: 'all things change, fall, perish, flee, die, decompose':

> et toute chose
> Change, tombe, périt, fuit, meurt, se décompose [...] (513).

The attempts by Metternich and other reactionaries to freeze into stasis the political tide of Europe or by half-hearted Liberals like Louis-Philippe and his supporters went—to Lamartine's mind—against the eternal and universal mutability.

When the political explosion Lamartine predicted did in fact begin around 1848, he demonstrated, as throughout his career, an ability to move with and even ahead of a changing political climate. What through twenty-first century hindsight seems a gradual and sincere move from bourgeois liberalism to a moderately leftist stance seemed to many of his contemporaries a drastic inconsistency to achieve political success. His success, however, was a theoretical one at best because it was Lamartine's

---

84 While my main source for the discussion of the politics of Lamartine is the poet himself, I also used Priscilla Robertson, *Revolutions of 1848: A Social History* (1952; New York: Torchbooks-Harper, 1960); further I compared my findings with those of William Fortescue, *Alphonse de Lamartine: A Political Biography* (London: Croom Helm; New York: St. Martin's, 1983).

85 Alphonse de Lamartine, *Oeuvres poétiques complètes de Lamartine*. Ed. Marius-François Guyard. (Paris: Pléiade-Gallimard, 1963). Throughout this entire book, all quotations from Lamartine's works will come from this edition.

sorrow to have been a statesman in a time of political conservatives and opportunists—the worst and most politically formidable among whom was Napoleon III. Lamartine was placed by the 1848 Revolution in such a strategic position that he could easily—like Louis Napoleon was to do—have set himself up as a dictator, thereby gaining the support of the Right, but he decided instead to share his power as the head of the Provisional Government with the leader of the Left, A.A. Ledru-Rollin; this decision lost him the crucial support of the wealthy. Alphonse de Lamartine's ability to assimilate change not only as a political thinker, but also, before that, as a Christian and a poet, objectifies more than anything else his Romantic *Weltanschauung*. Lamartine did not merely accept nineteenth-century historicism; he lived it and promoted it. Change is the dominant theme of his poetry and is essential to its innovations in style. If there were an artist of this culture who used the themes of his poetry as a complex adaptive mechanism for coping with the socio-political changes of his age, it was Lamartine, who was both an artist and a thinker of his age. William Fortescue, in his recent political biography of Lamartine emphases the poet-statesman's fear of the political violence and chaos of anarchy.[86] Lamartine, I think, felt that to move too fast in forwarding progressive measures would throw France into the political anarchy by a reaction from the Right or by excesses on the Left; while to move too slow would cause revolution of the Left. He sought in his political life to negotiate between these extremes of political development, for he saw that either extreme would throw society into chaos.

It is no surprise, then, that Lamartine's first collection of poems had a profound effect on the evolution of French poetry. *Evolution* is the operative word in not only politics but also poetics. Although the aesthetics of *The Poetical Meditations* when judged by a stylistic comparison with works by English and German Romantics of the same time demonstrates the same gradualism that characterized Lamartine the statesman, by contrast, it seemed to its first French readers a radical departure from the neoclassical sensibility that continued to dominate poetry under the Directorate, the Empire, and the early Restoration—indeed, it seemed so radical a departure that it was refused by the publishers to whom Lamartine first submitted it in 1817. Romantic elements acceptable in the prose of Staël and Chateaubriand were, until 1820, not palatable in verse—a more formalized realm because of its higher place in the Neoclassical hierarchy of genre. As always, Lamartine was attacked by his contemporaries as a radical, when he merely took the first and appropriately cautious step, as will emerge when we consider poems from this 1820 work.

Another poetic innovator, Victor Hugo—like Lamartine—was to make his own contributions to practical politics by serving in the French National Assembly (*Assemblée Nationale*), during the crucial years of the Revolution of 1848 to the take-over by Louis Bonaparte (his own *18–Brumaire*) in 1851. It was, however, earlier during the 1830 Revolution that he first crystallized his personal feelings and political thoughts on the tensions between permanence and change; he did this in book-length contribution to the Romantic lyric, with a powerful prose preface: *Autumn Leaves / Les Feuilles d'automne*. His 'Fantasy of a Passerby; speaking of a King' / 'Rêverie

86 Fortescue 44, 47–48, 50, 70, 91, 100–03, et passim.

d'un passant; à Propos d'un roi' (2: 20–22)[87] captures the sense of the old and threatened stasis by opening with the passing of the carriage of that most reactionary of all monarchs, the Bourbon King of Naples, on his way into the Louvre to visit the soon-to-fall French Bourbon, Charles X. Whether or not the date appended to the poem is the actual composition date, it is 30 June 1830—right on the brink of the July Revolution of that year. It is, however, certain that the poem does give one the feeling of what we would call today 'a breaking news story' and, before the publication of the *Autumn Leaves,* it was printed in *Revue de deux mondes* in June of 1831. The poem represents the reactions of the 'private thoughts of a poet as a citizen of the times'.[88] The open alexandrine has the casual air of reportage: 'A great noise of carriages and horses, the other day, carried the King of Naples to a court gala'.

> Voitures et chevaux à grand bruit, l'autre jour,
> Menaient le roi de Naple au gala de la cour (ll. 1–2).

A strong sense of temporality is established at the outset. In the rest of the poem, Hugo draws the paradox inherent in the confidence with which the Bourbons hold back the clock and fail to remember the Bastille and the earlier fall of their house to the people. Before, however, we examine more of the lyric, let me provide a political and historical context offered by Hugo himself.

The prose preface to *The Autumn Leaves* is intended to be as much a part of the collection as the lyrics themselves. Hugo claims that the genesis-date of the poems was the crucial year 1830. The preface reads as a manifesto of the politics of poetry. It provides a historical context in which to understand the rhetoric and the politics of his water-metaphorical strategies which we will examine later:

> Widely, here and there, over the face of Europe, entire populations are wiped out, deported en masse, or put in chains; Ireland has been entirely transformed into a cemetery, Italy into a penal colony, Siberia has been populated with the Polish nation; everywhere else, in the most peaceable of states, something worm-eaten which is disintegrating, and for the attentive ears, the soft sound in the process of making revolutions—still far down underground where the sappers are at work digging their subterrancan galleries underneath the kingdoms of Europe—all branching out from the great central revolution, the crater of which is Paris.

> Au dehors, çà et là, sur la face de l'Europe, des peuples tout entiers qu'on assassine, qu'on déporte en masse, ou qu'on met aux fers; l'Irlande dont on fait un cimetière, l'Italie dont on fait un bagne, Sibérie qu'on peuple avec la Pologne; partout d'ailleurs, dans les états même les plus paisibles, quelque chose de vermoulu qui se disloque, et pour les oreilles attentives, le bruit sourd que font les révolutions, encore enfouies dans la sape, en poussant sous tous les royaumes de l'Europe leurs galeries souterraines, ramifications de la grande révolution centrale dont le cratère est Paris (2: 5).

---

87 *Oeuvres complètes de Victor Hugo,* eds. Paul Meurice, Gustave Simon, and Cécile Daubray, 45 vols. (Paris: Imprimerie Nationale, 1904–52). Throughout this entire book, all quotations from Hugo's works will come from this edition.

88 J.C. Ireson, *Victor Hugo: A Companion to His Poetry.* 2 vols. Oxford: Clarendon, 1997–forthcoming) 1: 53.

For Hugo, the poet's study is secluded from the public clamor of voices; however, it is manifestly not in the ivory tower of Romantics, of the future Symbolists, and of formalistically minded Modernists.

Hugo sees the poet as above parties, as a freelance type of statesman, orator, rhetorician: 'Because the public forum is replete with Demosthenes, because the rostra are encumbered with Ciceros, because we have too many Mirabeaus is no reason for us to dispense with the services—in some obscure corner—of a poet'.

> Parce que la tribune aux harangues regorge de Démosthènes, parce que les rostres sont encombrés de Cicérons, parce que nous avons trop de Mirabeaux, ce n'est pas une raison pour que nous n'ayons pas, dans quelque coin obscur, un poëte.

Hugo goes on to delineate the type of audience the poet-as-politician reached—one that transcends all partisan concerns, yet is still a political one: 'For poetry not only addresses itself to a subject of a particular monarchy, to senators of a particular oligarchy, to the citizen of a particular republic, to the native of particular nation; poetry addresses itself to humankind—to the full-range of humankind'.

> Car la poésie ne s'adresse pas seulement au sujet de telle monarchie, au sénateur de telle oligarchie, au citoyen de telle république, au natif de telle nation; elle s'adresse à l'homme, à l'homme tout entier (7).

Hugo's poet is certainly not a politician nor is he precisely Shelley's 'legislator', but he bears all the qualities associated with a statesman. Much later, during the 1870s, Victor Hugo was legislator as well as radical statesman when 'he would be seen attending the Senate of the Third Republic purposely dressed like an old carpenter or bricklayer'.[89]

This conception of the poet as rhetorician and prophet will allow us to consider the implications of Hugo's use of a water-metaphor in 'Speaking of a King' and other poems from *Autumn Leaves* that we will later consider. The poem is a clear warning against those monarchists who follow in the extremely reactionary footsteps of the several ruling Bourbon Houses of Europe, especially Charles X of France, who with his Ordinance of 16 May 1830, went not only against the Constitution, but also against the popular will, by dissolving the Chamber of Deputies.

The climax of the poem perfectly sums up the way in which water motifs, so crucial to the other poems around it, can give expression to the political dialectic between stasis and dynamism:

> Each century, at its turn, whether it be an Age of Gold or of Iron, devoured (like a cape against which the sea rises) with its laws, customs, the monuments it builds—vain obstacles which seek in vain to stop the waves, with all that one sees and which one will never see again—disappears under the water which never recoils its flow. The ground forever disappears; the waves forever rise. Sorrow is for him who in the evening remains on the beach and fails to ask the fleeing fisherfolk why from beyond the horizon one hears a great roar! King hurry! re-enter the century where we are; quit the ancient shore!—make way for this human sea, or see if you wish to perish on the past century which the flood must cover!

89 Schenk 197; for a fuller coverage of Hugo's short term in the National Assembly during 1871, see Graham Robb, *Victor Hugo* (New York: Norton, 1997) 458–62.

Chaque siècle, à son tour, qu'il soit d'or ou de fer,
Dévoré comme un cap sur qui monte la mer,
Avec ses lois, ses mœurs, les monuments qu'il fonde,
Vains obstacles qui font à peine écumer l'onde,
Avec tout ce qu'on vit et qu'on ne verra plus,
Disparaît sous ce flot qui n'a pas de reflux.
Le sol toujours s'en va, le flot toujours s'élève.
Malheur à qui le soir s'attarde sur la grève,
Et ne demande pas au pêcheur qui s'enfuit
D'où vient qu'à l'horizon on entend ce grand bruit!
Rois, hâtez-vous! rentrez dans le siècle où nous sommes,
Quittez l'ancien rivage!—A cette mer des hommes
Faites place, ou voyez si vous voulez périr
Sur le siècle passé que son flot doit courir! (ll. 70–83).

The sea metaphor of these lines is more than just a passing figure of speech: it has revealed two different political attitudes to the *status quo,* and yet even these two attitudes fall under the same larger world view.

Hugo, as we shall see, will employ the sea for more than just a rhetorical metaphor for a political situation; but the Romantic feel and love for the sea, which reached a high point in the years between 1760 and 1870, is to be found in the above lines. The surprising emergent property is the mixture of hope with real fear of the change and its permanent place in the human condition, and it is to this point we will return again and again throughout the study.

I realize that certain Marxists will see the coping mechanism as escapist; I am of their opinion completely, at least in the case of *some* Romantics and throughout this study, in handling them I will gratefully rely on the approach of Raymond Williams and the Marxist scholarship since the 1980. Having said this, let me ask—what other political options did the Romantics have? I often get the feeling that Constable and Austen, for example, are being condemned for having died before *The Communist Manifesto* saw the light of day. The works of Marx and Engels have sharpened our diachronic sense; let us not get carried away with synchronic value judgments; it was the pre-dawn of a new era that Marx tells us was not begun until 1871. A capitalist in a feudal setting is a progressive in the march of history. And Romantic culture is a shifting mixture of feudalism and capitalism: it was the scene of a potential victory not to be claimed as decisive by the bourgeoisie. Seen in the light of what Marx and Engels really wrote about history, many important Romantics come off very well if we seek to judge them as progressives. What I have attempted to show is that the political Romantic sub-model contained key individuals who did have other political options than escape. *And* these options were determined by *class*. Byron, Lamartine, and Hugo had options that a Wordsworth, let alone a Keats, just did not have. I have spent a great amount of time discussing individuals who were struggling with their great cultural question: how much aesthetic, social, economic, and political change is enough, or how much is too much? Can we say that the appeal to a metaphor as a coping mechanism or as a way of expressing a comforting world view, amid an age of tensions and contradictions, was a cowardly escape mechanism for such active individuals? This was certainly not the case for Byron, Lamartine, and Hugo—and one might well add Staël and Shelley. And I also do not want to suggest that the water-cycle offered the only model for giving expression to this change; indeed, it is part of a whole rhetoric of cycles that filled the unquiet thoughts and restless dreams of the Romantics.

The Rhetorical Sub-Model: The Water-Cycle and the Other Cyclical Rhetorics of Romanticism

Although it is tangential to my focus, romantic political rhetoric parallels the poets in its appeal to water metaphors, and nowhere is this truer than among the progressives. Raymond Williams observes that the journalist William Cobbett 'saw and approved, in its infancy, the course of the labour movement, and he knew that it would not be beaten by laws: "Better call for a law to prevent those inconvenient things called spring-tides"'.[90] In his 'Sonnet: England in 1819', Shelley makes a similar appeal to water as a rhetorical metaphor to give expression to politics of the Peterloo Year of 1819, which Williams and others have found to be a landmark: he compares England to a muddy source. Whether as an extended metaphor, as that of Hugo and Lamartine, or a passing one as that of Cobbett and Shelley, water imagery was summoned as a polemic tool between 1760 and 1870, and to an unprecedented degree and range: above all else because it embodied change in a changing time. As I shall later show, the conservatives easily found the water-cycle as a symbol for permanence by recurrence, or the moderates, for permanence in change, particularly in the water-cycle—in short, embodying their contradictions within a paradox. The true radical could find a correspondence to his world view in the flux of a particular water phenomenon itself, isolated from the rest of the cycle. And the water-cycle takes energy from being only one component in a larger Romantic rhetoric of cycles—cycles of various types. In and by themselves, however, the signifier, the water-cycle, and the signified, world view, both constitute a subtle and flexible rhetoric. Indeed, the permanence and change embodied in the water-cycle and related symbols is a multiplicity in Romantic culture; the 'and' may signify two discrete, binary oppositions; however, it can just as well—and much more often—signify dialectically mixing, conjoining, and overlapping attributes—for example, permanence in change, an eternity of mutability, cyclic change, or the spiral as a unity.

It is not, however, my intention to focus entirely upon directly rhetorical and political metaphors in oratory nor even in the works of Lamartine and Hugo, where we have seen they abound. Such a rhetorical use of the image has its roots in Greco-Roman rhetoric, as well as poetry. Such isolated uses of water symbols, like the isolated occurrence of all such images, have their place in my study, but not as a point of emphasis. Such metaphors are specific and limited. And nothing I think better isolates a Romantic metaphor from the images and symbols of the early eighteenth century than a holistic and universal inclusiveness, interconnectedness. In Wordsworth's appendix to the preface to the second edition of *Lyrical Ballads,* he marks the continental divide between his poetic approach to natural phenomena and that of the very influential *Ossian,* which was fabricated (at least in great part) by James Macpherson (1736–96). Although it falls within the 1760s, having been written in the opening years of that decade, I still hesitate to term it Early Romantic; it is too transitional,

90 Williams 18.

too tentative in a way that the Ossianic sections of *Werther* are not; therefore, I will opt for John Canaday's term 'protoromantic'[91]—a term which is much to be preferred over 'preromantic', with its sense of *pre*-dating Romanticism and of not really being romantic at all. He begins by quoting from one of the prose-poetic sections from this collection's many 'epics', replete like much of *Ossian*, with water phenomena:

> Open this far-famed Book!—I have done so at random, and the beginning of the 'Epic Poem Temora', in eight Books presents itself. 'The blue waves of Ullin roll in light. The green hills are covered with day. Trees shake their dusky heads in the breeze. Gray torrents pour their noisy streams. Two green hills with aged oaks surround a narrow plain. The blue course of a stream is there. On its banks stood Ciarbar of Atha [...]' (*Prose Works* 3: 62–84).[92]

By the time of Wordsworth, Romanticism had evolved beyond the approaches of Macpherson. When *Ossian* sprung upon the literary scene, especially that across the Channel, it had an impact that neither Wordsworth nor we can re-feel. It has its own aptness and closeness to nature. 'Gray torrents', for example, is a happy recreation of streams tinted by the volcanic loam of Scotland, foam, and winter light; however, for Wordsworth, it lacked something quintessentially Romantic:

> I knew that the imagery was spurious. In nature everything is distinct, yet nothing defined into absolute, independent singleness. [E]verything [...] is in this manner defined, insulated, dislocated, deadened—yet nothing distinct (3: 77).

Such non-integrated and passing water images as those of Macpherson are not the fully developed water symbols demanded by Wordsworth and later Romantics to carry their philosophical thought: Macpherson's images lack the sense of interconnectedness demanded by their highly expressive rhetoric.

The task I have set for myself is further complicated: not just the verbal rhetoric of prose and poetry, the visual rhetoric of painting; but also, most forbidding of all, musical rhetoric. I have explored the theory and practice of musical rhetoric at some length, elsewhere.[93] Furthermore, I will set out the specific tools of analysis by which I will treat each of these rhetorics. And first and foremost it is important to consider that science has its own rhetoric also. I have already alluded to the fact that it has its own metaphors or models of thought with which to frame, or at least explain, its theories. Although more intricate than that of literature, the hydrological-cycle of science is a highly simplified model abstracted from a chaos of short-circuiting cycles. This model is the ground of all the rhetoric examined in this analysis.

For the sake of precision in handling Romantic water-cycle rhetoric throughout this study, the term *water image* is reserved for that which forms part of a descriptive

---

91 John Canaday, *The Lives of the Painters* (New York: Norton, 1969) 786.

92 *The Prose Works of William Wordsworth*. Eds. W.J.B. Owen and Jane Worthington Symyser. 3 vols. (Oxford: Clarendon, 1974). This will be cited as *Prose Works* throughout the text of this entire book.

93 Rodney Farnsworth, 'How the Other Half Sounds: An Historical Survey of Musical Rhetoric During the Baroque and After', *Rhetoric Society Quarterly* 20 (1990): 207–24; '"Hither, This Way": A Rhetorical-Musical Analysis of a Scene from Purcell's *King Arthur*', *The Musical Quarterly* 74 (1990): 83–97.

passage, and *water symbol* for that which is used expressively to represent something else. A further distinction is made between *closed* or *allegorical symbols* (*simile* also) and *open symbols*, often found in the form of a *metaphor*—but not all metaphors, however, are open. The former type allows many interpretations, the latter only one. It may be that my distinction of an open symbol is one of praxis and theory, and it has nothing to do with auctorial intentions. Now, I am not one for disregarding auctorial intentions, but it has been noted again and again over the last decades that where author's intentions are ambiguous, critics' intentions fill the void. While I refuse to reject all interpretations, I will place emphasis upon the rich and appealing shadows of symbolic doubt.

The reason the closed symbol has only one meaning is that it forms part of a consistent pattern; it is usually an element within a narrative. The most common closed, allegorical symbol treated in this study is the river as a symbol of life: the river's flowing from its spring to the sea represents human life in its course from birth to death. The life's passing and the river's flowing constitute an allegorical narrative, and the symbols which form the elements of the narrative—spring, waterfall, estuary, sea—have no meaning beyond their allegorical context.

While I am more interested in the open symbols in a metaphoric form, allegory is an important method among Romantic painters and poets, and one not to be ignored; indeed, it becomes very interesting when the particular dynamics of the cycle or natural phenomenon being used as a closed symbol has qualities which deconstruct the one-for-one allegorical significance. Similarly, I find of greatest interest those open symbols where the most apt of several interpretations is undercut also by the nature of the phenomenon used as signifier; where a metaphor representing order contains elements suggesting chaos; where another is used for permanence but really suggests change. In some cases, these inconsistencies may be conscious or unconscious on the authors'/painters' part; where these are part of a larger pattern, I suspect intentionality—indeed, in the case of several important Romantics, I suspect a real appreciation for the resonance set up by the inner tension between the qualities of the signifier and those of the signified.

Coleridge made the perfect distinction between allegory and symbols. Allegory, for Coleridge, has a lower valuation:

> Now an Allegory is but a translation of abstract notions into a picture-language which is itself nothing but an abstraction from objects of the senses; the principal being more worthless even than its phantom proxy, both alike unsubstantial, and the former shapeless to boot.

Symbol has a larger complexity, of many meanings:

> On the other hand a Symbol [...] is characterized by a translucence of the Special in the Individual or of the General in the Especial or of the Universal in the General. Above all by the translucence of the Eternal through and in the Temporal. It always partakes of the Reality, which it renders intelligible; and while it enunciates the whole, abides itself as a living part in that Unity, of which it is the representative (30).

The meaning of the water-cycle in allegory is set; obviously, the cycle as a symbol could take on all sorts of meaning.

In painting, the question of symbolic meaning is even more complicated than in poetry, for a poet can state a direct relationship via a simile (example: my love is like a white waterfall) or indirectly by juxtaposing the metaphor with the concept represented (example: He stood by the foaming waterfalls, thinking of his unrequited passion). While it is true that both rely on semiotic equivalencies—one verbal, the other visual—representing the thing itself, they have their own peculiar relationship to that represented thing. While the sign of the painting and drawing lie visually closer to what is represented than the verbal, poetry has the potential to aurally approximate natural sounds to capture the sounds of water in motion. While music also has this aural capability, poetry has over music the same simile or metaphor advantage it has over painting. Even when the composer has added a descriptive title of 'By the Waterfall' or has a full program describing the natural sounds to be imitated in the course of the musical 'narrative', there may not be any way of knowing with absolute certainty that the sound an instrument makes is meant by the composer to be a sign for the rush of water, say. Thus, the relation between symbol and meaning in music or painting is usually perforce open and rarely is the relation closed and one-to-one. Symbols used in a series of paintings which form a narrative in their recurrence offer a context on which to base allegorical interpretations. Moreover, there were several important approaches to landscape painting, which were prevalent during Romantic culture. They can be arranged conveniently along a gamut from the mimetic to the highly symbolic: (1) topographical; (2) factual; (3) ideal; (4) symbolic; (5) allegorical.

(1) The *topographical* approach to landscape painting seeks to reproduce a particular locus with what might best be termed 'photographic' accuracy. Indeed, the use of the word 'photographic' is to the point because these topographical paintings were often the eighteenth- and early nineteenth-century equivalent of the tourist snapshots and postcards of today. For this topographical landscape, the French painter J.-P. de Loutherbourg's *Falls at Schaffhausen* offers a paradigm.

(2) The 'landscape of fact'[94] is one in which the artist has made an attempt at accurately rendering natural phenomena which can be observed at many different, widely distant loci. The observations thus gathered by the artist are 'compiled' into a composite landscape. Constable's *Landscape: Noon* (*The Hay-Wain*) exemplifies this kind of landscape painting. This accurate rendering can be looked at semiotically as a set of signs consisting of lines and brushstrokes that constitute the basic *facture*—the application of the paint or other medium to the canvas. In the case of *Hay-Wain*, the essential characteristic is the tiny stroke and flecks of white or whitish-yellow paint added to suggest movement under the intermittent, but bright sunlight of midday.

94 Concerning 'landscape of fact', see Kenneth Clark, *Landscape into Art*, rev. ed. (New York: Harper, 1976) 33–71. While this term, like all terms employed by Clark, is used in a broader historical context ranging from Pompeii to Matisse, his concepts have been useful in formulating my own. Other works which I found useful in writing this section on landscape are the following: Barzun 218–19; Elizabeth Wheeler Manwaring, *Italian Landscape in Eighteenth Century England* (New York: Oxford UP, 1925) passim; Morse Peckham, *Beyond the Tragic Vision: Quest for Identity in the Nineteenth Century* (New York: Braziller, 1962) 129–45; Ulrich Weisstein, 'Romanticism: Transcendentalist Games or wechselseitige Erhellung der Künste?' *Colloquia Germanica* 1 (1968): 47–69.

(3) Painters, as well as poets, construct the *ideal* landscape as a composite vision combining factual material based on their own observation of nature with aesthetic influences based on their knowledge of earlier landscape artists such as Claude Lorrain and Salvator Rosa. In other words, ideal landscape is a blending of features derived from actual observations with those influences gained through that body of conventions constituting the history of art. The *Bay of Baiae* with its strange juxtaposition of waterfalls with a placid sea offers a paradigm for this type of landscape. Turner, in his large-scale historical and mythological canvases, combines what he has learned in his color sketches made from nature with what can only be described as landscapes which exist nowhere on earth, but are a strange melding of aspects of actual English and Italian landscapes with ideal 'Italian' landscapes of seventeenth- and eighteenth-century painting. Water is almost as crucial to the ideal landscape as the human figure. According to John Canaday, the American painter Thomas Cole and his patron Robert Gilmor agreed that in order for a landscape to be ideal it needed 'a pool, a river, or water of some kind'.[95]

(4) The term *symbolic,* as used with regard to poetry, will be reserved for that representation which gives expression to a generalized world view for the purpose of this study, the notion of 'permanence in change' and the corollary cyclic articulations. While getting at this symbolic meaning is difficult enough in poetry, it is even more so in painting, which is a non-verbal art. In order to avoid misinterpreting a symbol or, worse, attributing symbolic meaning where there is none, I have employed the writings of painters or those of members of their circles whenever possible in my analysis. Moreover, before making any conclusions, I will cite supporting verbal evidence for my interpretation of these symbols.

(5) Finally, there are those landscapes that form part of a series and have a discernible *allegorical* pattern. Friedrich's paintings evidence this type of landscape. The one-for-one relationship between the meaning being symbolized and the object symbolizing it can be gathered indirectly from Friedrich's own written interpretations of *The Teschen Altar* and indirectly from his close friend G.H. von Schubert's writings on an early version of *The Cycle of Life's Ages / Lebensalter-Zyklus*. Such interpretations can then be applied to various Friedrich paintings, whether these form part of a series or not. In addition, within the paintings there are relationships between the symbols that constitute the entire composition. For example, take *The Cycle of Life's Ages*: first, there is the juxtaposition of symbols (such as children next to a rivulet, youths next to a small stream, adults next to a navigable river, old men next to the sea); second, times of day and seasons are related to water and life symbols, for example—spring buds, early morning light, morning, children, and rivulet. Cole's *The Voyage of Life* series (1840) works much the same way. It remains to compare this American work with its German counterpart and to demonstrate that the water-cycle in both offers a quintessentially romantic symbol. This linking—used as it is to convey human life allegorically—is an instance of the expressive connection of parts of the water-cycle with seasons and times of the day discussed by Frye.

95 Canaday 1036–37.

The water-cycle, as used in the arts, is even more complicated by being related to a whole rhetoric of cycles found in the Romantic disciplines. This involves cycles that seem, at first glance, to have nothing to do with the water-cycle.

Music has its own rhetoric of cycles: on the circle or cycle of keys arranged in fifths—conceptualized on the basis of Rameau's theoretical writings, developed by the Classical composers Haydn, Mozart, and Beethoven, and daringly pushed to extremes by Liszt and other Romantics: it is the basic harmonic paradigm for the musical language of Classical-Romanticism.[96] This cycle is so important to our understanding of the works of Liszt and others in the chapters that follow that I should introduce it here (capitals denote major keys and small case, their relative minors):

F♯ (d♯) / G♭ (e♭)
B (g♯)/D♭ (b♭) C♯ (a♯)/D♭ (b♭)
E (c♯) A♭ (f)
A (f♯) E♭ (c)
D (b) B♭ (g)
G (e) F(d)
C (a)

Such symbolism is, of course, tricky to read in a musical work: there is always the danger of reading into it something that is not there. It is not just fortuitous that Liszt includes a cyclical symbolism in a work that takes a poem by Victor Hugo as its programmatic basis. In music, as with painting, verbal texts offer useful interpretive ballasts.

Romantic philosophers took to cycle and related metaphors with an eagerness not seen since the ancient world. To these philosophers, the cycle symbols are what have earlier been described as the ground of their world view and not mere figures of speech; similarly, what I am seeking is that very fine line beyond which the passing rhetorical metaphor—however expressive it may be—can be distinguished from an extended metaphor that is expressive of the world view of an age: a way of embodying the hopes and fears of a culture. Let me offer an example from Herder's *Yet Another Philosophy of History Concerning the Development of the Human Race / Auch eine Philosophie der Geschichte zur Bildung der Menschheit* (1774). He poses the following question on cultural progress: 'Does this mean that there can be no manifest progress and development in a higher sense than what was believed?'

> Sollte es nicht offenbaren Fortgang und Entwicklung aber in einem höhern Sinne geben, als mans gewähnet hat?

He answers the question with two metaphors of cosmic implications: the first is the organic metaphor frequently treated by twentieth-century scholars; the second is the

---

96 Cf. Charles Rosen, *The Classical Style: Haydn, Mozart, Beethoven* (New York: Norton, 1972) 23–25. You might notice that the way I picture the cycle of fifths is different from Rosen or most other musicologists; while I keep the sequence of the keys intact and do not materially change things, I place C Major on the bottom and the flatted keys on the left; this is just the way, for some reason, I conceptualize the cycle in my mind—in spite of great resistance from my past harmony and composition teachers.

water-cycle: 'Can you see this river flow away: how it sprang from a tiny source, swells, stops there, starts here, ever meanders and sinks farther and deeper—but always remains water! a river! a drop! Nothing but a drop, until it plunges into the sea—might it not be the same as with the human race?'

> Siehest du diesen Strom fortschwimmen: wie er aus einer kleinen Quelle entsprang, wächst, dort abreißt, hier ansetzt, sich immer schlängelt und weiter und tiefer bohret—bleibt aber immer Wasser! Strom! Tropfe! immer nur Tropfe, bis er ins Meer stürzt—wenns so mit dem menschlichen Geschlechte wäre? (2: 36).[97]

Already, cosmic in proportions, the water-cycle is here used to convey the entire course of human progress—seen in the new eighteenth-century way, in the Voltairean form of cultural historicism, clearly evolutionary with optimistic implications.

Coleridge, in *The Statesman's Manual,* combines the cycle with permanence of process to define reason working within the imagination, which he views as

> that reconciling and mediatory power, which incorporating the Reason in Images of the Sense, and organizing (as it were) the flux of the Senses by the permanence and self-cycling energies of the Reason, gives birth to a system of symbols, harmonious in themselves, and consubstantial with the truths, of which they are the *conductors* (Coleridge's emphasis).[98]

If, then, the ideal is for symbols to be consubstantial with the truths they convey, then can one imagine a more apt symbol, or 'system of symbols', for permanence in change than the hydrological-cycle? This cycle, as employed by the Romantics, can be no better described than as a 'reconciling and mediatory power' to deal with a world in 'flux'. Notice the language of the Eucharist that Coleridge uses. If one were so inclined, one might be able to view the Romantic water-cycle model 'consubstantial' with and 'conductor' of 'truths' as states of grace; however, I prefer 'coping mechanism'.

The cycle can take on a sense of progression by becoming a spiral. Although it is not central to this study, it is important to consider the rhetoric of cycles implicit in the philosophy of the forerunner of Romanticism, Gianbattista Vico, and the great Romantic philosopher Hegel. Vico has been taken by Romantics, as well as later commentators, as a key source of their cyclic grounds of thinking. In *The New Science / La Scienza Nuova* (Third Edition, 1744), he sets forth, in book four, the three stages that politics, customs, and humanities pass through in 'The Course the Nations Run' / 'Del corso che fanno le nazioni', as he titles the book. Book five follows the three stages through the later cycle of the same nations: 'The Recourse of Human Institutions that the Nations Take When They Rise Again' / 'Del ricorso delle cose umane nel risurgere che fanno le nazioni'.[99] It would seem that in the

---

97 Johann Gottfried Herder, *Werke in Zwei Bänden,* ed. Karl-Gustav Gerold, 2 vols. (Munich: Hanser, 1972) . Citations are all in the text.

98 Samuel Taylor Coleridge, *Lay Sermons,* ed. R.J. White (London: Routledge; Princeton: Princeton UP). Citations are all in the text.

99 Giambattista Vico, *Opere,* ed. Pasquale Soccio (Milan: Garzanti, 1983) 519, 569. It is true that Vico is not precise on whether he is describing a repetitive cycle or a spiral inclusive of changes, but the standard interpretation is that Vico's analyses indicate a spiral; for an example of this standard reading, see the classic study—J. B. Bury, *The Idea of Progress: An Inquiry into its Growth and Origin* (1935; New York: Dover, 1955).

tradition of Vico, Saint-Simon posited a cyclical history revolving from periods of equilibrium (the Middle Ages) and periods of breakdown and reformation (Saint-Simon's own time).[100] M.H. Abrams, in his *Natural Supernaturalism,* found just such a cycle-as-spiral in the writings of Schiller, where 'the concept of a fall, or historical decline, has been fused with the idea of progress so as to give history a spiral form; for the unity with himself to which man will return is not only as good as the unity he has forfeited; it is far better'.[101]

The cycle rhetoric of the High Romantic philosophers and its influence upon German and English literature has also been carefully traced by Abrams. Fichte and Schelling, as well as Novalis, found a source of optimism in this cyclic return: 'When the history of mankind is envisioned as this circular design manifested in time, the Christian view of the lost and future paradise assumes the form of unity, unity lost, and unity-to-be-regained'. The problem, here, is the vision of a lack of progress in this return; indeed, Abrams had already pointed out the dangers involved in the intricate combinations of linear views of history, which have a beginning, middle, and end, with cyclical ones.[102]

This paradoxical blend of the line and the cycle is crucial in many Romantic artifacts using the water-cycle. Once again, I must look outside humanities for a useable model in science, one closely related to hydrology and the allied meteorology, even sharing with them the scientist Hutton: geology. Stephen Jay Gould's examination of this era of geology offers one of the best critiques of dichotomies and of cyclic rhetoric, not only as it appears in the history of scientific rhetoric, but as it occurs in any history. On dichotomies and the binary reductionism it implies, Gould writes:

> We often try to cram our complex world into the confines of what human reason can grasp, by collapsing the hyperspace of true conceptual complexity into a single line, and then labeling the ends of the line with names construed as polar opposites—so that all richness reduces to a single dimension and contrast of supposed opposites. All these dichotomies are false [...] because they can capture but a fraction of actual diversity.

He continues: 'if we must dichotomize, time's arrow and time's cycle is the most fruitful contrast for understanding the major issues underlying the greatest transformation that geology has or could contribute to human thought—the discovery of deep time'. This 'deep time' is one that opts for complexity, for tension among the opposites: 'Something deep in our tradition requires, for intelligibility itself, both the arrow of historical uniqueness and the cycle of timeless immanence'.[103]

In dealing with the two great geologists writing amid the Romantic milieu, Hutton and Lyell, Gould writes: 'We need to recapture the dichotomy of their day—time's arrow versus time's cycle—to grasp their deeper divergence in different attitudes toward the meaning of time's cycle'. Hutton was entirely bound up in the eternal recurrence of the cycle so that he had no sense of historical progression. Charles

---

100 Bury 284; for Saint-Simon's place in the Vico tradition, see Edmund Wilson, *To the Finland Station: A Study in the Writing and Acting of History* (1940; New York: Farrar, 1972) 97.

101 Abrams, *Natural Supernaturalism* 207.

102 Abrams, *Natural Supernaturalism* 179, 181, 34–37, respectively.

103 Stephen Jay Gould, *Time's Arrow/Time's Cycle: Myth and Metaphor* (Cambridge: Harvard, 1987) 191, 200, respectively.

Lyell, contemporary with High Romantic philosophers and artists, saw the cycle less rigidly and more creatively; he felt less bound to the eternal return; therefore, more able to combine the sense of permanence-in-change with the sense of history.[104] Gould finds even more interesting tensions in an earlier work, Thomas Burnet's *Sacred Theory,* and I will draw from Gould's analysis of it some conclusions about the Judeo-Christian usages of the cycle in Romantic culture.

The spiral offers the best way to combine the change of history with its larger share of continuity and recurrence in patterns, and if the spiral is upwards, progress is implied—indeed, the Romantics used the spiral for its combination of recurrence in the gross with progress in crucial elements. In Hegel, as well as Goethe and Coleridge, Abrams finds the vision of the cycle as an upward spiral, which 'rotates along a third, a vertical dimension, to close where it had begun, but on a higher plane of value.' He says, 'It thus fuses the idea of the circular return with the idea of linear progress, to describe a distinctive figure of Romantic thought and imagination—the ascending circle, or spiral'. As for Coleridge, Abrams writes that mental and physical life 'moves along a circular and ascending course'.[105]

The Romantic rhetoric of cycles brings us back to the sub-model treated in the previous section: the political sub-model. One of the basic truths of complexity theory is that divisions interconnect.

Concerning the cycle rhetoric of Hegel, Abrams sums it up as not just moving spatially upwards but also outwards taking in more of the universe; it spirals 'upward and widening outward until they constitute one great *Kreis von Kreisen* that comprehends everything' by returning to the beginning.[106] While the cycle implies change, any sort of personal or social initiative of change is discouraged by a belief that it may well be futile. Sir Isaiah Berlin summed up this anti-change aspect of Hegel's system: 'When a man in the name of some subjective idea attempts to destroy a tradition instead of modifying it, he opposes the laws of history, attempts the impossible, and thereby reveals his own irrationality'.[107] This may have been the saving grace of Hegel's system in the eyes of the Prussian authorities and other forces of reaction in Metternichian Europe—perhaps part of the key to the flourishing of this system in this milieu.

This reactionary-political tendency to look for some hope of stasis within the ceaseless change of the cycle has been detected in Keats' 'To Autumn' by Jerome McGann. The cycle in question here is that of the seasons—a continuity in spite of change, where every winter is supposed to be like every other winter, and so on through the cycle to the fall. McGann writes:

> 'To Autumn' asks us to believe—to willingly suspend our disbelief—that all autumns are the same. We must imagine them to be universally, the 'season of mists and mellow fruitfulness'. But Keats asks us to believe this because he knows, as we know, that it is not true. Such an autumn of perfect harvests and luxurious agricultural abundance is an autumn in the mind.

---

104 Gould, *Time's Arrow* 151–52.

105 Abrams, *Natural Supernaturalism* 184, 271, 266, respectively.

106 Abrams, *Natural Supernaturalism* 227.

107 Berlin, *Karl Marx* 58.

I might also add that to separate autumn from the winter is to extract it altogether from the cycle of the seasons, which negates it as a season. Of course, the eternal return of a pristine and absolutely undifferentiated season, as McGann points out, never happens. Keats' poem is an exercise in unreality as removed from the cycle of seasons as surely as it is removed from human history:

> This is [...] very negation of the negation itself, wherein all events are far removed from the Terror, King Ludd, Peterloo, the Six Acts, and the recurrent financial crises of the Regency and where humanity escapes the inconsequence of George IV, the absurd Prince Regent, the contemptible Wellington.[108]

The cycle can allow for reconciliation to the actuality of the world and its bias for change, but it can never allow escapism. If cycle rhetoric is nothing else, it is inclusive not exclusive.

The concept of the cycle is never simple: it is always gathering layers of meaning connected to reconfigurations of the cycle form itself. Like all aspects of rhetoric, the cycle can historically evolve with human culture to express its changing value system. The cycle of one discipline (the natural science of astronomy, say) can become the source for another discipline (the social science of politics). Take, then, the word 'revolution', which only came into its own during this time: this word finds its source in cycle rhetoric:

> Only with the French Revolution did the concept of 'revolution' take on its modern meaning. This etymological fact signals a larger truth about the grouping of events in world history. For many centuries prior to the 1500s, 'revolution' referred to astronomical cycles. During the period of the English upheavals from 1640 to 1688, 'revolution' was used to refer to fundamental political changes, yet still retained the sense of cycling back to a previous state of affairs. By the time of the French Revolution, however, 'revolution' came to connote a sudden, fundamental, and innovative departure in a nation's social and political life—and the term has retained this connotation down to the present day.[109]

Such is the power of cycle rhetoric, as a great whole, to give expression to the great changes in human history.

The whole is only one of the many manifestations of the protean cycle that—it should be remembered—need not be taken whole, but also can be viewed as dismantled into its component quarters, which can be further fragmented. Only one of the paradigms—perhaps, the controlling one that potentially linked the rest—in the rich grammar at the command of Romantic water rhetoric, the water-cycle conveys both the change to something new and the continuity, even permanence, of that motive process. In short, the cycle is so powerful that it can signify, implicitly and beyond the level of polemic engagement on specifics: the basic assumptions that continued to drive the Encyclopedists, the American Revolutionaries, their French allies, and the French revolutions of 1789, 1830, 1848, and 1871. The Liberal and Radical

---

108 McGann, 'Keats and the Historical Method,' Kroeber and Ruoff 439–64, qtd. from 455–56, 458. See Jonathan Bate's ecological take on this poem: 'Living with the Weather', *Studies in Romanticism* 35 (1996): 431–48.

109 Theda Skocpal and Mayer Kestnbaum, 'Mars Unshackled: The French Revolution in World Historical Perspective', *The French Revolution and the Birth of Modernity,* ed. Ferenc Fehér (Berkeley: U of California P, 1990) 13.

movements, unlike their forerunners since the Middle Ages, conceived 'Revolution' to be something different and on a higher plane than a mere rebellion against an intolerable present and a restoration of a partly mythic past: the Roman Republic of the American and French Republics; the lost representative government found among the Anglo-Saxons by the Whigs; that found among the Franks by Montesquieu and his 1789 successors. This cosmic conception of an eternal return to a pristine past was to find in the water-cycle its cosmic symbol.

What is true of the French Revolution(s) is perhaps even truer of Napoleon. He surprised people around him by enforcing his power with a series of unexpected changes: diverting the vector of the Revolutionary Army from the Rhine to Italy; then, diverting it even further to Egypt, of all places. Then came the Brumaire and the Emperorship. He was ever creative in inventing grand gestures that usually wrought great changes and always-great surprises. Georges Lefebvre, a key historian of the first French Revolution, wrote that the people around him never had an 'inkling of the invincible romanticism of Napoleonic imagination!'[110] He was constantly bombarding his contemporaries with yet another abrupt change of tack. It is crucial to note, however, that change was framed by the cycle's permanent return—it was a return to the past. His change of direction was variously presented, depending on the audience. For the liberals, he introduced his change as a return to the pristine goals of the Revolution before the corruption of the Jacobins, the Thermadorians, or the Directory. For the Pope and the aristocrats, it was a return to pre-Revolutionary order and values. For the Emperor himself, it was a return to the Consulship, then Emperorship of Ancient Rome—now of course, not the Republic appealed to by the 1790s revolutionaries, but the Empire. Like Augustus, the person who created the basis on which Roman Emperorship was constructed to look like a restoration of the Republic after the Civil Wars,[111] Napoleon was careful to present revolution under the aspect of an eternal return to one of these various past goals: to the original goals of the Revolution, to the Ancient Regime, or to Augustus.[112]

In concluding my discussion on the rhetorical sub-model, I dare affirm that the Romantic climate of opinion was cycle/spiral directed and that the water-cycle contributes to Romantic rhetoric, one model, among others. One of the prime reasons for this foregoing section was a desire to demonstrate with a larger context that I do not claim too much for my water-cycle as a cyclical model embodying an intellectual adaptive mechanism for dealing with change. It only represents one important voice in a veritable fugue of cycle rhetorics within Romantic Culture.

Conclusion

The water-cycle model, by containing so many sub-models, which are in turn parts of a greater scientific, aesthetic, political, or rhetorical whole does offer a breadth of vision and a chance to derive artifacts from an extensive range of fields. For all its

110 Georges Lefebvre, *The Directory*, trans. Robert Baldick (New York: Vintage-Random, 1967) 221.

111 Ronald Syme, *The Roman Revolution* (Oxford: Oxford UP, 1960) 313–28.

112 Felix Markham, *Napoleon* (New York: New American, 1963) 132–33.

significant appeals to the natural and social sciences, my enquiry remains a study centering on the humanities and more specifically on literature. By seeking, however, to contextualize its artifacts, this study marks a departure from the literary studies that have almost entirely focused on the important formal aspects of recurring images and symbols: most of the studies on the water-cycle have been traditional motif-studies covering a single author or motif-histories covering vast stretches of literary history (see the bibliographical essay at the end of this study). These studies have sought to connect an author's work to another of his or to those of other authors of the same country. Only rarely and timidly have these studies been taken across national boundaries. Oddest of all, the connections with contemporaneous hydrological and related sciences, like meteorology, were rarely made. However, thc larger question of context is rarely posed: why this complex of images or symbols at this particular time? why in Europe and its political and/or cultural colonies? Standing on the shoulder of these earlier Positivist collectors of factals and insights, I hope to see a bit further and perhaps to place water-cycle motifs in as full a context as I can. I will seek to explore direct influences among artists (by which term I always mean to includc writers) that have interested traditional comparatists; the seemingly unconnected parallel employment of a metaphor for similar purposes that most interests me; other such affinities and analogies that Hegel would have attributed to his *Zeitgeist*; the effect of science upon artists, and in some cases, the impact of artists within fields of science; the correlations between the social and political contexts of these artists, and in quite a few significant instances their own efforts at politics. Out of this multiplicity clustered around the water-cycle, I will seek to create one possible vision of change: one that seeks the agency of mediation between order and chaos.

CHAPTER ONE: SOURCE

In its place at the beginning of the cycle, at least as it is conceived by the artistic mind, as well as by this study, the source is disadvantaged compared to the placement of the other quarters. As interesting as each quarter is when viewed in isolation from the other quarters, the discussion of them is only enlivened by their interconnection—especially by their relation to the fourth, the linking sky quarter, where contemporaneous discoveries in science come into play. The primary mode of this first chapter is, therefore, presentational not relational; it may seem at points to border on cataloguing, but I trust that, when seen under the aspect of the whole, it will not prove so.

Toward a Romantic Model

Of the two periods immediately preceding the Romantic, the Baroque comes closer than the Neoclassical Enlightenment to offering a source of imagery comparable to that of Romanticism. According to Jean Rousset, employment of water imagery is characteristic of the Baroque, at least in France. The water is usually in a state of turbulence—a quality it shares with that of Romanticism. In the works of seventeenth-century French poets, Rousset detects an 'active presence of water in the landscape [...] and most often of water in motion, relishing in its own motion, as a symbol of flow and of flight, as a center of a metamorphosis, as a creator of swelling, sinuous, or winding forms [...] as [...] a universe which is a theatre of inconstancy, of mobility, of illusory reflections [...]'.[1] Thus, the elemental features of water offer to Rousset an emblem for the Baroque *Weltanschauung,* which entailed an obsession with change in its various manifestations. In another study not only of Baroque poetry but also of landscape architecture, Rousset indicates the ways in which water dominated the scene and how it fits into the artificial world of the garden, at least in France: 'When it was not rivers or other natural water-formations, it was the grottos, fountains, and spurting jets of water—all the artificial formations of water—that gave life and movement to the parks'.[2] Water, then, is important for our understanding of the style and meaning of the key arts of the Baroque period.

On the subject of cloud images and symbols (among other corollary phenomena) in the seventeenth century, Rousset writes the following:

1 My translation of the following: 'présence active de l'eau dans le paysage [...] et le plus souvent de l'eau en mouvement, goûteé dans son mouvement même, comme symbole d'écoulement et de fuite, comme centre de métamorphose, comme créatrice de formes gonflées, sinueuses ou tournoyantes [...] comme [...] d'un univers qui est un spectacle d'inconstance, de mobilité, de reflets illusoires; tel est, semble-t-il, le sens de la 'rêverie' que l'eau provoque en eux'. Jean Rousset, 'L'Eau et le soleil dans le paysage des poètes', *Cahiers de l'Association Internationale des Etudes Françaises* 6 (1954): 49–55, qtd. from 52.

2 'Quand ce n'étaient pas les rivières et les eaux naturelles, c'étaient les grottes, les fontaines et les jets d'eau, les eaux artificielies sous toutes leurs formes, qui donnaient vie et mouvement aux parcs'. Jean Rousset, *La Littérature de l'âge Baroque en France: Circe et le paon* (Paris: Corti, 1953) 146; also see passim 142–57, 161–65.

> These poets are so eager to employ all the images of movement and metamorphosis that one is not surprised to see them carried away with delight towards the most changeable of substances: the cloud, the rainbow, the play of twilight, and of sunlight—not at all interested in a romantic taste for half-shadow and for nocturnal mystery, they on the contrary loved brightness and the light—but they sought to take hold of the shifting mixture of colors, of nuance ranges of color, of fleeting forms in the act of coming undone and of recomposing themselves.[3]

Rousset's analysis is an old-fashioned structuralism analysis that fails to seek socio-political explanations for this form. Are there such explanations?

Theodore Rabb's expansive cultural study treats how Baroque culture met its crisis of changes, which filled the century between the beginning of the Reformation and the height of the Baroque: the pillage of Rome by the largely Protestant army of Emperor Charles V, the French Civil Wars, the Thirty Years War, the Puritan struggles with Charles I. He sees in the works of Rubens and other Baroque artists a rhetoric on behalf of the dynastic monarchs who were bringing stability to Europe:

> There was one final conspicuous feature of the cultural reaction to the upheavals set in motion in the early sixteenth century. After 1600 the efforts to re-establish a sense of stability and confidence entered a stage of ever more intense striving, of epic enterprises. Again the most immediate evidence comes from painting, where the birth of the Baroque epitomized the belief that grandeur and immensity could subdue uncertainty. Gone was the unsettling vision of Mannerism; in its stead arose a direct, powerful, and unashamed affirmation of emotion, often encouraged by Counter-Reformation Catholicism, that was the visual equivalent of religious and political authoritarianism.[4]

Beginning somewhere after 1650, especially in France, there was a feeling that stability was achieved; the grandeur and immensity of the earlier half of the century were combined with a Neoclassical sense of order affirming that stability.

Could one perhaps see in, for example, Louis XIV's great canals, pools, and fountains—paradoxical controls, by architecture and sculpture, of a mutable substance—the perfect emblems of the stasis and power over change that he sought to culminate over France? It is possible; this subject bears consideration by scholars with a vision larger than that of Rousset. And yet whatever the limits to Rousset's critical vision might be, it will still be useful for the analyses in the chapters which follow to keep in mind this Baroque predilection for 'images of movement', such as clouds, because a similar fascination for images and symbols with just this quality is to be discovered among the Romantic poets, painters, and composers to be examined.

---

3 'Ces poètes à la recherche de toutes les images du mouvement et de la métamorphose, on ne s'étonne pas de les voir se porter avec prédilection vers la matière de toutes la plus changeante: le nuage, l'arc-en-ciel, les jeux du crépuscule et du soleil—non point par goût romantique de la pénombre et du mystère nocturne, ils aiment au contraire l'éclat et la lumière—, mais pour s'attacher à capter des mélanges mouvants de couleurs, des dégradés, des formes fuyantes en train de se défaire et de se recomposer'. Rousset, *L'Age Baroque* 129; also see passim 129–33. See *L'Intérieur et l'extérieur: Essais sur la poésie et sur le théâtre au XVIIe siecle* (Paris: Corti, 1968) 183–93, 197–235. Also see Erik Michaelsson, 'L'Eau, centre de métaphores et de métamorphoses dans la littérature française de la première moitié du XVIIe siècle: Le Miroir de l'eau et le déluge'. *Orbis Litterarum* 14 (1959): 121–73.

4 Theodore K. Rabb, *The Stuggle for Stability in Early Modern Europe* (New York: Oxford UP, 1975) 58.

As the eighteenth century moved further from the Age of Louis XIV after his death in 1715, the landscape ideal became more natural, freer. The artificial fountains of Rousset's description were eventually to fall out of favor in landscape gardening and have little if any importance in landscape painting. Such artificially driven fountains occur only in the works of a minority among the poets studied in this chapter: the German Romantics—Novalis and Eichendorff—employ jets of water so unreal and sinister that they seem to demand a supernatural force driving them; the later generation of English Romantics—Keats and Shelley—use fountains that combine the supernatural with a disembodiment to the point of their being hardly more than metaphors for abstractions. The Romantic usage of these unnatural sources is very specialized and self-conscious. In French such an artificial fountain is usually termed *fontaine* with a *jet-d'eau*; in German, *Springbrunnen*, or less often *Springquell*, with a *Wasserstrahl.*[5] Although the evidence I will present in this chapter may not warrant any definitive conclusions, it would seem that these artificial sources held little interest for the poets active between 1760 and 1870 and that lack of interest sets these off from poets before them. Particularly in the 1760s and 1770s, according to Christopher Hussey, such artificial sources fell out of favor with the landscape architects and poets, who branded them as 'unnatural', hence undesirable.[6] Such artificial fountains are entirely man-made and, therefore, not part of the water-cycle; examples of these types of fountains are touched on in this study for the sole purpose of contrasting them to those that have natural origins.

The natural sources can be somewhat arbitrarily subdivided into the following categories:

(1) One type of source is untrammeled by man and forms the source of a stream—indeed, the headwater of a river central to the poem. In English, the preferred word for such a wild and untouched source is *spring*; in French, *source*; in German, *Quelle*; *fuente* in Spanish. Since my emphasis is on works which employ all or a considerable part of the water-cycle, this chapter foregrounds streams which each have their head-spring high in mountainous landscape, spilling precipitously from the source.

(2) Related to these mountainous sources are the run-offs from glaciers and fields of ice. Often included with all three natural sources are small brooks lying close to their source. On the basis of the above categories, the important distinction to be made is that natural fountains release a *brook* with a tranquil flow; whereas, glacier run-offs usually gives rise to torrents. Among the Early Romantics there is a great preference for the mountainous sources. The third type, however, was preferred by the Neoclassicists:

(3) Although it is still a natural source and, therefore, part of the cycle, the water from this type of fountain is collected into a man-made basin or retaining wall for

---

5 For a more extensive discussion of the different words in German for fountains and springs see Ursula Wiegers, 'Der Brunnen in der deutschen Dichtung: Eine motivgeschichtliche Untersuchung', diss., Bonn, 1957, 1–6. For another national study of water motifs see Anneliese Bach, 'Das Motif des Brunnens in der deutschen Lyrik vom 17. Jhdt. bis zur Gegenwart', *Germanisch-Romanische Monatsschrift* 37 (1957): 213–31.

6 Christopher Hussey, *The Picturesque: Studies in a Point of View* (1972); rpt. Hamden, Conn.: Archon, 1967) 139–40, 179.

holding drinking water. In English, the standard word used to denote this type of source is *fountain*; in French, *fontaine*; in German, *Brunnen*; however, in Spanish, no such distinction is made, and *fuente* is used for all types of sources—artificial and natural. The Neoclassicist fountain is usually the architecturally improved natural source; during our period, both Meléndez Valdez and Goethe gravitated toward this usage. The concept of Neoclassicism can be applied to water images in painting less easily than it can in literature. For if one takes the sort of noble Greeks and Romans, posing before columns and pediments—who derive in part from the seventeenth-century Neoclassicism of Nicolas Poussin's mythic and historical canvases and that culminated in the eighteenth-century form of Jacques-Louis David—as a paradigm for Neoclassicism, then one is left with waterless interiors and exteriors; that is to say, one will find there no water images to use as points of comparisons. It is true, however, that the seventeenth century offers the great example of Poussin's landscape vision of placid streams and still pools and that the eighteenth century offers some followers of Poussin, and indeed one of considerable ability: Richard Wilson. Establishing paradigms for fountain images and symbols in the early and middle eighteenth century is a relatively simple matter. The topographical drawing by Richard Wilson, 'A Fountain',[7] from his 1752 Italian sketchbook, is presumably a sketch made *in situ*. It is merely a hastily conceived and executed piece that represents an actual locus with peasant women drawing water. The fountain is rendered as a descriptive image for no discernible symbolic purpose, which I will show runs against the grain of most visual artists between the 1760s and the 1870s.

The ordering of the treatment of fountains and springs in visual arts follows that of poetry: first, I shall discuss natural springs that have been improved into fountains by manmade structures; then, springs which, like those found in poetry, are located in high mountains; finally, the springs which are employed in the larger context of the water-cycle. The painting and other visual arts offer a varied range of sources to rival poetry; however, music does not. To achieve the full sense of a water-source, music seems to need the verbal and scenic resources offered by opera; therefore, most musical sources come from there. I will, however, discuss a piano piece by Franz Liszt that comes close to rendering the sense of a source by musical elements—helped greatly, of course, by a descriptive title.

The (Romantic) Neoclassical Source Itself

Johann Wolfgang Goethe, the creator of such a diversity of landscape visions and of their symbolic expressiveness, offers throughout his poetic career a series of fountain images and symbols that form a complex rhetoric, which is used in turn by him to dialectically oppose permanence with change—then to resolve them into the paradox of permanence in change. In the course of doing this, however, Goethe uses these images and especially the symbols for expressing the full complexity of individual human psychology, as well as of the social-political tensions and contradictions of

---

7 'A Fountain' from *Studies & Designs by Ri. Wilson, done at Rome in ye Year 1752*, British Museum, London; illus. in *Painter and Poet: Studies in Literary Relations of English Painting* (Cambridge: Harvard UP, 1938) 125.

his time. My study will spend a great amount of space considering *The Sorrows of the Young Werther / Die Leiden des jungen Werthers* (1774; 4: 382–511)[8]; however, I feel this emphasis can best be justified with the aid of the first great comparatist—Germaine de Staël. Moreover, according to a recent scholar, Staël is the first to unify 'scattered, local and half-formed agendas' and give them 'a single name, *Romanticism*'.[9] Most importantly for my study, she stands as one of the pioneers in relating arts to their larger social and political context.

Can the emphasis I place on *Werther* as a representative work be justified? Writing in 1800, Staël sees *Werther* as the one work composed by a German that can be held as equal to the best ever produced in world literature (1: 273–74). Deprived as Germans were then of directing their energies toward 'representative governments' ('gouvernements représentatifs'), Staël sees that the intellects of German writers had been forced to turn 'toward the contemplation of nature and the examination of their own psyches' / 'vers la contemplation de la nature et l'examen d'eux-mêmes'. Using these as her theoretical springboards, she goes on to write of this short novel that 'It conveys with the greatest possible force the harm a bad social system can do to an energetic mind; a sort one encounters more in Germany than elsewhere'.

> Il représente dans toute sa force le mal que peut faire un mauvais ordre social à un esprit énergique; il se rencontre plus souvent en Allemagne que partout ailleurs (273).

As strange as it may seem, the decline and fall of Werther is conveyed throughout the novel by an interlocking appeal to water symbols.

It will seem odd to call attention to a Neoclassical fountain in the series of Storm and Stress letters that make up the bulk of *Werther*; however, in a plethora of its Early Romantic imagery which we will consider later, Goethe includes a long descriptive passage concerning a rustic fountain. In a letter dated 12 May 1771, Werther has just left behind a disappointed admirer, 'poor Leonore' / 'arme Leonore' (4: 382), has not yet met Charlotte, by whom it will be his turn to be left unrequited, and has not yet become resentful of the gilded cage of his upper middle-class idleness and uselessness. He feels that he is participating in a fresh and yet peaceful life in a new location. It is surely no accident, then, that among the first of many passages describing local scenes there is one of a tranquilly flowing fountain. I will be arguing that this fountain—like all the other water phenomena in the novel—offers an analogue to Werther's mood; or perhaps, his description of the fountain might be rather colored by his mood and by his subjective perception—indeed, distortion, of the external world, as the persona (Werther) wonders in the opening lines: 'I don't know whether deluding spirits hover about this spot, or whether it is the warmth of a heavenly fantasy pervading my heart that lends everything about me a paradisiacal aura'.

---

8 Johann Wolfgang von Goethe, *Gedenkausgabe der Werke, Briefe und Gespräche*, ed. Ernst Beutler et al. 24 vols. (Zürich: Artemis, 1949). Throughout the text of this entire book, all quotations from Goethe's works—unless otherwise specified in the notes—are from this edition.

9 John Claiborne Isbell, *The Birth of European Romanticism: Truth and Propaganda in Staël's 'De l'Allemagne', 1810–1813* (Cambridge: Cambridge UP, 1994) 4.

Ich weiß nicht, ob täuschende Geister um diese Gegend schweben, oder ob die warme himmlische Phantasie in meinem Herzen ist, die mir alles rings umher so paradiesisch macht.

These fantasies of his heart—as Werther calls them—seem, in part, to dictate his description of the fountain.

The feelings evoked in him are described both in purely objective terms and in allusions to fairy tales or to things mysterious, which hint more at Early Romanticism than signify the classical *genii loci*:

Just at the entrance of the village, there is a fountain—a fountain that enchants me like Melusine and her sisters.—You go down a little knoll and you find yourself before a vault; from there, about twenty steps lead down to where the clearest water flows from marble rocks. The little wall above that forms an enclosure, the tall trees that shades the space around it, the coolness of the place—all this has something so attractive, something so awe-inspiring about it.

Da ist gleich vor dem Orte ein Brunnen, ein Brunnen, an den ich gebannt bin wie Melusine mit ihren Schwestern.—Du gehst einen kleinen Hügel hinunter und findest Dich vor einem Gewölbe, da wohl zwanzig Stufen hinabgehen, wo unten das klarste Wasser aus Marmorfelsen quillt. Die kleine Mauer, die oben umher die Einfassung macht, die hohen Bäume, die den Platz rings umher bedecken, die Kühle des Ortes, das hat alles so was Anzügliches, was Schauerliches.

The hand of man is everywhere evident in the description; the original, natural spring has been 'improved' with steps and wall; I contend in this study that landscapes that celebrate cultivation and utilitarian improvements partake of the Neoclassical aesthetic; Romantic, of the wild. Nature, here, is not wild but idyllic.

Contributing to the idyll is Goethe's early evocation of Biblical and Homeric times—not surprising when one considers that the persona is under the influence of Homer's *Odyssey* during this portion of the novel: 'Not a day goes by but that I spend an hour there. The young women from the town come there to fetch water—that most innocent and necessary occupation, which in ancient times the daughters of kings themselves performed'.

Es vergeht kein Tag, daß ich nicht eine Stunde da sitze. Da kommen dann die Mädchen aus der Stadt und holen Wasser, das harmloseste Geschäft und das nötigste, das ehemals die Töchter der Könige selbst verrichteten.

The Homeric aura here must be kept in mind, when later in the novel, Werther is in a radically different state of mind and under the influence of a different sort of 'epic poet'—Ossian. As there are Homeric waters, there will prove to be Ossianic ones.

In the closing passage of this letter, the themes are restated; first comes the poet's subjective perception of the manifestations of the patriarchal past: 'As I sit there, the Patriarchal ideal with the greatest vividness springs to life around me: our ancestors socializing and courting around the fountain':

Wenn ich da sitze, so lebt die patriarchalische Idee so lebhaft um mich, wie sie alle, die Altväter, am Brunnen Bekanntschaft machen und freien;

then comes the aura of fairy tales or of ancient times when places and objects were conceived of as having spirits innate to them: 'and how benevolent spirits hover around the fountain and springs',

und wie um die Brunnen und Quellen wohltätige Geister schweben.

Finally, the persona points out the more sensate, objective effects of the fountain: 'Oh! the person who is unable to feel this with me must never have refreshed himself by the coolness of a fountain after a hard day of summer hiking'.

> Oh, der muß nie nach einer schweren Sommertagswanderung sich an des Brunnene Kühle gelabt haben, der das nicht mitempfinden kann (4: 385).

After all the foregoing patriarchal exaltation that precedes this last line, this might seem a bit mundane, flat; but it may well be that the antique is sustained by Goethe's intent that we take this as a reference to Horace's famous thirteenth of *Third Book of Odes,* where the poet sacrifices a kid-goat and sings in honor of the cold refreshment offered by the Bandusian spring.

These antique and patriarchal overtones appear again almost a decade and a half later in Goethe's *Hermann and Dorothea / Hermann und Dorothea,* which also draws on the *Odyssey*; however, even more so: it is a contemporaneous story crafted in the form and style of Classical epics. It is not a coincidence that a fountain similar to that depicted in *Werther* not only appears, but is the central location and, indeed, symbol for the lovers' courtship. Here Hermann stops and waits as the doctor and the parson enter an encampment, where there is an immigrant group of which Dorothea is a member, in order to find out if she is worthy to be Hermann's bride. As in *Werther,* the natural water-source has been improved, as it were, with steps and walls for practical reasons and social occasions; but it is notable that the improvements to the latter source must have been slight because it is called *Quelle* instead of *Brunnen.* In the epic, moreover, these social sides, especially courtship—which is only referred to in passing in *Werther*—become central to the structure of the plot, characters, and symbols of *Hermann and Dorothea.* The scene of the courtship is introduced:

> There before the village was a place of fun and sport for peasants and burghers who lived nearby—a wide, verdant meadow, spreading beneath the green sward, shaded by the dignified dark-foliage of sublime linden trees, which had been rooted there for centuries. Under the trees was a fountain, shallowly dug. If one descended its steps, one encountered stone benches ringing the spring, which gushed perennially and purely and which was enclosed by a low wall—to make drawing water easier'.

> Von dem würdigen Dunkel erhabener Linden umschattet,
> Die Jahrhunderte schon an dieser Stelle gewurzelt,
> War mit Rasen bedeckt ein weiter grünender Anger
> Vor dem Dorfe, den Bauern und nahen Städtern ein Lustort.
> Flachgegraben befand sich unter den Bäumen ein Brunnen.
> Stieg man die Stufen hinab, so zeigten sich steinerne Bänke,
> Rings um die Quelle gesetzt, die immer lebendig hervorquoll,
> Reinlich, mit niedriger Mauer gefaßt, zu schöpfen bequemlich ('Polyhymnia' 5: 151–58).

Hermann chooses this spot to wait while friends of the family go to find out if Dorothea is 'worthy of marriage' ('Ob das Mädchen auch wert der Hand sei', l. 162). Like Hermann, however, who is convinced of the fact ('Zwar ich glaub es', l. 163), the reader might well have no doubts about Dorothea because the fountain of the preceding lines ('Reinlich' / 'Purely' in l. 158 above) may well be offered up by the poet as an objective correlative for her purity—indeed, for her virginity.

In the next canto when the narrative returns to Hermann, he is waiting by this fountain, and he appropriately daydreams of Dorothea. In the reflection upon its water, he sees 'the lovely form of the young woman taking shape' / 'die liebliche Bildung des Madchens' on the surface of the water ('Erato' 7: 6); the 'Bildung' here carries the sense of formation or creation, which implies the creative formation of Dorothea's image on the water—a sort of effect one might find in a modern science-fiction novel. The key for this study is that it materializes quite naturally on the fountain's reflective surface. Only after does Dorothea appear incarnate, as it were. It would seem that not only hers but also Hermann's purity and innocence are represented by the fountain. This also seems likely if one remembers that Hermann—before meeting Dorothea—has been shy with women. Finally, if this were true, it seems symbolically appropriate that the reason which Dorothea gives for coming is that the fountains of the town have been abused and dirtied by overuse, and that the water of this particular fountain remains pure:

> In order for you to know right away why I came here to draw water where the spring flows unceasingly and purely, I will tell you this: negligent people have sullied all the water within the village, with horses and oxen wading through even the springs which provide water for the inhabitants. And thus, with washing and cleaning, they have also fouled all the troughs of the village and polluted all the fountains, and all the fountains are polluted; for each person thinks only of himself and of the immediate fulfillment of present needs and not of what will follow'.

> Daß ihr aber sogleich vernehmet, warum ich gekommen,
> Hier zu schöpfen, wo rein und unablässig der Quell fließt,
> Sag ich euch dies: es haben die unvorsichtigen Menschen
> Alles Wasser getrübt im Dorfe, mit Pferden und Ochsen
> Gleich durchwatend den Quell, der Wasser bringt den Bewohnern.
> Und so haben sie auch mit Waschen und Reinigen alle
> Tröge des Dorfes beschmutzt und alle Brunnen besudelt;
> Denn ein jeglicher denkt nur, sich selbst und das nächste Bedürfnis
> Schnell zu befriedgen und rasch, und nicht des Folgenden denkt er (ll. 28–36).

Since the poet seems to have already suggested a symbolic connection between, on the one hand, his heroine and the fountain and, on the other, the earnest moral tone of the last two lines, it becomes more and more certain that something beyond the sullying of pure water is implied: it is the consequences of a promiscuity that Dorothea is morally far above.

And what a powerful expression of (proto)ecological thinking do these lines offer! What better description is there, even in our own age, of the psychology of polluting? Goethe perfectly captures thinking in the short term, answering immediate needs, and refusing to see one's environmental actions in their larger spatial and temporal context. He shows, with incisive clarity, that it is a selfishness in which the individual hurts himself along with the group.

I think it is crucial for us to pause in our examination of this, the complex nexus of signification in the central moment of *Hermann and Dorothea,* and to ask ourselves: why has Goethe freighted this meeting at a source and the act of drawing pure water from it with so much symbolic weight and, not only in this work, but in the earlier *Werther,* with its many Neoclassical allusions, as we would expect from a work that seeks to adapt the Homeric epic to a modern subject? We must remember that the moment and place are set vividly in time of war. Dorothea is a war immigrant, escaping from the French Revolutionary Army. It is an irony, not an uncharacteristic complexity of the period, that Goethe's epic *Hermann and Dorothea* is an exalted appeal to the same Neoclassicism that Jacques-Louis David has employed to glorify the exploits of that army and its revolution and that André Chenier has used to convey not only its *pro* but later its *contra*. Those who yearned for the changes brought by the French Revolution and those who feared them both paradoxically seek the sense of tradition, even permanence, suggested by Neoclassical trappings. In times of convulsive change, the essential revolutionary would seek justification for the changes as a cyclic return to a past state; the essential conservative would hope for a return to the immediate past; both, then, would look to the past as an escape from the present. I believe in both cases there is a certain fear involved in any vision of pure change: it must be somehow softened by a matrix. And the cycle contains both the change and the permanence.

We, however, need not look to France for parallels. In 1797, during which *Hermann and Dorothea* was finished as well as published, Friedrich did a series of watercolors uncharacteristically Neoclassical in subject and style: a series of exactly the sort of improved fountain Goethe had in his mind—surely, everywhere before his eyes in rural Europe. The watercolors, like Wilson's 'A Fountain' already examined in the opening of this chapter, were done *in situ*; however, unlike Wilson's purely topographical drawing, Friedrich's watercolors employ fountains for highly symbolic purposes, similar to those of Goethe's novels. While the 'Spring in the Frederiksberger Garden' / 'Quelle im Frederiksberger Garten', improved by a spring-house, shares the simplest of utilitarian purposes of the fountains in Goethe's novel and epic, two of these watercolors, the 'Emilia's Kilde' / 'Emilias Kilde' and the 'Luise Spring in Frederiksdahl' / 'Luisenquelle in Frederiksdahl', consist of stone monuments constructed over the tops of natural sources. It is not clear whether the fountains have been improved, at least in part, for utilitarian purposes; although the rustic quality of the covering structure would seem to indicate that the 'Spring in the Fredriksberger Garden' fits in that category, yet this need not be the case because the picturesque qualities of its rustic look might have been aesthetically valued for its own sake. It is clear, however, that their unity of style, their creation in the same year, and the uncharacteristic contribution they make to Friedrich's oeuvre suggest that the three must be read as a series. In his inscriptions for the two covered by stone monuments, the artist used the same phrase, 'drawn in accordance with nature'.[10] And yet more is involved.

Clearly the artist's choice of these subjects was based on something more complex: the symbolism of the springs' having been made into fountains for the purpose of each being a monument for a dead person was almost certainly the aspect which drew the interest of Friedrich, who loved allegory. Concerning the 'Emilias Kilde', William Vaughan writes: 'The monument in this watercolour, erected over the source of a stream by Count Ernst Schimmelmann in memory of his first wife, was designed by Abildgaard. It was a favourite motif among Copenhagen landscapists of

10 Original: 'nach der Natur gezeichnet', qtd. in *Caspar David Friedrich*, ed. Werner Hoffmann, Hamburger Kunsthalle Catalogue (Munich: Prestel, 1974) 114–15.

the time'.[11] There being funeral structures, combined with the constant flow of the natural springs beneath, suggests that the fountains are meant as symbols for the immortality of the soul; however, even a constant flow is still change. In the works of Friedrich, Goethe, and several of their contemporaries, the specific image of water is forced to bear symbolic signification that runs counter to its physical qualities. The painter, in these instances, does not take the easy way of representing form; his signifier must be perched at the edges of a chaos of implied contradictions; the signifier is fraught with the implication of flux and reflux and, yet, is made to symbolize permanence. The signifier, at once, represents and deconstructs. In spite of or because of its contradictions, this symbol is simultaneously more philosophically powerful and aesthetically focused. This particular expressive use of water symbols will run the full gamut of the water-cycle. It might also be noted that in the 'Emilia's Kilde' there are stone benches, as there are around the fountain in *Werther*.

'Emilia's Kilde' and 'Luise Spring in Frederiksdahl' are most interesting because of their Neoclassical architecture—again very uncharacteristic in a painter who prefers rendering Gothic ruins. The style of the 1797 drawings, in addition, has the linear severity that one would have to look for in the Neoclassicist David, or J.-A.-D. Ingres, or Asmus Jakob Carstens. The formidable Neoclassical fountains by Friedrich and Goethe, their frequent recurrence, and their symbolic density—adding to their dating from the 1790s—all make me suspect that something larger is at stake: they are German and conservative responses to the cosmic changes of the French Revolution. Somewhat like the way that Werther sought to quench his overheated emotionalism in a spring blessed by *genii loci,* who suggest the ancient world of Homer and Horace, as well as the Bible, Friedrich and the Goethe of the 1790s both sought out isolated images of permanence in the midst of a superheated *Zeitgeist*. Goethe's epic is a vision of two young people clutching at the continuity involved in that basic building block of organized society: the family growing out of the rituals of courtship and marriage. And anyone who has followed the conservative political rhetoric in modern France or the USA, for example, knows that the myth of the family is a key aspect in the hope of a 'Great Political Return' today, as it was then. The eternal return of courtship in the generation of humankind is symbolized by the fountain.

The poet weights every word and gesture of this ritual of courtship at the fountain with significance. Even the simplicity of the gestures of drawing water seems to belie this deeper significance. Perhaps, their mutual, simultaneous gesture is a symbol for courtship; its rhetoric—particularly its pleonasms—creates the archaic quality of ritual repetition; yet paradoxically, there is a strong sense of the mundane, too: 'She stooped over to draw water; and he took the other vessel and stooped over'.

> Sie beugte sich über, zu schöpfen;
> Und er faßte den anderen Krug, und beugte sich über (ll. 39–40).

Their indirect encounter by reflections—indeed, it takes place against the 'heavens'—seems part of this aura of courtship ritual: 'And they saw their two images,

11 William Vaughan, *German Romantic Painting,* 2nd. ed. (New Haven: Yale UP, 1994) 78.

rocking, reflected in the blue of the sky, and they each nodded and gave a friendly greeting to the mirrored image of the other'.

Und sie sahen gespiegelt ihr Bild in der Blaue des Himmels
Schwanken, und nickten sich zu, und grüßten sich freundlich im Spiegel (41–42).

Even the quivering motion made by the reflection on the water seems to offer yet another objective correlative to the lovers' emotions. Thus, in contrast to the previous verbal formality, the nod and the greeting between the lovers are transmitted to each other by a fluxive medium, which appears to add a tremor of emotion to these gestures. Hermann asks for and accepts a drink of water from Dorothea: 'Then the merry boy said: "Let me have a drink!" She hands him the vessel. Then they both rested, leaning against their pitchers, friendly and close'.

Laß mich trinken, sagte darauf der heitere Jüngling;
Und sie reicht' ihm den Krug. Dann ruhten sie beide, vertraulich
Auf die Gefäße gelehnt [...] (ll. 43–44).

The act of the offering of water is a ritual act; if not perhaps eucharistic, it clearly is a communion of sorts. In the rites of betrothal—which for Hermann, this is just that—the giving of a beaker of pure spring water is still performed in many areas of Central Europe (something similar is brilliantly captured, in the next century, during the great scene at the end of Hugo von Hofmannthal's libretto for Richard Strauss' *Arabella.*)

A comparison between this stately and simple proffering of water in this epic from the Weimar period with a frenetic and ambiguous one in *Werther* demonstrates how the developed fountain becomes more clearly a part of Goethe's (Neo)Classical-Romantic strategy. A young child brings Lotte a glass of water. This happens during a walk in which Werther, Lotte, and two of the latter's younger sisters come upon the fountain which has, and takes on, an extraordinary meaning for the persona:

After walking for an hour and a half, we headed back towards town—to the fountain which is so precious to me and is now a thousand times more precious. Lotte sat down on the little wall, and we stood before her. I looked about me, alas, and vividly recalled that time when my heart had been so alone. 'Beloved fountain', I said, 'since then I have not rested in your coolness, have sometimes not even glanced in your direction when passing by in a hurry'.

Nach einem Wege von anderthalb Stunden kamen wir gegen die Stadt zurück, an den Brunnen, der mir so wert und nun tausendmal werter ist. Lotte setzte sich aufs Mäuerchen, wir standen vor ihr. Ich sah umher, ach, und die Zeit, da mein Herz so allein war, lebte wieder vor mir auf.—Lieber Brunnen, sagte ich, seither hab' ich nicht mehr an deiner Kühle geruht, hab' in eilendem Vorübergehn dich manchmal nicht angesehn.

Goethe, here, makes it clear how freighted with meaning the place has already been and how much more it soon will be.

Then Werther notices the child's act of carrying a glass to Lotte and immediately comes to a full realization of something he has as yet been only half aware: the nature of his feelings for Lotte. Both the juxtaposition and the parallel construction of the following sentences—describing, as they do, the action and the reaction—make the connection between the two events unmistakable: '—I looked down and saw that Malchen was mounting the steps while being very preoccupied with a glass of water.—I looked at Lotte and deeply realized how much she really means to me'.

—Ich blickte hinab und sah, daß Malchen mit einem Glase Wasser sehr beschäftigt heraufstieg.—Ich sah Lotten an und fühlte alles, was ich an ihr habe (413).

The gesture of carrying a glass of pure water—symbolic, as it is, of the child's innocent, sisterly love for Lotte—becomes distorted, by some neurotic sort of displacement within the consciousness of the persona, into a symbol for sexual passion; Werther gives vent to his passion by kissing the child; judging by the intensity of their reaction, both Malchen and Lotte sense this neurotic undercurrent; the child's innocent gesture has been corrupted by Werther's passionate one that has been displaced from Lotte unto the child, and the former must receive ablution from the fresh spring:

Just then, Malchen came with a glass. Marianne wanted to take it from her: 'No!' cried the child with the sweetest expression, 'Dear Lotte, you must have the first sip'. I was so enchanted by the frankness, by the goodness with which she uttered this, that I could not give expression to my feelings other than by picking up the child up from the ground and kissing her so passionately that she began at once to scream and weep. 'You did a bad thing', said Lotte. I was taken aback. 'Come, Malchen', continued Lotte, taking her hand and leading her down the steps, 'wash yourself in the fresh spring—hurry—hurry so that it won't do you any harm'. So there I stood and watched: with what assiduity the little one rubbed her cheeks with her tiny wet hands; with what faith that the miraculous spring would wash away all impurities and spare her from the disgrace of growing an ugly beard; how Lotte said that she had washed enough and how the child continued to wash assiduously, as if too much were better than too little.—I can assure you, Wilhelm, that I have never attended a baptism with greater respect—and when Lotte came up, I felt a desire to fall on my knees before her, as before a prophet who has forgiven the sins of a nation through consecration.

Indem kommt Malchen mit einem Glase. Marianne wollt' es ihr abnehmen: Nein! rief das Kind mit dem süßesten Ausdrucke, nein, Lottchen, du sollst zuerst trinken!—Ich ward über die Wahrheit, über die Güte, womit sie das ausrief, so entzückt, daß ich meine Empfindung mit nichts ausdrücken konnte, als ich nahm das Kind von der Erde und küßte es lebhaft, das sogleich zu schreien und zu weinen anfing.—Sie haben übel getan, sagte Lotte.—Ich war betroffen.—Komm, Malchen, fuhr sie fort, indem sie es bei der Hand nahm und die Stufen hinabführte, da wasche dich aus der frischen Quelle, geschwind, geschwind da tut's nichts.—Wie ich so dastand und zusah, mit welcher Emsigkeit das Kleine mit seinen nassen Händchen die Backen rieb, mit welchem Glauben, daß durch die Wunderquelle alle Verunreinigung abgespült und die Schmach abgetan würde, einen häßlichen Bart zu kriegen; wie Lotte sagte, es ist genug, und das Kind doch immer eifrig fortwusch, als wenn viel mehr täte als wenig.—Ich sage dir, Wilhelm, ich habe mit mehr Respekt nie einer Taufhandlung beigewohnt—und als Lotte heraufkam, hätte ich mich gern vor ihr niedergeworfen wie vor einem Propheten, der die Schulden einer Nation weggeweiht hat (413–14).

The rhetoric of a protestant sermon—leading ineluctably up to the similes of an Old-Testament prophet and a New-Testament baptism, added to an earlier communion—offers us Judeo-Christian images to place along side the Old Testament and the Homeric ones in order to yet further certify the spirit of the place; the passage culminates on a prophet who could be Jeremiah or John the Baptist, but just as likely is Goethe's Storm and Stress hero—Mohammed; but whichever, it demonstrates the high moral symbolism involved and the extreme complexity involved in the seemly mundane ritual performed at the side of the source, in Goethe and (he makes one feel) in his age. It further shows how Goethe employs his water images in the novel as objective correlatives for the fluctuations in Werther's psyche.

The fountain itself and the act of drinking spring water from it, thus, in the two works, strongly relate to love and courtship; but, since marriage is to be the culmination of events in *Hermann and Dorothea,* the domestic aspects of the fountain carry much more weight than they do in *Werther*. In one letter of the novel, the persona describes how he breaks class barriers by helping a young servant girl place a heavy pitcher of water on her head (386); however, nothing more is involved.

In the epic, however, Dorothea's final act of fetching and ladling out water to the people whom she has been aiding is a symbol of the domestic duties which she will be taking on (see especially ll. 114–16). Before discussing and contextualizing this important element further, it is crucial—with our feminist hindsight of today—to examine another very different action on Dorothea's part. This action, moreover, establishes Goethe's hatred of a disorder brought upon by revolution. In the canto (6) called after Clio, the muse of history, an old judge among the immigrants, talking in general terms of the valor of his immigrant band, then focused on a young woman (Dorothea), in order to tell her history in very heroic terms:

> And above all there was one fine deed, which I would here like to mention, that was performed by a bravehearted young woman, a noble and excellent maiden, who had stayed behind at the large farm alone with young girls; for the men had marched against the foreigners. They were attacked at the farm by a bunch of plundering rabble—who forced their way to the women's quarters at once, and there set eyes on this maiden—who had grown into a beautiful woman—and on the charming young girls—little more than children they were still. Then these deserters were seized by passionate lust; with brutish frenzy they attacked the girls—who stood trembling with fear—and the bravehearted maiden. But she immediately snatched the sword from the scabbard of one of them, and struck him down with a powerful blow; he fell at her feet, bleeding. Then with virile blows, she valiantly freed the girls; she wounded four more of the marauders; but those escaped death. Then she barred the doors of the farmhouse and waited for help, armed.

> Und so laßt mich vor allen der schönen Tat noch erwähnen,
> Die hochherzig ein Mädchen vollbrachte, die treffliche Jungfrau,
> Die auf dem großen Gehöft allein mit den Mädchen zurückblieb;
> Denn es waren die Männer auch gegen die Fremden gezogen.
> Da überfiel den Hof ein Trupp verlaufnen Gesindels,
> Plündernd, und drängte sogleich sich in die Zimmer der Frauen.
> Sie erblickten das Bild der schön erwachsenen Jungfrau
> Und die lieblichen Mädchen, noch eher Kinder zu heißen.
> Da ergriff sie wilde Begier; sie stürmten gefühllos
> Auf die zitternde Schar und aufs hochherzige Mädchen.
> Aber sie riß dem einen sogleich von der Seite den Säbel,
> Hieb ihn nieder gewaltig; er stürzt' ihr blutend zu Füßen.
> Dann mit männlichen Streichen befreite sie tapfer die Mädchen,
> Traf noch viere der Räuber: doch die entflohen dem Tode.
> Dann verschloß sie den Hof, und harrte der Hülfe, bewaffnet ('Klio' 6: 104–18).

A violent action, perhaps, but certainly warranted in preventing a multiple rape and in time of war. This event—told in a Homeric flashback—is the single heroic act of this epic, this example of that most heroic genre. In both the deed and its exceptional position within the epic, Dorothea takes her place beside Virgil's Camilla.

It is Dorothea's heroic deed in the realm of history and in defense of herself and the little girls that, I think, ennobles the nurturing action of the bearing and ladling out of water. Also we should briefly consider how it fits in the general emphasis on the engagement in Goethe's other late works: *Faust, Part Two,* where Faust turns from the wild individualism of his earlier age, to participation in the Emperor's court and to his attempts at a Holland-like Utopia, and much more importantly *Wilhelm Meisters Wanderjahre / Wilhelm Meister's Journeyman Years,* which has a hero that makes a similar turn and which culminates in the attempts to form a Utopia in America. But even more than these, the water-bearing passage from *Hermann and Dorothea* refutes the validity of a Goethe as the Olympian standing above the French Revolution of 1789 and embodies this engagement with people as strongly Liberal (in the nineteenth-century sense) as any of the bourgeois revolutionaries; but, his was a purely evolutionary Liberalism holding on to many remnants of Feudalism. Goethe's later characters—indeed all his characters with the exception of *Urfaust* and *Werther*—are always engaged and socially committed.

Dorothea's parting words to these people, which follow the act of water bearing and ladling, clarify this deeper meaning:

> Then the young woman addressed them with an earnest expression, saying: 'Friends, this may well be the last time that I raise a vessel to your mouth—that I wet your lips with water: however, when from this time on a drink refreshes you on a hot day, when in the shade you enjoy peace and the clear spring, then think of me and of my friendly services, which I performed more out of love than out of kinship duty'.

> Da versetzte das Mädchen mit ernsten Blicken und sagte:
> Freunde, dieses ist wohl das letztemal, daß ich den Krug euch
> Führe zum Munde, daß ich die Lippen mit Wasser euch netze:
> Aber wenn euch fortan am heißen Tage der Trunk labt,
> Wenn ihr im Schatten der Ruh und der reinen Quellen genießet,
> Dann gedenket auch mein und meines freundlichen Dienstes,
> Den ich aus Liebe mehr als aus Verwandtschaft geleistet ('Erato', 7: 145–51).

The language has a feel of Luther's translation of the Bible—e.g., 'Da versetzte', 'und sagte', as well as the abundant use of anaphora. The fountain ritual, then, is central to expressing the glorification of the subservient place of women in the Patriarchal structure of the eighteen-century German bourgeoisie. Dorothea's heroic exploits, nevertheless, brings home the fact that she has chosen this social place in spite of or because of an inner strength. As a woman colleague of mine expressed it: 'Dorothea assumes the subservient task, as she says, out of love and respect rather than out of weakness and that such humility and sacrifice from one so strong adds a special nobility to the act'.[12] I, for my part, think it is the historical context of war and immigration that makes the gesture noble, especially, in light of Dorothea's earlier deed of defending herself and others.

Referred to in passing in *Werther,* the image of young maidens bearing water from the fountain has become central to Goethe's epic: the persona of the novel

12 Elaine Stuart, letter to author, 20 Aug. 1997.

merely evokes the patriarchal times of ancient Greece and of the Middle East, whereas these become central to *Hermann and Dorothea* (see especially 'Polyhymnia' 5: 225–7).[13] And yet although the fountain appears passingly in *Werther*, I do think it says a very important thing about Goethe's Storm and Stress: even during this Early Romantic outpouring of material that would prove to be the material upon which later Romantics would build, Goethe demonstrates that the material of his own Neoclassicism was there in force, albeit as yet a secondary force in the whole. Indeed, its presence in Werther suggests Goethe's lifelong attraction to things Neoclassical—the other side of the tension that balanced his aesthetic state. The victory of the Protoromantic material of *Ossian* over the elements of *Odyssey*, as I will demonstrate at length later, was a Pyrrhic one in Goethe's larger oeuvre. The link between the novel and the later epic clearly indicates a larger one between Storm and Stress and Weimar Classicism—just as it will later prove to contextualize them with Romantic culture.

And yet I feel that the difference goes even beyond one of style. I have sought to unveil in *Werther* the contrast between the placid fountain and streams of the first part and the vision of storm and flood of the second, to be examined later. This stark binary opposition in symbolism exemplifies the simplification of this novel compared to the epic *Hermann and Dorothea*. The novel, especially but not entirely the letters or first-person portion, is concentrated on the mood oscillations of the mind of a suicidal individual—even social considerations of class are confined within this focus. The epic draws back to offer a comprehensive view of a real historical time and a turning point in two lives, which promises a future. If one were to insist upon today's feminist viewpoint, the Patriarchal courtship-at-the-well thematics of the fountain promise to bind the heroine to what could be seen as an ignoble life of servitude—the giving of the drink of water to Hermann is an earnest of this bondage. This, however, judged by the dreadful standards of the time is not the worst of bondages possible for the immigrant Dorothea, which is further balanced in part by the heroic presentation of a woman actively defending herself and others from rape. The real effects of revolution and war upon the populace is shown with a certain proto-Realism. Above all, the (proto)ecological insights are equally if not superior to anything accomplished in England at the time. The particular of the fountain is given universal significance in exactly the way that Georg Lukacs describes in his description of the quality of Totality / *Totalität*.[14] Lukacs feels that the epic's ideality, more than its reality, lifts the mundane details into cosmic signification.[15]

Neoclassicism is perfectly fused with Romanticism to create a nominally Persian vision: the *West-Easterly Divan* / *West-östlicher Divan* (3: 163–243). In a work that is permeated by the same patriarchal aura ('Patriarchenluft' 287) as the fountain in *Werther* and as the whole of *Hermann and Dorothea*, Goethe not only makes significant use of the fountain symbol but also roots this symbol firmly in its

---

13 Ernst Loeb finds a similar significance in the scene, 'Am Brunnen' in *Faust I* (ll. 3544–86); *Die Symbolik des Wasserzyklus bei Goethe* (Paderborn: Schöningh, 1967) 97. This scene from *Faust* is one of the many occurrences of water symbolism that had to be omitted from this study, in spite of their significance.

14 Georg Lukacs, *Ästhetik*, 2 vols. (Neuwied: Luchterhand, 1963) 1: 237–38.

15 Lukacs, *Goethe und seine Zeit* (Bern: Franke, 1947) 92–94.

time—that of the Napoleonic sequel to the French Revolution: 'North and West and South are splintering, thrones are cracking, kingdom are trembling—make your escape into the pure Orient in order to partake of the air of the Patriarchs! Amid loving and drinking and singing, let Khiser's fountain renew your youth'.

Nord und West und Süd zersplittern,
Throne bersten, Reiche zittern
Flüchte du, im reinen Osten
Patriarchenluft zu kosten,
Unter Lieben, Trinken, Singen
Soll dich Chisers Quell verjüngen (ll. 1–6).

The emphasis, as in the earlier novel and epic, is on *loving*; however, Goethe's symbolic usage has undergone a profound change to parallel his philosophical transformation. The symbols are made to resolve permanence and change into a perfect tension and harmony.

In Goethe's 'At the Edge of the Joyous Fountain' / 'An des lustgen Brunnens Rand' (359), the first persona, Suleika, says, in effect, that Hatem loves her so strongly that he has written her name on the water, and it has remained for her to see. By this she means that she is so much in love with him and so keenly aware of his love for her that she is able to imagine this miracle. Paradoxically, the metaphor of water, a highly mutable element, becomes a representation of Suleika's and Hatem's immutable love for each other, their awareness of the other's love, and its signification, if not perhaps communication: 'At the edge of the merry fountain, which was playing with its threads of water, I didn't know what was holding me fast; however, there was my name, lightly drawn by your hand. I looked down, feeling benevolent towards you'.

SULEIKA. An des lustgen Brunnens Rand,
Der in Wasserfäden spielt,
Wußt ich nicht, was fest mich hielt;
Doch da war von deiner Hand
Meine Chiffer leis gezogen.
Nieder blickt ich, dir gewogen (ll. 1–6).

This fantastic emblem of communication—that of writing upon the transient surface of water—offers the paradoxical tension between the mutable and the eternal; this tension is embodied by water phenomena and can be traced throughout Goethe's oeuvre. I will attempt to do so. In the works of this poet, just as was seen in the funereal fountain paintings of Friedrich, the specific image of water is forced to bear symbolic meaning that runs counter to its qualities. It is no accident that the resulting symbols in poets and painters are at once more philosophically powerful and aesthetically focused. This fountain symbol dispenses with the descriptive imagery of those encountered in *Werther* and *Hermann and Dorothea*: the receptacle has—save its rim—disappeared; only the water which is contained remains. This fact can be attributed only in part to the difference of genres: *Werther*—although it is in prose—is generically closer to 'At the Edge of a Playing Fountain' than the omniscient epic since each has a lyric persona; in addition, Werther's objective description of the fountain is mixed with subjective impressions; later, in the scene with Malchen and Lotte by the same fountain, the persona causes what has been a descriptive image to be transformed into a symbolic representation of his passion for Lotte. This again suggests that *West-Easterly Divan* is a fusion of the Romantic leanings of *Werther* and the more austerely Neoclassicism of *Hermann and Dorothea*.

In this lyric of the later Goethe, however, the descriptiveness has been peeled away, leaving only the symbolic representation of the speaker's/lover's emotion, as the second stanza demonstrates even more clearly than the first; indeed, not only the water of the fountain, but another aspect of objective nature is stamped by the subjective perception of the lovers: 'Here, at the end of the canal in the well-appointed central alley I look up again and once more I see the letters of my name finely drawn: "Remain! Remain benevolent toward me!"'

Hier, am Ende des Kanals
Der gereihten Hauptallee,
Blick ich wieder in die Höh,
Und da seh ich abermals
Meine Lettern fein gezogen:
Bleibe! bleibe mir gewogen! (ll. 7–12).

In this stanza, the descriptive detail, uncharacteristic as we have seen of the somewhat rarified even abstract style of the *Divan*, offers objective correlatives of the emotionally ordered and stable sort of love that the mature Goethe had come to prize over the years and works since *Werther*.

The lover's reply follows; but, just as the question of Dorothea's purity was answered by the symbolism of the fountain, so too is the question of Hatem's fidelity; indeed, any further or more direct verbalizing of this faithfulness would be superfluous. The poet-as-persona has given it a subtle twist: the symbols of movement are melded into the cyclical symbol of coming and going—of Alpha and Omega: 'HATEM: May the waters—leaping, moving—may the cypresses confess this to you: "From Suleika to Suleika is my coming and going"'.

HATEM. Möge Wasser, springend, wallend,
Die Zypressen dir gestehn:
Von Suleika zu Suleika
Ist mein Kommen und mein Gehn (ll. 13–16).

Thus a lyric in which symbols of movement and change have been converted into symbols of constancy and permanence culminate in one of the only symbols that can paradoxically include both permanence and change: the cycle.

These passages from the three works of Goethe demonstrate a complexity that traditional *Motifgeschichten* rarely even broaches at all: the fact that water motives recur is far less important than the variety of stylistic guises in which they can appear among works of a great and long-lived author, as well as between the works of two lesser authors each working in a different style. Let me review: *Werther* oscillates between descriptive and symbolism; the epic of his middle-age holds the two in perfect balance; the lyric, molded as it is with the concise density of Goethe's later style, privileges the symbol and its experience. The final point to be gained is a supreme demonstration, indeed ultimate proof, of the extremely flexible rhetoric offered by water phenomena as employed during the (Neo)Classical-Romantic period.

Few of the poets treated in this study share with Goethe this recurrent and complex usage of the fountain symbol. One is tempted to speculate that Goethe's usage of the fountain is one aspect of his brand of (Neo)Classicism—found, if in an evolving form, amid the stress and storm of *Werther*. It is no surprise that Schiller—also working himself out of a bout with Early Romanticism as well as the co-founder of Weimar Classicism—employs the fountain symbol in one of his rare uses of water phenomena—in 'The Fountain of Rejuvenation' / 'Quelle der Verjüngung' (1:283).[16] Schiller, here, employs the fountain to represent the idea that poets transcend death in the poetic process: 'Believe me, it is no fairytale, the fountain of youth runs truly and eternally—you ask where? In the art of poetry'.

> Glaubt mir, es ist kein Mährchen, die Quelle der Jugend, sie rinnet
> Wirklich und immer, ihr fragt wo? In der dichtenden Kunst.

Schiller employs the cyclic dynamism of the fountain and its concomitant embodiment of permanence and change in a manner that so closely resembles Goethe that one cannot help seeing this older poet's hand or his influence in it.

So rare and *sui generis,* indeed, is Goethe's usage of the fountain that to find anything comparable, it is necessary to turn as far afield as Spanish literature: of all the poets under examination only Juan Meléndez Valdés made similar use of the natural source and its qualities. This parallel is significant, but not unexpected. First, Meléndez falls but a few years short of being Goethe's exact contemporary. Second, like Goethe, he had an ambiguous position relative to Romanticism: on the one hand, he considered himself a Neoclassicist and is generally regarded as the greatest of the Spanish Neoclassic poets; on the other hand, he came under the influence of Rousseau and, therefore, anticipated certain aspects of the Romantic usage of nature imagery. This strange blend of Rousseauist and Neoclassic treatments of nature is nowhere more apparent than in one of Meléndez' *Anacreonic Odes / Odas anacreónticas* entitled 'To a Fountain' / 'A una fuente' (7–9).[17] This lyric, according to R. Merritt Cox, establishes 'the close relationship between man and nature. [...] The poet is not objective in his presentation of nature, but internalizes his reactions created from the former sensations'.[18] Although correct in part, Cox has overstated his case: Cox bases his reading of the poem in great part on the word 'espíritu' / 'soul' or 'spirit': 'Oh, smiling little fountain, how my spirit takes delight!—my eyes become enchanted in your crystalline mirror!'

> ¡Oh! ¡cómo en tus cristales,
> Fuentecilla risueña,
> Mi espíritu se goza,
> Mis ojos se embelesan! (ll. 1–4).

My reading of the lines would center, rather, on the word 'cristales'—which in this context can mean 'crystal', 'glass', 'mirror', or 'lens'. Meléndez juxtaposes the fluid

16 *Schillers Werke: Nationalausgabe*, ed. Julius Petersen, Friedrich Beissner et al. (Weimar: Böhlau, 1943–[not yet completed]). All citations are in the text.

17 Juan Meléndez Valdes, *Poésias*, ed. Pedro Salina, Clásicos Castellanos 64 (Madrid: Espasa-Calpe, 1941) 7–9.

18 R. Merritt Cox, *Juan Meléndez Valdés* (New York: Twayne, 1974) 103.

with the permanent. Like the fountains of Goethe's *Werther* and *Hermann and Dorothea,* these lines from Meléndez' lengthy lyric are primarily devoted to such passages which can be taken as objective description. Like the above passages from *Werther,* the objective and the subjective subtly intertwine, and there is no fully developed symbolic meaning in Meléndez' lyric. Whereas the fountains of *Hermann and Dorothea* and 'At the Edge of the Joyous Fountain' are symbols, the fountains in *Werther* and Meléndez' lyric are not; rather, they are images which have a cheering effect upon the persona's feelings or correspond to feelings already within the persona. Above all, both Goethe's and Meléndez' fountains are pure and tranquil in their flow. Of the Spanish poet, William E. Colford writes, 'There is much mention of water in Meléndez' landscapes. [...] His vocabulary in describing water and its surroundings is uniformly pleasant. The pools are always limpid, the waters crystal-clear'.[19]

These (Neo)Classical-Romantic fountains and springs of Goethe, Friedrich, Schiller, and Meléndez have common traits: they carry the weight of tradition amid scenes of unrestrained individual passion or cataclysmic social change; they can be rendered with objective clarity or stand pure in splendid isolation from their surroundings; the fountains or improved springs are surrounded by Neoclassical architecture. These four creators embody permanence in change, with emphasis on the permanence. All four are complex enough, even paradoxical enough to fit perfectly the Postmodernist punctuation I have adapted from the musicologist Blume: if the term (Neo)Classic-Romantic fails to fit their contemporaries, it certainly fits them.

The Romantic Natural-Supernatural Source

Like the fountains in Early Romantic and (Neo)Classical-Romantic works, the fountain in the Romantic novel is a natural spring which has been improved by man-made structures of some sort. If, however, I am correct in my Homeric reading and Friedrich visual-parallels, the architectural style in *Werther* and *Hermann and Dorothea* is Neoclassical. In Romantic ones, they are often Gothic in architecture. Even Lord Byron's *Don Juan,* canto 13, for all its Neoclassical elements, has a natural fountain covered by a Gothic edifice:

> Amidst the court a Gothic fountain play'd,
>   Symmetrical, but deck'd with carvings quaint—
> Strange faces, like to men in masquerade,
>   And here perhaps a monster, there a Saint.
> The spring gush'd through grim mouths, of granite made,
>   And sparkled into basins, where it spent
> Its little torrent in a thousand bubbles,
> Like man's vain glory, and his vainer troubles (canto 13: stanza 65 [5: 544]).

The simile in the final lines suggest eternal change. My reading of Byron's stanza tends to focus, rightly or wrongly, upon the gushing, sparkling, and bubbling action of the spring; it is this absolute delight in motion for its own sake that sets the

19 William E. Colford, *Juan Meléndez Valdés: A Study in the Transition from Neo-Classicism to Romanticism in Spanish Poetry* (New York: Hispanic Institute, 1942) 209.

Byronic approach to water phenomena off from either the Goethean or (Neo)Classical-Romantic approach. There is no urge to see the springs gushing forth to be part of some eternal or recurrent whole. Motion and change is all.

Franz Liszt, in his series *Years of Pilgrimage / Années de pèlerinage*, created a rare and surprisingly effective vision of the gushing, sparkling, and bubbling actions of a spring in his 'Beside the Spring' / 'Au bord d'une source'; it is a surpassing achievement because, on the one hand, the spring is an ongoing action of the flow of different water and a repetition of the same process of flowing out. Since excessive repetition is a threat to musical form, a balance must be struck. Liszt achieves a vision of a spring by a use of the rondo form in music; he oscillates between two essentially differing thematic sections. The first theme group depends on the reiteration of notes, but not of the same exact notes, rather of a sequence of notes having the same intervalic distance between them; this sequence is placed at all the registers of the scale—so the actual notes change. This delightful tension between the same intervalic relationship and the varied notes creates the paradoxical sense of sameness and difference. This first theme group seems intended to represent the recurrence of the bubbling forth of the spring. A second theme group essentially consists of half-step (chromatic) runs up and down the scale; perhaps this represents that other aspect of a free-running source: the rivulet that it creates and that runs off from it. Liszt's unusual time signature of 12/8—unlike the more common 3/4 time, say—would tend to de-emphasize any accentual beats within the measure and allow for a sense of flow; this is the same time signature Beethoven used for the brook movement in the Pastoral Symphony. Something occurs in the course of this which is hard to interpret in the context of this work, but which I will treat later in a different work by the same composer: Liszt chromatically moves in passing harmony out of the key of the piece A♭ all around the circle or cycle of keys to the absolutely opposite key of E—where the harmony remains fully established for two measures before passing back to A♭. Is it pushing interpretation to say that this extraordinary move suggests the larger water-cycle? Yes, however tempting it may be, such an interpretation seems untenable here; however, later I will treat a long work by Liszt where it is much less so.

Close musical as well as literary Romantic parallels to Liszt's and to Byron's fountains actually exist in a Scottish novel and an Italian opera freely adapted from it.

The source that is employed as an emblem of sorrowful change is the spring in Scott's *The Bride of Lammermoor* (1819). It has a Gothic structure erected to protect its waters from falling leaves and other debris, and it is described as 'a plentiful and pellucid fountain, which had been once covered in, screened and decorated with architectural ornaments of a Gothic character'. Scott's fountain is in a wild, rarely visited spot. Unlike often-frequented and well-kept fountains in Goethe's works, the spring here has been allowed to return to its wild state:

> But now the vault which had covered it being broken down and riven, and the Gothic font ruined and demolished, the stream burst forth from the recess of the earth in open day, and winded its way among the broken sculpture and moss-grown stones which lay in confusion around its source (57).

Unlike the fountain in *Werther* and *Hermann and Dorothea,* Scott's fountain offers no sense of the utilitarian center of ongoing communal processes; unlike the *courtship* of Hermann and Dorothea, the *tryst* of Scott's lovers has no part in the eternal continuity of life through courtship, marriage, and reproduction. The fountain, the locus of tryst, in *Bride of Lammermoor* promotes no such continuity. The love of Edgar Ravenswood for Lucy Ashton leads to the immediate end of the former family and the eventual dying out of the latter. Their love is antisocial and destructive. At the time of the novel, the fountain is no longer a center of social life: it might even be seen as the antisocial life—because it is here where the heroine meets the outcast hero.

It is not just a question of the architectural improvements that separate these High Romantic natural sources from those of the (Neo)Classicists-Romantic: they also have a supernatural element. This supernatural element of course had been touched on as we have seen in the Early-Romantic, Storm and Stress novel *Werther,* specifically in the references to the fairy tale of Melusine and her sisters, as well as the 'benevolent spirits' of the fountain—elements totally omitted from the severely Neoclassical *Hermann and Dorothea.* Melusine introduces one of those water sprites that are so important to German High Romanticism. It must of course be remembered that nymphs of fountains have mythic origin that go back at least to the Greeks and Romans. Byron includes, in his fourth canto (2: 120–86) of *Childe Harold's Pilgrimage,* a powerful and significant reference to one of the most famous of these source nymphs, Egeria the beloved and the teacher of Rome's great lawgiver. The passage comes in the midst of Byron's extremely pessimistic vision of the cycle of empires: running, as it does, from the first freedom to the pride of glory; then, to wealth, vice, and military competition; and finally, to barbarism. He returns to the beginning of Rome with the all 'too brief' reign of Numa and writes upon the still extant fountain where he is supposed to have met Egeria. In a very complex piece of psychology, Byron explains that Egeria and the divine inspiration she represents are an emanation of a temperament:

> Egeria! sweet creation of some heart
> Which found no mortal resting-place so fair
> As thine ideal breast! whate'er thou art
> Or wert,—a young Aurora of the air,
> The nympholepsy of some fond despair;
> Or, it might be a beauty of the earth,
> Who found a more than common votary there
> Too much adoring; whatsoe'er thy birth,
> Thou wert a beautiful thought, and softly bodied forth (canto 4: stanza 115).

Byron moves from Egeria, who signifies a thought emanating from a human imagination, to the water emanating from the spring—water which bursts the confines of the broken architectural embellishments:

> The mosses of thy fountain still are sprinkled
> With thine Elysian water-drops; the face
> Of thy cave-guarded spring, with years unwrinkled,
> Reflects the meek-eyed genius of the place,
> Whose green, wild margin now no more erase
> Art's works; nor must the delicate waters sleep,

Prisoned in marble, bubbling from the base
Of the cleft statue, with a gentle leap
The rill runs o'er, and round, fern, flowers, and ivy creep,

Fantastically tangled [...] (116–17).

The retaining wall once built for utilitarian purposes and its aesthetic embellishment are ruined, and the spring water pours out to nurture a verdant futility. At the end of Byron's meditation, the fountain of Egeria—so important in the first legendary days of Rome—becomes the symbol of the lost loves in the youth of civilizations and of each human being:

Alas! our young affections run to waste,
Or water but the desart [sic]; whence arise
But weeds of dark luxuriance, tares of haste,
Rank at the core, though tempting to the eyes [...] (120).

The source symbol, which usually carries a sense of youthful beginnings, seems crafted by the poet to contrast paradoxically with the sense of water flowing out merely to nourish the wastes of a civilization's decline and death.

The symbolism of the ruined structure of the fountain in Byron is also central to the fountain in Scott's *Bride of Lammermoor*. The sense of building at the height of glory and the ruin to follow is even stronger in the Scott passages. This sheltering structure had been added to the spring not for utilitarian purposes but rather as a monument to a dead beloved; it is, then, a funeral memorial constructed over a natural source like we observed in Friedrich's 'Emilia's Kilde' and 'Luise Spring in Frederiksdahl':

in memory of his Naiad, he had previously ornamented the fountain in which she appeared to reside, and secured its waters from profanation or pollution, by the small vaulted building of which the fragments still remained scattered around it. From this period the house of Ravenswood was supposed to have dated its decay (58).[20]

This Naiad, a water sprite or nymph of a spring, and her beloved come to a tragic end, which forebodes the destiny of the hero and heroine of the novel. The couple of the novel use the fountain as a trysting place as the earlier couple of the legend are said to have done. Yet this link with the past, even when reinforced by a reference to the classical myth of Egeria and Numa, is tenuous and lacks the sonorous resonances of either Byron's or Goethe's allusions to the patriarchal past; indeed, Scott goes on to rationalize away the legend:

Such was the generally received legend, which some, who would seem wiser than the vulgar, explained, as obscurely intimating the fate of a beautiful maid of plebeian rank, the mistress of this Raymond, whom he slew in a fit of jealousy, and whose blood was mingled with the waters of the locked fountain, as it was commonly called.

Not all the supernatural aspects, however, are rejected:

---

20 Sir Walter Scott, *The Bride of Lammermoor*, ed. Fiona Robertson (Oxford: Oxford UP, 1991). All citations are in the text.

All however agreed, that the spot was fatal to the Ravenswood family; and that to drink of the waters of the well, or even approach its brink, was [...] ominous to a descendant of that house [....] (58–59).

Later Lucy Ashton and the Lord of Ravenswood set up a tryst at Mermaiden's Fountain, as it is called, by which she sits in such a way that 'might have suggested the idea of the murdered Nymph of the Fountain' (205). The special nature of thc place is stated by Lucy, with a certain auctorial self-consciousness that suggests intertextual references to Gothic Romances—but the rather straightforward picture of nature naturing in the environs also suggests, at least according to my reading of the text, *Werther* (a novel by an author much read and imitated by Scott[21]):

'I like this spot', said Lucy at length [...] 'the bubbling murmur of the clear fountain, the waving of the trees, the profusion of grass and wild-flowers, that rise among the ruins, make it like a scene in romance. I think, too, I have heard it is a spot connected with the legendary lore which I love so well' (206).

It is at this point in the novel that the two plight their troth.

Later while riding past, Ravenswood sees a figure over the fountain. At first, he thinks it is Lucy; however, it turns out to be the ghost of Alice, an old woman—retainer of the Ravenswoods—allowed by Ashton to live in a cottage in the forest (245–46). Alice had earlier warned Ravenwood and Lucy that their relationship was cursed; it turns out that the old woman had died moments before her ghosts appearance to Ravenwood, and this is her last warning. It is interesting but not unexpected that, when Savatore Cammarano adapted this novel into an opera libretto for Gaetano Donizetti, he ignored the naiad element of the old legend about the fountain and emphasized instead the motif about the young woman's death at the hands of a jealous lover. Only her ghost is retained for Lucy's opening aria: 'In silence, darkness reigned amid deepest night, a pallid ray of the dismal moon struck the fountain [...] when suddenly a low groan broke through the air; and here, here, on this fountain's rim, ah, the ghost revealed itself to me! Ah!'

Reganava nel silenzio
alta la notte e bruna,
colpia la fonte un pallido
raggio di tetra luna [...]
quando un sommesso gemito
fra l'aure udir si fe';
ed ecco, ecco, su quel margin,
l'ombra mostrarsi a me, ah!

The fountain itself is given a significant musical emphasis according to the operatic standards of the time. During the early nineteenth century, there was much opera and almost no concert music in Italy, so operas often contained long mini-concerti. The scene at the fountain, scene two of act one, opens with one such concertino between orchestra and harp. I have no doubt but that Donizetti meant the long elaborate runs on the harp, as well as lightly plucked strings (*pizzicato*) to represent the rivulet run-off from the fountain. The runs move rapidly (in sixteenth and thirty-second

21 G.H. Needler, *Goethe and Scott* (Toronto: Oxford UP, 1950); for the possible influence of *Werther* on Scott's Waverley novels, see esp. 39–42.

notes) in scales or up-and-down figures, moving in wide stretches of an octave and often more. During the aria proper, it is the clarinet that does the rapidly running figures skipping intervals (*arpeggios*).

Close operatic parallels to the Egeria element in the fountains of Scott and of Byron are found in the German High literature[22] and operas, where they appear as Ondine legends. Indeed, Scott's fountain has more parallels with the High German Romantics than with the Goethe of Weimar Classicism. The sinister aspect of Scott's fountain bears greater similarities to the passages by Novalis and Eichendorff to be treated next.

Artificial or Supernatural Fountains

Unlike the sources of Scott, of Friedrich, of Liszt, of Meléndez, and those of Goethe that we have considered so far—all five of whom use natural springs whence flows real water—the following passages from the works of Novalis and Eichendorff contain fountains which are artificial or supernatural: either ones with water jets or those that give forth not water but strange flaming substances. The fountains of Weimar Classicism or the equally hybrid Neoclassicism of Meléndez display none of pyrotechnics that will be observed in the passages below taken from the High Romantics. For Schiller, as for Meléndez and Goethe, the objective nature of a tranquil fountain contains enough qualities for manifold symbolic employment, and their fountains remain natural sources. It seems significant that this most artificial and fantastic Romanticism—to which both the naturalist as well as the classicist in Goethe took offense—relished these unnatural / supernatural *sources*. To my mind, nothing underlines so clearly the distinctiveness of the Weimar Classicism of Goethe and Schiller from the High Romanticism of their younger, German contemporaries as the preternatural fountains of the latter—set, as they are, in either the landscape of dreams or a place that is truly underpinned by the supernatural. It also links the German works more to English High Romanticism in its less naturalistic, more fantastic forms: that of the second generation of Romantic poets, especially Shelley and Keats.

It is the artificial *Springbrunnen* which fill the enchanted, artificial garden of evil in Eichendorff's 'The Marble Statue' / 'Das Marmorbild'. As Fredric Jameson writes his own reading of 'a determinate and signifying *absence* in the text' of another prose work by Eichendorff, 'it is the possibility of such a reading which in turn allows us to grasp the text as a socially symbolic act, as the ideological—but formal and immanent—response to a historical dilemma'. Jameson finds in the other work a similar aristocratic world apart described in Eichendorff's Romantically lush style, signifying 'survival in Germany of a quasi-feudal power structure' in Eichendorff's time; this critic goes on to comment: 'under the spell of this wondrous text, the French Revolution proves to be an illusion, and the grisly class conflict of decades of Napoleonic world war fades into the mere stuff of bad dreams'.[23] As Theodor

22 Andrews, M.T. 'The Water Symbol in German Romanticism Culminating in Fouqués *Undine*', diss., U of Washington, 1969, 22.

23 Fredric Jameson, *The Political Unconscious: Narrative as a Socially Symbolic Act* (Ithaca and New York: Cornell UP, 1981) 137, 138–39, 150, respectively.

Adorno has made clear, Eichendorff is politically complex: 'His superiority to all the reactionaries who are claiming him today is shown by the fact that like the great philosophy of his time he understood the necessity of the revolution he was terrified of'.[24] The ambivalence that Eichendorff feels about the past is expressed in the unnatural, uncanny quality of the fountains and their garden setting in 'The Marble Statue'. Eichendorff's fountain symbol negotiates between the need for orderly change and the fear of a chaos caused by too much change, too quickly.

The hero of 'The Marble Statue', Florio, passes through an alley of trees into a hidden garden:

> Countless fountains splashed monotonously in the great solitude—playing with gilded balls. Looking between the trees, one saw in the distance a splendid palace glimmering with tall, slender pillars. No one could be seen near it; a deep stillness prevailed all around. Only now and then a nightingale would awake and sing as if almost sobbing in slumber. Florio gazed with astonishment at the trees, fountain, and flowers. It seemed to him as if all this had been lost and forgotten for ages, and the stream of days were flowing over it with light, limpid waves; as if under it all, only the garden lay there bound and bewitched and dreaming of the bygone existence.

> Unzählige Springbrunnen plätscherten, mit vergoldeten Kugeln spielend, einförmig in der großen Einsamkeit. Zwischen den Bäumen hindurch sah man in der Ferne einen prächtigen Palast mit hohen, schlanken Säulen hereinschimmern. Kein Mensch war ringsum zu sehen, tiefe Stille herrschte überall. Nur hin und wieder erwachte manchmal eine Nachtigall und sang wie im Schlummer fast schluchzend. Florio betrachtete verwundert Bäume, Brunnen und Blumen, denn es war ihm, als sei das alles lange versunken, und über ihm ginge der Strom der Tage mit leichten, klaren Wellen, und unten läge nur der Garten gebunden und verzaubert und träumte von dem vergangenen Leben[25] (344).

The uncanny vision of the uncounted fountains 'playing with gilded balls in their spray' can only be explained that these balls have been suspended by the water jets from the fountain. Perhaps not necessarily supernatural, this imagery does suggest a decadent refinement, an element of excess. This is the Catholic conservative dreaming of a bygone Renaissance Italy—it is a daydream. In his description of the landscape, it is made a dream with even the nightingales poised between waking and slumber. All is silence and stasis (appropriate to the marble statue of the title), or what is very like these forms of stillness—the constant reiteration of sound and motion, by seemingly infinite repetition of forms and processes. This is a Metternichian pipe dream from which all the challenge and fear of change is abolished or frozen into harmlessness.

Later when the hero returns to the garden, the *Springbrunnen* and their *Wasserstrahlen* are described again: 'From time to time jets of water from the numerous fountains, as if rejoicing, leaped clear over the tops of the shrubbery, sparkling brightly in the gold of twilight'.

> Von Zeit zu Zeit schwangen sich Wasserstrahlen von den vielen Springbrunnen wie jauchzend bis über die Wipfel der Gebüsche, hell im Abendgolde funkelnd (358).

---

24 Theodor Adorno, 'In Memory of Eichendorff', *Notes to Literature,* 2 vols., ed. Rolf Tiedemann, trans. Shierry Weber Nicholsen (New York: Columbia UP, 1991) 1: 55–79.

25 *Eichendorffs Werke,* ed. Richard Dietze (Leipzig and Vienna: Bibliographisches Institut, n.d.). All citations are in the text.

Here again the fountains are artificial, and this seems to be closely connected to the evil and enchanted world of the garden. As Seidlin writes, 'the fountains complete the picture of vacuity and futility, a monstrous *perpetuum mobile,* a motion continuously sinking back into itself, a senseless playing with shimmering toys, up and down, rising and falling, incessantly'. Thus the fountain represents 'the automatic motion spending and replenishing itself by inane circling—this is the garden of sin, sin as a garden'.[26] There is in the fountain symbols of Eichendorff a deeply negative and troubling side—the sense of endless repetition, without any progressive development; by implication, behind his fountain lies that larger symbol of the water-cycle symbol.

Goethe hated the High Romantics of Germany; so, I think, it is no accident that when, in *Faust, Part Two,* he has his devil contemplate building a little chateau for his own pleasure, Mephistopheles includes waterjets, indeed, rather obscene or at least earthy ones: 'And water jets of every kind [...] at all sides, there wee squirts hiss and piss by the thousands'.

> Und Wasserststrahlen aller Art [...]
> [...] an den Seiten,
> Da zischts und pißts in tausend Kleinigkeiten (ll. 10132–66).

Since unnatural sources appear with such infrequency in Goethe and are the preference of his nihilistic devil, one cannot but be tempted to read this as a dismissal of artificial fountains and the sorts of poets, drunk on the supernatural, who employ them. The reason for this particular dislike is the inveterate optimist's—Goethe's—dislike of the pessimistic use that the cycle symbol is put to by Eichendorff and, as we will soon consider, Novalis.

The English poets of the second generation prefer the supernatural fountain, but they demonstrate that fountains do not necessarily have to be used for such a negative vision of continual change: Keats and Shelley see the overflowing fountain as a symbol of poetical and mythological tradition from which every poet takes his inspiration from his forerunners; there is, furthermore, a sense of the larger water-cycle.

Keats uses this fountain symbol in the introduction to *Endymion* (65–168).[27] This is one of his rare uses of water imagery; very unusual, considering the rich uses of this imagery in Shelley's works—a poet with whom Keats name has been traditionally and somewhat facilely linked. This paucity of water in Keats' poetry is due, I think, to this poet's overpowering yearning for permanence and his morbid fear of change. Water is much too fluid—far too emblematic of the unstable and changeful. When Keats sought to embody change, he sought it in symbols that could more readily mediate a changeful world and his desire for permanence—the seasons or human beauty, for example. In his fountain passage from *Endymion,* Keats opens

26 Oskar Seidlin, 'Eichendorff's Symbolic Landscape', *PMLA*, 72 (1957): 645–61; qtd. from 650.

27 *The Poetical Works of John Keats*, ed. H.W. Garrod, 2nd ed. (Oxford: Clarendon, 1958) 65–168. In his discussion of the water-cycle, Frye makes particular use of this work: 'We find a great number of symbols from phases one and two in Keats' *Endymion*, and of symbols from phases three and four in *The Waste Land*' (*Anatomy*, 160).

book one with these lines containing the famous concept of the permanence of a work of art:

> A THING of beauty is a joy for ever:
> Its loveliness increases; it will never
> Pass into nothingness; but still will keep
> A bower quiet for us, and a sleep
> Full of sweet dreams, and health, and quiet breathing (ll. 1–5).

After a catalogue of supporting images, he culminates his rhetoric with the eternal source—as much a rain shower as a fountain—but from the gods:

> All lovely tales that we have heard or read:
> An endless fountain of immortal drink,
> Pouring unto us from the heaven's brink (ll. 22–24).

While to put too fine a point on the details would do damage to the tone, it certainly does suggest a sort of eternal source—divine, but only just the other side of Paradisiacal borders, falling to earth as a shower; this use of showers to symbolize poetic gifts will be seen to be a major usage in the lyrics of Keats' friend Shelley.

In the following lines taken from his long and, admittedly, complex poem, *Adonais: An Elegy on the Death of John Keats* (2: 385–405),[28] Shelley uses the same symbol in order to express an optimistic concept, just as Keats does, but the characteristics of his fountain more resembles those in the passage of Novalis to be examined next, for like the latter, Shelley found the clear and tranquilly splashing water to be insufficiently dynamic:

> Dust to the dust! but the pure spirit shall flow
> Back to the burning fountain whence it came,
> A portion of the Eternal, which must glow
> Through time and change, unquenchably the same [...] (ll. 338–41).

Thus, for Shelley, the cyclic dynamism implicit in the fountain does not suffice as a symbol of the eternally recurring process of poetry: his fountain must be burning and glowing.[29]

In Shelley's last great work, *The Triumph of Life* (4: 167–85), there is a fountain that has as sinister a context as anything envisioned by the German High Romantics. It is particularly valuable here because it has received one of the most profound readings that any Romantic poem has benefited from in recent decades: Paul de Man has examined its Romantic rhetoric. After considering it in light of Shelley's

---

28 *The Complete Works of Percy Bysshe Shelley,* eds. Roger Ingpen and Walter E. Peck, The Julian Edition, 10 vols. (London: Benn, 1927). Throughout the text of this entire book, all quotations from Shelley's works will come from this edition.

29 While I feel that the water-cycle is clearly implied in the lines, 'shall flow / Back to the burning fountain whence it came', this phenomenon has melded with another, 'the burning fountain'. It has been brought to my attention that the fiery aspect to Shelley's fountain is probably derived from his view of molten lava in the caldera of Mt. Vesuvius, described by the poet as a 'fountain of liquid fire', in a letter to Peacock dated 22 December 1818. The poet's idea, however, of a flowing back to the fountain clearly derives not from the volcano imagery but rather from the water-cycle.

fountain image, I will use it to treat the rhetoric in Novalis' *Heinrich von Ofterdingen*—this will be useful for comparison (as well as contrast) because in both works the fountains appear unexpectedly in dream visions of sorts. Paul de Man writes of a 'positional act', something similar in effect to the unexpected juxtapositioning sought by the Surrealists—the famous umbrella on an operating table.[30]

Paul de Man shows how figures appear throughout Shelley's *The Triumph of Life* in a 'brusque and unmotivated' manner—in a way that De Man describes as 'a positional act'.[31] The fountain and stream imagery in this poem is more unexpected precisely because it has a naturalistic quality that clashes with the supernaturalistic appearances of everything else in the poem; in fact, it resembles the realistic born springs that will be examined later in this chapter:

> Under a mountain, which from unknown time
> Had yawned into a cavern, high and deep;
> And from it came a gentle rivulet,
> Whose water, like clear air, in its calm sweep
> Bent the soft grass, and kept for ever wet
> The stems of the sweet flowers, and filled the grove
> With sounds [...] (ll. 312–18).

For De Man, this fountain and the stream it engenders jointly constitute Shelley's symbol for the gradual loss of consciousness; but the critic gives considerable emphasis to the natural 'semi-synaesthesia' of the metaphor in these lines:

> The property of the river that the poem singles out is its sound; the oblivious spell emanates from the repetitive rhythm of the water, which articulates a random noise into a definite pattern. Water, which has no shape of itself, is molded into the shape by its contact with the earth, just as in the scene of the water washing away the tracks, it generates the very possibility of structure, pattern, form, or shape by way of the disappearance of shape into shapelessness. The repetition of the erasures rhythmically articulates what is in fact a disarticulation, and the poem seems to be shaped by the undoing of shapes. But since this pattern does not fully correspond to what it covers up, it leaves the trace which allows one to call this ambivalent shaping a forgetting.[32]

It may well be the very shapelessness of water, described so eloquently by De Man, that makes this element such a challenge for not only Romantic poets, but also for the painters and musicians as well. Water's fluidity paradoxically is for them palpably described as neither shape nor shapelessness. This, however, is only part of water's fascination for them. We will see that it is not only this fluidity of water described here, but also its sounds, as De Man also indicates, that fascinates. The refluxing and re-sounding quality will correlate—as I shall show throughout this study—to 'erasures' not only of outward phenomena, but also erasures of, or rather, within human consciousness. In Romantic poets, each erasure is never total as in a Zen or a Tao state, but rather a part of the consciousness—the part that might be called the static, the 'random noise' of the mind, to allow the imagination to work.

---

30 Paul de Man, 'Shelley Disfigured', *The Rhetoric of Romanticism* (New York: Columbia UP, 1984) 93–123; qtd. from 117–18.

31 De Man 117.

32 De Man 107.

In works of Shelley, Goethe, or Rousseau, we will see that the water itself—for all its correspondence to subjectivity or its absence—remains actual, realistic water.

The Triumph of Life—however much in other respects it is a bona fide dream vision—represents the norm for Shelley's treatment of water: one that lies close to the natural actuality of water phenomenon. The juxtaposition in this poem of this actuality next to the supernatural manifestations of a dream vision only emphasizes the quasi-scientific accuracy with which he sought to render water. In this regard, its contrast with Novalis could not be greater.

In Novalis' *Heinrich von Ofterdingen,* there is a significant instance of this later type of artificial fountain being juxtaposed with a natural spring (131–32)[33]; both occur early in the novel, during the course of the hero's dream. The first of these is an artificial fountain (*Springquell*); the second, a natural spring (*Quelle*). According to M.T. Andrews, the first of these is a symbol for 'erotic, lustful sensations'.[34] Andrews' reading of the fountain symbol is supported by the fact that it, unlike the spring, is hidden away in a cave within a high cliff and by the comparison of the water to the body of young women:

> As he entered the cavern, he noticed a mighty beam of light, which was jetting up to the ceiling of the cavern as if coming out of a spring and fragmented there into countless sparks which gathered below into a large basin. The beam glistened like flaming gold. Not the slightest sound could he hear; a holy silence surrounded the glorious spectacle. He approached the basin, which surged and trembled in endless colors. The walls of the cave were all coated with this fluid, which was not hot but cool, the fluid on the walls emitted nothing but a dim bluish light. He dipped his hand into the basin and wet his lips. It was as though a spirit wafted through him, and he felt deeply refreshed and strengthened. An irresistible longing to bathe seized him; he undressed and entered the basin. It seemed to him as if a sunset cloud was enveloping him, a heavenly sensation flooded his heart; countless thoughts struggled with voluptuous delight, to interbreed within him; new and as yet not before seen images intermingled and palpable forms appeared around him, and every wave of the beautiful element nestled up to him like a tender bosom. The waves appeared to be attractive young women dissolved in liquid form, which immediately embodied themselves before him.

> Wie er hineintrat, ward er einen mächtigen Strahl gewahr, der wie aus einem Springquell bis an die Decke des Gewölbes stieg, und oben in unzählige Funken zerstäubte, die sich unten in einem großen Becken sammelten; der Strahl glänzte wie entzündetes Gold; nicht das mindeste Geräusch war zu hören, eine heilige Stille umgab das herrliche Schauspiel. Er näherte sich dem Becken, das mit unendlichen Farben wogte und zitterte. Die Wände der Höhle waren mit dieser Flüssigkeit überzogen, die nicht heiß, sondern kühl war, und an den Wänden nur ein mattes, bläuliches Licht von sich warf. Er tauchte seine Hand in das Becken und benetzte seine Lippen. Es war, als durchdränge ihn ein geistiger Hauch, und er fühlte sich innigst gestärkt und erfrischt. Ein unwiderstehliches Verlangen ergriff ihn sich zu baden, er entkleidete sich und stieg in das Becken. Es dünkte ihn, als umflösse ihn eine Wolke des Abendrots; eine himmlische Empfindung überströmte sein Inneres; mit inniger Wollust strebten unzählbare Gedanken in ihm sich zu vermischen; neue, niegesehene Bilder entstanden, die auch ineinanderflossen und zu sichtbaren Wesen um ihn wurden, und jede Welle des lieblichen Elements schmiegte sich wie ein zarter Busen an ihn. Die Flut schien eine Auflösung reizender Mädchen, die an dem Jünglinge sich augenblicklich verkörperten.

---

33 *Novalis Werke*, ed. Gerhard Schul (Munich: Beck, 1969). All citations are in the text.

34 Andrews 22.

The unformed quality, lacking in any erotic focus, suggests the nightmarish wet-dream of an adolescent. This sort of sexuality is not, for Novalis, full-formed, adult love. This and the spring in the next passage to be discussed, when thought of sequentially, offer a sort of sexual *Bildung* in the psyche of the novel's hero.

In the second 'illumination' ('Erleuchtung') or scene of his dream, the hero sees a somewhat more natural but still uncanny, spring ('Quelle'): this, according to Andrews, symbolizes 'divine love'.[35] The unnaturally jetting fountain ('Springquell') considered above is associated with lustful, sensual sexuality; on the other hand, this natural spring is closely connected with the pure, sinless love that Heinrich will have for his heroine Mathilde, who is symbolized by the 'light-blue flower' / 'lichtblaue Blume' which grows next to the spring: 'He found himself on a soft lawn next to a spring'.

> Er fand sich auf einem weichen Rasen am Rande einer Quelle [...].

But even it fails to behave in an entirely natural way:

> which jetted up into the air and seemed to consume itself there. Dark blue cliffs with multicolored veins arose further on; the daylight around him was brighter and milder than ordinary daylight, and the sky was dark blue and entirely clear. What, however, most powerfully attracted him was a tall, pale-blue flower, which grew beside the spring and touched him with its broad glistening leaves.

> die in die Luft hinausquoll und sich darin zu verzehren schien. Dunkelblaue Felsen mit bunten Adern erhoben sich in einiger Entfernung; das Tageslicht das ihn umgab, war heller und milder als das gewöhnliche, der Himmel war schwarzblau und völlig rein. Was ihn aber mit voller Macht anzog, war eine hohe lichtblaue Blume, die zunächst an der Quelle stand, und ihn mit ihren breiten, glänzenden Blättern berührte.

For the purposes of this study, the important thing to notice is that the word for an artificial fountain or jet of water, *Springquell,* is used for the first source. It is not a source of water but of some strange flaming gold and sparkling fluid. There is something artificial, occult, and sinister about this unnatural fountain. On the other hand, the word used to denote the second source is simply 'spring' (*Quelle*) denoting its natural character. Most sinister, I think, is the abrupt and surprising way that Novalis throws this cave and fountain into his narrative and shrouds it with supernatural manifestations; he does so in a way that suggests 'a positional act.' The full effect is that of the disorientation of surreality, of dreams. It is a phantasmagoric light-and-dark show of revelations that reveal nothing but a greater mystery. De Man might better be describing Novalis' fountain vision than (as he does) Shelley's when he penned the following: 'This play of veiling and unveiling is, of course, altogether tantalizing [...] a highly erotic experience; it is like glimmering light because it cannot be decided whether it reveals or hides'.[36] Interestingly enough, the first, unnatural source is described in exhaustive detail, whereas the poet denotes the second, natural one merely with one word, 'Quelle'. In Neoclassical works, the natural spring was the subject of elaborate descriptive passages. In *Heinrich von*

35 Andrews 23.

36 De Man 106.

Ofterdingen, the natural attributes of the source are reduced, making in the end a vaguely delineated symbol; its supernatural trappings are more important than the sight and sound of its flow. *The Triumph of Life,* on the other hand, lies closer to the Neoclassical works in that Shelley retains actuality in the fountain's flow with all the verbal means available; indeed the verbal strategies are key to the deeper signification of the metaphor. Shelley's use of the cave-born spring lies close to that of the poets to be treated next.

Since my larger topic is so rooted in the actual facts of nature as we can know them (i.e., as phenomena rather than noumena), I have devoted an unusually considerable space to both Novalis and the Eichendorff fountain passages simply because they establish the exemplars for the source as created predominantly from Romantic-style elements. Separating them both from those composed from Neoclassical elements or from Realistic elements, the Novalis and Eichendorff fountains are distinguished by the non-natural and—when compared to Scott's fountain passage—by extremes.

Source in the Mountains[37]

The poets to be treated in this section employ water phenomena evidencing considerable movement and especially seizing upon moments of cataclysmic change. This love for water in motion they share with the Baroque poets, of whom Rousset writes: 'No landscape is without water: it is a constant fact as much in the landscape gardening as in the speculations of the theoreticians and in the dream of the poets: and the preference is for a white, living water, rushing and splashing, a water in movement'.[38] What the water jets of Versailles and the artificial fountains of Rome were to the Baroque poets, torrential mountain springs were to the Early Romantic poets. The demand of the latter for natural water sources drew them to springs which are located high in the mountains and—at least, in the works of the three poets to be considered in this section—which are at once springs and falls.

Chronologically, the first of these poets is Friedrich Leopold von Stolberg. In the mid-1770s, he composed 'The Rock Stream: Immortal Youth' / 'Der Felsenstrom: Unsterblicher Jüngling' (1: 104–06),[39] which opens with the persona addressing and then describing a cataract as it spills from a cleft in a rock cliff: 'Immortal youth! You stream forth from the rocky cleft. No mortal has viewed the cradle of

37 While Marjorie Hope Nicholson does not treat water images and symbols as such and while she treats only English poets, I owe much of my initial impetus for this section to her quotations from the romantic poets in the Introduction and Epilogue to *Mountain Gloom and Mountain Glory: The Development of the Aesthetic of the Infinite* (New York: Norton, 1963) 8–17, 375–93.

38 My translation of: 'Pas de paysage sans eau: c'est une vérité constante dans la pratique des jardins aussi bien que dans la spéculation des théoriciens et dans la rêve des poètes; et, de préférence, une eau vive, courante ou jaillissante, une eau en mouvement'. Rousset, *Littérature de l'âge Baroque* 143 et passim.

39 *Gesammelte Werke der Brüder Christian und Friedrich Leopold Grafen zu Stolberg*, Authorized ed. (Hamburg: Perthes, 1827). Throughout the text of this entire book, all quotations from Stolberg's works will come from this edition.

the mighty one; it is given to no ear to hear your babbling nobility within the gushing source'.

Unsterblicher Jüngling!
Du strömest hervor
Aus der Felsenkluft.
Kein Sterblicher sah
Die Wiege des Starken;
Es hörte kein Ohr
Das Lallen des Edlen im sprudelnden Quell (ll. 1–7).

The lines that follow (ll. 8–46, which will be analyzed in subsequent chapters) inconsistently combine the aspects of the water phenomena, and the symbolic meanings which the poet attaches to them are also inconsistent. In the opening lines, the poet uses a spring/waterfall and its aspect of seeming to changelessly pour forth and employs it for a symbol of an 'immortal youth'. On the other hand, in the remaining stages of the poem, the spring/waterfall is shown in its destructive mode and then it is treated as the source of a river which in turn flows into a lake; that is to say, it is treated there as a phenomenon subject to time and change; further, it is employed as a symbol for the stagnation of adulthood and the human subjection to death. The inconsistency in these two different symbolic meanings is clear: something which is always youthful does not grow old and die.

Stolberg's poem was not only important in its contribution to Storm and Stress and German High Romanticism, but it also had a direct influence on one of Coleridge's most sorely neglected lyrics, 'On a Cataract, from a Cavern near the Summit of a Mountain Precipice' (1: 303–09), written almost two decades after Stolberg composed his. The Coleridge work cannot, by the broadest use of the term, be called a translation; at most, it appropriates from the Stolberg poem a few key images from the opening stanza of the German poem. Unlike Stolberg, Coleridge devotes his entire poem to the eternal youthfulness of the spring/waterfall. The fact that it is a source for a stream is irrelevant symbolically because the source per se is the symbol:

Unperishing Youth!
Thou leapest from forth
The cell of thy hidden nativity;
Never mortal saw
The cradle of the strong one;
Never mortal heard
The gathering of his voices;
The deep-murmured charm of the son of the rock,
That is lisp'd evermore at his slumberless fountain (ll. 1–9).

In this more expanded form, the simple allegorical symbol of the Stolberg poem takes on added dimension by representing the abstract concept of continuity-in-change (i.e., 'ceaseless renewing'); also, the sense of religious mystery found in Stolberg's lines concerning the covert birth of the spring (ll. 4–5) is expanded in the Coleridge poem:

There's a cloud at the *portal*, a spray-woven *veil*
At the *shrine* of his *ceaseless renewing*;

It embosoms the roses of dawn,
It entangles the shafts of the noon,
And into the bed of its stillness
The moonshine sinks down as in slumber,
That the son of the rock, that *the nursling of heaven*
May be born *in a holy twilight*!

The wild goat *in awe*
Looks up and beholds
Above thee the cliff inaccessible [...] (the emphasis is mine; ll. 10–20).

All of which supports the interpretation that the persona is speaking of the supernatural, the unchanging processes of universal life, which in its turn contains within it the changes of individual life and death—of birth and destruction:

Thou at once full-born
Madd'nest in thy joyance,
Whirlest, shatter'st, splitt'st,
Life invulnerable (ll. 21–24).

This idea of a spring born far within a cave, so that the source is hidden, will prove a rich one in Romantic Culture. It is interesting that Coleridge has ended his poem at this point; as I will show, the later part of Stolberg's lyric introduces Storm and Stress ideology relative to Germany of the 1770s and not to England of the 1790s.

In the very last part of *Faust, Part Two* (5: 289–526) that Goethe finished in 1831, Goethe has what I take to be a parody of the Romantic rock-born spring/falls; I say parody because it comes as part of Mephistopheles' arsenal to help Faust win the Emperor's war and gain his lowlands by the sea for his development project. Faust describes how rock springs spurt forth to wash away the enemy troops attempting a flanking movement over mountain paths: 'I see a trickle there already. At many a spot in the dry and bare rock the abundant, rapid spring develops':

Dort fängt es schon zu rieseln an.
An mancher trocknen, kahlen Felsenstelle
Entwickelt sich die volle, rasche Quelle [...] (ll. 10, 719–21).

These little gushing springs form into a giant waterfall, sweeping all the rival emperor's army before it; the sense of the flux and reflux of the rushing water coming from all sides is created by the constant reiteration of the conjunction, *und*/*and*:

Already one stream rushes down powerfully to join another stream, from out of gorges they come back double in size and form a grand outwardly curved arc; all of a sudden it settles in a rocky part and roars and foams to this side and that side, and then gradually throws itself into the valley. What is the use of bold, hero-like resistance when the powerful wave streams to wash them all away? I myself shudder at such a fierce surging'.

Schon rauscht ein Bach zu Bächen mächtig nieder,
Aus Schluchten kehren sie gedoppelt wieder,
Ein Strom nun wirft den Bogenstrahl;
Auf einmal legt er sich in flache Felsenbreite
Und rauscht und schäumt nach der und jener Seite,
Und stufenweise wirft er sich ins Tal.
Was hilft ein tapfres, heldenmäßiges Stemmen?

Die mächtige Woge strömt, sie wegzuschwemmen.
Mir schaudert selbst vor solchem wilden Schwall (II. 10, 725–33).

In part, this moment in *Faust* reflects Goethe's horror at the extremes of colossal change. Taking Mephistopheles' magic as a metaphor—in its sixteenth-century guise—for the science and technology of Goethe's own time, we can understand the horror that Faust feels at this victory. These mountain water sources constitute weapons of mass destruction. In terms of the water-cycle, Goethe returns again to the old Storm and Stress symbolism of rock-born waterfalls, the joining of streams into larger, and the sheer downward, collective force of water coming out of the mountains; these elements are not really modified by being combined into the magic and science-technology symbolism. Later, we will see how this magic as signifier of science and technology—both in its positive and negative effects—receives its culminating treatment in act five of *Faust, Part Two*.

The German-English parallels and influences run even deeper. In order to fully understand the symbolic meaning of 'On a Cataract', it is necessary to look at selected passages from another poem by Coleridge that employs the symbol of a rock-born spring/waterfall, 'Hymn before Sun-Rise, in the Vale of Chamouni [sic]' (2: 376–80)[40]; this poem is loosely based on a fine German poem, which falls outside the range of this study because of its very marginalized use of water phenomena—indeed, it is the emphasis on the latter that is Coleridge's true contribution to his own version.[41] Above, in dealing with 'On a Cataract', I rely entirely on the poem itself; whereas, with 'Hymn before Sun-Rise', I am aided by the information contained in Coleridge's prose introduction to the poem. Here, the intellectual undercurrents of the poem are clearly set forth: the imagery of this poem has been chosen to show God manifest in his nature: 'Who *would* be, who *could* be an Atheist in this valley of wonders!' (2: 377, fn.; italics are in the original). The spring/waterfall offers a symbol for continuity in change, and this in turn symbolizes God—that greatest of all mythologizations for eternal continuity. Here, as in 'On a Cataract', the waterfall springs from the caves or crevasses beneath the melting glaciers and is 'invulnerable life' as compared to 'Life invulnerable' ('On a Cataract', l. 25). These streams of water are being destroyed, but by means of this destruction, there is created an enduring and beautiful vision of foam:

And you, ye five wild torrents fiercely glad!
Who called you forth from night and utter death,
From dark and icy caverns *called you forth,*
Down those precipitous, black, jagged rocks,
For ever shattered and the same for ever?
Who gave you your invulnerable life,

---

40 *The Complete Poetical Works of Samuel Taylor Coleridge*, ed. Ernest Hartley Coleridge (Oxford: Clarendon, 1912).

41 In 'Hymn before Sun-Rise', there again was a German inspiration for this lyric of Coleridge: Friedrike Brun's 'Chamouny [sic] beym Sonnenaufgange'. Since the water images of the German are inconsequential, an analysis of Brun's fine poem—even in light of that of Coleridge—would contribute nothing to this study; for a copy of her poem, see Coleridge's *Poetical Works*, 2: 1131. Concerning Coleridge's poem and its relation to Brun's poem, see Adrien Bonjour, *Coleridge's 'Hymn before Sunrise'* (Lausanne: Concorde, 1942).

Your strength, your speed, your fury, and your joy,
Unceasing thunder and eternal foam? (my emphasis; ll. 39–46).

The language of the above lines (especially ll. 40–46) suggests the concept of Resurrection. He employs a symbol of potential permanence in change, a metaphor for the potential, yet-unlocked energy of Apocalypse and Resurrection.[42] Even within the context of the poem, Coleridge has his persona clearly set forth his ideas; for his answer to his questions is placed in all-capital letters:

Ye Ice-falls! ye that from the mountain's brow
Adown enormous ravines slope amain—
Torrents, methinks, that heard a mighty voice,
And stopped at once amid their maddest plunge!
Motionless torrents! silent cataracts!
Who made you glorious as the Gates of Heaven
Beneath the keen full moon? Who bade the sun
Clothe you with rainbows? Who, with living flowers
Of loveliest blue, spread garlands at your feet?—
  GOD! let the torrents, like a shout of nations,
  Answer! and let the ice-plains echo, GOD!
GOD! sing ye meadow-streams with gladsome voice!
Ye pine-groves, with your soft and soul-like sounds!
And they too have a voice, yon piles of snow,
  And in their perilous fall shall thunder, GOD! (emphasis is Coleridge's, ll. 58–63).

Thus it is this deified continuity which pervades a landscape that might otherwise superficially seem to be a terrifying vision of change; here is yet another use of natural phenomenon which seems to run counter to what it signifies. The description of the spring/waterfall is objective and yet also colored by a concept of God which the poet's vision finds innate within the landscape. The reiteration seems to me to be an intellectual embarrassment and suggests some underlying spiritual crisis: why else would Coleridge feel the need to protest so much.

Mary Shelley, similarly, has her character Frankenstein (1818)[43] see in cataract-pervaded landscape around Mont Blanc an emblem of Divinity:

> The weight upon my spirit was sensibly lightened as I plunged yet deeper in the ravine of Arve. The immense mountains and precipices that overhung me on every side—the sound of the river raging among the rocks, and the dashing of the waterfalls around, spoke of a power mighty as Omnipotence—and I ceased to fear, or to bend before any being less almighty than that which had created and ruled the elements, here displayed in their terrific guise (91).[44]

---

42 It seems rather strange to me that with all the poem's apocalyptic visioning, M.H. Abrams omits this from consideration in *Natural Supernaturalism*.

43 Although they did not really treat the passage in the Mont Blanc locus of the novel, two works, in particular, that helped me negotiate the difficulties of this novel are Anne K. Mellor, 'Possessing Nature: The Female in *Frankenstein*', *Romanticism and Feminism*, ed. Anne K. Mellor (Bloomington: Indiana UP, 1988) 220–32; and Margaret Homans, 'Bearing Demons: Frankenstein's Circumvention of the Maternal', *Mary Shelley's Frankenstein*, ed. Harold Bloom (New York: Chelsea House, 1987) 133–53.

44 Mary Shelley, *Frankenstein, or The Modern Prometheus*, rev. ed., ed. Maurice Hindle (Harmondsworth: Penguin, 1992) 91. Throughout the text of this entire study, all references to this work by Mary Shelley will be from this edition.

Yet, as I have tried to show again and again throughout this study, the symbol is determined, in part, by the particular intellectual inflections given by each artist—let alone, for the moment, each reader/viewer. The scene makes Frankenstein think of a more traditional form of divinity, much closer at least rhetorically to Coleridge than that held by her free-thinking husband. Mary Shelley clearly kept her own metaphysical council. I think juxtaposing her prose vision of the Arve with Percy Shelley's 'Mont Blanc' reminds one just how anti-humanistic and traditionally religious *Frankenstein* is: its message is that mankind has no right to tamper with God's nature. Indeed, everything—river from melting ice or rivers of ice itself—is an emblem of the power of God's nature violated by Frankenstein:

> Immense glaciers approached the road; I heard the rumbling thunder of the falling avalanche and marked the smoke of its passage. Mont Blanc, the supreme and magnificent Mont Blanc, raised itself from the surrounding *aiguilles*, and its tremendous *dôme* overlooked the valley (92).

This scene—where not just water cascades but great sheets of mountain ice—represents the creator, whose power Frankenstein the Monster's 'creator' has usurped; but it, perhaps, more cogently represents the hidden, latent power of nature that the inventor-scientist has unleashed in the form of his Monster—and that his colleagues have further potential to let out. It is after all in this magnificent Alpine scenery that Mary Shelley has set the C/creator's confrontation with his C/creature. The water symbol offers a gauge by which to measure all the radical mood shifts of Frankenstein during this episode. At first, he feels his mood lighten, but after he is weighted down with despair and throws himself 'on the grass, weighted down with horror and despair'.

More and more, however, nature soothes him, but not without reminding him of her power:

> The abrupt sides of vast mountains were before me; the icy wall of the glacier overhung me; a few shattered pines were scattered around; and the solemn silence of this glorious presence-chamber of imperial Nature was broken only by the brawling waves or the fall of some vast fragment, the thunder sound of the avalanche or the cracking, reverberated along the mountains, of accumulated ice, which, through the silent workings of immutable laws, was ever and anon rent and torn, as if it had been but a plaything in their hands (93).

Shelley enfolds her concept of N/nature in figures of speech that give it a primacy over a more orthodox conception of God. There is a very powerful feeling here that we are dealing with something closer to an ancient fertility deity whose power the male Frankenstein would have usurped in creating life without a woman. Mary Shelley's deity may be atavistic, but at least it is not Patriarchal.

The glacial ice is subject to the same mutability as the sea quarter of the water-cycle, so much so that it would be good to bear this passage in mind when we consider the Polar Sea episodes of the novel in chapter three:

> The surface is very uneven, rising like the waves of a troubled sea, descending low, and interspersed by rifts that sink deep. The field of ice is almost a league in width, but I spent nearly two hours in crossing it (95).

Also, as I will discuss in chapter four, the link quarter of the water-cycle is appealed to in this Mer de Glace section. Mary Shelley's choice of setting this key scene of her great novel along the great Mer de Grace glacier reminds us of the powerful impact that the Mont Blanc, Chamonix area had upon the Romantic minds. Both flowing ice, as well as rushing water, amid the world of ice and rock gave Romantics a perfect vehicle to express permanence, change, and the Power behind them. Wordsworth too fell under the spell of this alpine region.

Not only the description of the overhanging cliff fraught with waterfalls, but above all the staging of a key moment of epiphany is to be found in the crossing-the-Alps 'spot of time' in the 1805 version of *The Prelude.*[45] Wordsworth describes the landscape with mountain waterfalls in the Alps and clearly states that they are symbols of the E/eternal—that is, symbols of God or, at least, of the eternal processes of nature; yet, paradoxically, they are symbols of cataclysmic change, representing the greatest one of all—'the great Apocalypse' of *Revelations*. Here, the signifier's dual qualities of permanence and change are consciously crafted by Wordsworth to signify respectively the eternal and the end-time, which would also be equally split and clashing in nature:

> The stationary blasts of water-falls,
> And every where along the hollow rent
> Winds thwarting winds, bewilder'd and forlorn,
> The torrents shooting from the clear blue sky,
> The rocks that mutter'd close upon our ears,
> Black drizzling crags that spake by the way-side
> As if a voice were in them, the sick sight
> And giddy prospect of the raving stream,
> The unfetter'd clouds, and region of the Heavens,
> Tumult and peace, the darkness and the light
> Were all like workings of one mind, the features
> Of the same face, blossoms upon one tree,
> Characters of the great Apocalypse,
> The types and symbols of Eternity,
> Of first and last, and midst, and without end (6: 558–72).[46]

The poet seems to desire a certain ambiguity between the waterfall-stream quarter and the rain-cloud quarter when he describes the 'drizzling crags' and the 'torrents shooting from the clear blue sky'. The processes convey change—thus they have a beginning, a middle, and an end—yet they are permanent, 'without end'. As M.H. Abrams comments that here 'the Scriptural Apocalypse is assimilated to an apocalypse of nature; its written characters are natural objects, which are read as types and symbols of permanence in change' (107). I agree with Abrams that, at least in this instance, Wordsworth has for once not allowed his metaphysical elements to lessen the actualism of the physical reality of water phenomena in the alpine region near Chamonix—Supernaturalism does not here get the better of Naturalism.

In understanding the importance of the falls springing out or down from the rocks to convey the sense of permanence in change, it is significant to examine the

---

45 See Kenneth MacLean's useful study, 'The Water Symbol in *The Prelude* (1805–06)', *University of Toronto Quarterly* 13 (1947–48): 372–89.

46 William Wordsworth, *The Prelude or Growth of a Poet's Mind*, ed. Ernest de Selincourt, 2nd ed, rev. by H. Darbishire (Oxford: Clarendon, 1959). Throughout the text of this entire study, all references to the 1805 and 1850 versions of *The Prelude* will be from this edition.

rhetoric, to distinguish image and symbol, and to observe that the same Alpine scene of the above passage was described in another earlier poem by Wordsworth. In his *Descriptive Sketches* (*PW* 1: 42–91)—which, as their name implies, still remain in great part within the tradition of eighteenth-century descriptive poetry—the image of the falls is buried among other landscape imagery and has no symbolic purpose:

> —Thy torrents shooting from the clear-blue sky [...] (ll. 113).[47]

The waterfall is one of the many landscape images that is part of the imagery list, which constitutes this early poem. The falls image in line 113, however, is the same as that in line 561 of the 1805 version of *The Prelude* and would seem to be the kernel from which the latter's waterfalls-symbol grew. It is, nevertheless, certain that the complexity of this symbol would never have grown had not the seeming failure of the French Revolution compelled Wordsworth to escape from his fear of change. This fear maneuvered Wordsworth into a world view where one change cyclically swallows up another change. Change can thereby be safely viewed by a poet inclining to conservatism: he can view it as a Burkean *revolving* to the past.

In order to better understand the above passage from the 1805 version, it is useful to examine some other passages in the same version which illumine the concept of the 'workings of one mind' in line 567 and the use of the workings of water phenomena to symbolize it. Later in the Alps section of book six, there is a passage (6: 658–72) in which the persona says he must 'break off [...] the record of these wanderings' and must forego the temptations of descriptive verse when images which might fail as symbols to convey changeful, metamorphic qualities of his main theme—that of the 'Growth of a Poet's Mind'. As he makes clear in a later passage, the speaker privileges the subject over the object in the act of perception. The subjective portion of the poet's mind—which is symbolized by the second stream—is added to by the objective portion—the first stream. The objective portion of the mind is constructed from perceptions of the external world separate from the subjective, but 'kindred':

> Finally whate'er
> I saw, or heard, or felt, was but a stream
> That flow'd into a kindred stream, a gale
> That help'd me forwards, did administer
> To grandeur and to tenderness, to the one
> Directly, but to tender thoughts by means
> Less often instantaneous in effect;
> Conducted me to these along a path
> Which in the main was more circuitous (1805 version, book 6: ll. 672–80).

Similarly, in book two, Wordsworth uses the image of a river with its various tributary springs to represent the mind and its contributive experiences:

> Who knows the individual hour in which
> His habits were first sown, even as a seed,
> Who that shall point, as with a wand, and say,

---

47 The passage qtd. from *Descriptive Sketches* is in the original 1793 version not the 1849 revision, which is printed opposite the original version.

'This portion of the river of my mind
Came from yon fountain?' (1805, 2: 211–15).

In all these passages from *The Prelude,* Wordsworth presents a vision perched on the edge of chaos—indeed, superficially, it seems to be chaos itself; however, the vision is suffused with symbols of mind, of order.

In his 'Mont Blanc: Lines Written in the Vale of Chamouni [sic]' (1: 229–33), Percy Shelley, like Wordsworth in the passage just examined, makes significant use of the spring/waterfall as a symbol of human thought and some gnosis of a higher order; it, too, is set in the context of seeming chaos:

The everlasting universe of things
Flows through the mind, and rolls its rapid waves,
Now dark—now glittering—now reflecting gloom—
Now lending splendour, where from secret springs
The source of human thought its tribute brings
Of waters,—with a sound but half its own,
Such as a feeble brook will oft assume
In the wild woods, among the mountains lone,
Where waterfalls around it leap for ever,
Where woods and winds contend, and a vast river
Over its rocks ceaselessly bursts and raves (ll. 1–11).

Here Shelley takes the mountain landscape as a symbol for the human mind. He divides water imagery into two types: (1) the 'rapid waves', 'waterfalls', and 'a vast river' which represent the external world of all things including the transient and the Platonic 'Everlasting universe of things'; (2) then, the 'secret springs' which are 'The source of human thought' forming, in turn, a tributary in the form of 'a feeble brook', which is often passively affected by the external and assumes 'a sound but half its own'.

In the next section, the Arve is described as springing from its sources in 'ice gulphs' and evidencing destructive power, as in Stolberg and Coleridge, in the forms of rapids and waterfalls. The 'Power' of transient and, by implication, of transcendent things is symbolized ('in the likeness of the Arve'):

Thus thou, Ravine of Arve—dark, deep Ravine—
Thou many-coloured, many-voiced vale,
Over whose pines, and crags, and caverns sail
Fast cloud shadows and sunbeams: awful scene,
Where Power in likeness of the Arve comes down
From the ice gulphs that gird his secret throne,
Bursting thro' these dark mountains like the flame
Of lightning thro' the tempest [...] (ll. 12–19).

In these lines and those that follow them, Percy Shelley seeks and finds, in this varied landscape of change, a compendium of natural processes to represent the evanescent workings of the human mind (and perhaps divine mind, in the Platonic sense). Among the manifold meanings contained in the above reference to 'Power', Shelley might well have had in the back of his mind the concept of water as a technological source of power. Chamonix and these very waterfalls that inspired Shelley's multivalent conception of power fit into the science of thermodynamics—the study of the motive power of heat—and into the biography of one of its great pioneers, James Prescott Joule of Manchester: 'He married at the age of thirty-nine, chose Chamonix for his wedding trip, and took with him a long thermometer to the Alpine waterfalls at various altitudes, in the hope of finding the temperature increasing with the motion. The trip held disappointment. There was too much spray'.[48] The heat driving the steam engines of his native Manchester was on Joule's mind in the Alps. Mundane paradigms cannot but help shape even the tourist's mentality. Shelley and Joule were predisposed by their water-wheel and stream-driven world to think of the physics of power and energy in the presence of the waters of Chamonix.

But unlike Joule, Shelley is, for once, not being a scientist; but, in this fairly early poem, very metaphysical. Shelley's 'eternity' is, as in Wordsworth and Coleridge, represented as 'ceaseless motion' and 'unresting sound':

> Thy caverns echoing to the Arve's commotion,
> A loud, lone sound no other sound can tame;
> Thou art pervaded with that ceaseless motion,
> Thou art the path of that unresting sound [...] (ll. 30–33).

The seat of the Eternal power that drives all this transient energy is the mountain. As Earl R. Wasserman has written: 'We learn that the Arve is a likeness of the Power [i.e., not the Power itself], that the Power is inaccessible, and, at the climax, that Power is symbolized by the summit of Mont Blanc and governs both mind and matter'.[49] The climatic lines that Wasserman refers to are:

> Far, far above, piercing the infinite sky,
> Mont Blanc appears,—still, snowy, and serene—
> Its subject mountains their unearthly forms
> Pile around it, ice and rock; broad vales between
> Of frozen floods, unfathomable deeps,
> Blue as the overhanging heaven, that spread
> And wind among the accumulated steeps [...] (ll. 60–68).

The phrase 'the frozen floods' conveys with its oxymoronic structure the sense of permanence in change; on a purely scientific level, it conveys a sense of the potential untapped energy of mountain ice fields and glaciers, just as we have found Coleridge doing: 'Motionless torrents! silent cataracts!' Ice then is one source of the actual water-cycle and, with its combination of potential energy locked in a steady-state, becomes a symbol for Eternal Power that offers the permanent source behind the transient workings of the natural world.

---

48 Charles Coulton Gillispie, *The Edge of Objectivity: An Essay in the History of Scientific Ideas*, pb. ed. (Princeton: Princeton UP, 1973) 371.

49 Earl Wasserman, 'Mont Blanc' in George M. Ridenour, *Shelley* (Englewood Cliffs: Prentice, 1956) 69–102, qtd. from 81. I still prefer Wasserman's reading of the poem but I feel my reading is supported by Jerrold E. Hogle's view that the poem takes us 'beyond polarity' to reveal nature as 'the continuous swerving between separating differences', *Shelley's Process: Radical Transference and the Development of His Major Works* (New York: Oxford UP, 1988) 85; William Keach's application of an evolved, higher formalism is useful, *Shelley's Style* (New York: Methuen, 1984) 194–200. Of recent essays, I prefer Francis Ferguson, 'Shelley's 'Mont Blanc': What the Mountain Said', *Romantic Poetry: Recent Revisionary Criticism*, ed. Karl Kroeber and Gene W. Ruoff (New Brunswick: Rutgers UP, 1993) 335–44.

The Eternal Entity of Coleridge, of Wordsworth, and of each of the Shelleys differ from one other, and yet all found their expressive vehicle in the same high-mountain landscape; I think this is so because these symbolize, at their lowest common denominator, the permanent substance underlying change. The chaos of shattered boulders and precarious ice becomes a symbol of transcendent order. Crucial too is the fact that these landscapes contain, for symbolic purposes, water phenomena that are strikingly similar to those found in the Stolberg lyric—a waterfall welling forth and falling from rocks and/or ice—which share a sense of dynamism. The works discussed in the section that follows are not falls but sources which are set in mountains—if perhaps not the high, alpine type.

In at least four works, Friedrich focuses on unimproved springs in mountains, and—as with the poetic works we have just considered—these paintings are accompanied by symbols of eternity; while they are Christian, they are radical because of their wild, natural context. In *Cross in the Mountains / Kreuz im Gebirge,* a spring and its brook dominate the lower center of the canvas. Flowing, as this spring does, from the foot of a cross, with the Gothic spires of a church looming above, the spring clearly represents the source of faith. The iconography is the same in the *Cross in the Forest / Kreuz im Walde,* with the exception that there one finds no church and the spring issues into a small waterfall. Börsch-Supan interprets that work in the following terms: 'The cross with the sun above the spring indicates the relationship of the beginning and end of earthly existence with Christ, and the promise of eternal life through Christ'.[50] Whereas the connection of the springs in these two paintings with the rest of the cycle is only intimated, the title of *Source of the Elbe / Elb' Quelle* makes the cycle-context clear. The latter work is also interesting because it uses a real locus, but in this instance for symbolic purposes. Seen in the context of the *Cross in the Mountains* and *Cross in the Forest,* the spring in the *Source of the Elbe* represents either the origins of life or the sources of faith, which the human figure in the painting contemplates. The water derives ultimately from the mountains in the hazy background—an arrangement that may offer a symbol for the poorly understood mystery of the signified metaphysical sources and origins.

Birth and Infancy: The Spring in the Water-Cycle

Springs, when viewed as sources for streams, are thereby placed in the context of the water-cycle. Indeed, springs are here being considered conveniently as one quarter of the whole cycle. In the first work to be discussed, the poet treats only three of the quarters—the spring, stream, and ocean. The process he describes is bounded at either end, as is the thing which this process represents: the three main stages of human life—birth and youth, middle age, old age and death. Written in the early 1770s, Goethe's 'Mohammed's Song' or—less literal but more accurate—'Song in

---

50 Helmet Börsch-Supan, Catalogue Notes, *Friedrich* (London: Tate Gallery, 1972), 51–97; qtd. from 66. Concerning this spring symbol, Robert Rosenblum writes, 'A river source (which almost seems to flow like blood, from the base of the cross)' symbolizes 'the coming of new life from the cross', *Modern Painting and the Northern Romantic Tradition: Friedrich to Rothko* (New York: Harper, 1975) 31.

Praise of Mohammed' / 'Mahomets Gesang' (1: 304–05) begins with the spring as symbol for birth of the Prophet, whose life will be described throughout the poem by means of direct, allegorical usage of the river—that is, the prophet's life is comparable to the course of the river.[51] Thus, the symbol has a closed, one-to-one relationship with the thing symbolized and no further meanings are suggested. Moreover, although there is much objective description of the river, Goethe is describing an unspecified river which is important only as a symbol.

Since the lyric falls within the pale of the Storm and Stress movement, it is not surprising that the 'birth' of the river/prophet is the tumultuous one of a rock-born spring which immediately gives forth into a small, rapid mountain stream. In these short lines the Germanic way of compounding words not only contributes to the density of ideas or sensations, but also allows for the springing rhythms of youthful exhilaration and the gushing of the water, an effect I have approximated in my translation at some cost to the English syntax. Also, I have sought to convey the ambivalent sense of what is at once a spring and a young prophet in rendering the pronouns: 'See the rock-born spring, joy-bright, like a star's gaze; above the clouds, kindly spirits have nurtured his/its youth in thickets amid the crags. Youth-fresh he/it dances out of the clouds, down upon the marble crags and rejoices again to the heavens'.

Seht den Felsenquell,
Freudehell,
Wie ein Sternenblick;
über Wolken
Nährten seine Jugend
Gute Geister
Zwischen Klippen in Gebüsch.

Jünglingfrisch
Tanzt er aus der Wolke
Auf die Marmorfelsen nieder,
Jauchzet wieder
Nach dem Himmel (ll. 1–12).

Similarly, in the passage that follows, the overall effect of the language is in keeping with the revolutionary sentiments of the Storm and Stress movement: the river/prophet moves with the steps of a young leader who calls to his brother streams which join his motion/movement: 'Down through the clefts in the mountain peaks, he chases after multi-colored pebbles, and with the steps of a young leader, he carries away his brother springs with him'.

---

51 This is one of the few poems that Ernest Loeb and I each treat. As he points out, the rest of the cycle—the cloud quarter—is in the poem and the entire cycle is implied: 'In "Mahomets Gesang" "sehen" wir zunächst [...] die zükünftige Bestimmung des Felsenquells von guten Geistern "über Wolken" voherbestimmt: ein bedeutsamer Hinweis auf die schicksalhafte Ursprungs- und Endbestimmung, von der auch das Elementare grundsätzlich nicht ausgeschlossen, deren bildhaft-einprägsamster Ausdruck vielmehr der Kreislauf des Wassers ist. Der Wolke ent-"tanzend" weiß der junge Quell sich im Abwärtstaumeln dennoch dem himmlischen Ursprung verbunden, zu dem er grüßend emporjauchzt' (102).

Durch die Gipfelgänge
Jagt er bunten Kieseln nach,
Und mit frühem Führertritt
Reißt er seine Bruderquellen
Mit sich fort (ll. 13–17).

Here and hereafter, I am going to keep to the personification in translating the pronouns. As with the Stolberg work, there is an exuberance in the use of the water symbol in Goethe's poem; however, there are qualitative differences between the two lyrics. Whereas the rock-born spring is inconsistently employed as a symbol, here the anthropomorphic usage is sustained by means of such terms as 'Führertritt', 'reißt', and 'Bruderquellen'—anthropomorphic to such a degree that Ruskin's 'pathetic fallacy' comes immediately to mind. For to say that a stream of water moves with 'steps' or 'strides' is to violate any sort of correspondence between the symbolized and the symbol: A river's movement has no qualities which are analogous to human walking.

In a sonnet written four decades later, 'Powerful Overwhelming' / 'Mächtiges Überraschen' (1: 269), Goethe employs this same motif of birth and rebellious youth expressed by the spring/stream symbol (but as I shall later show, it is for an ironic purpose): 'A stream gushes out of a cloud-encompassed cavern, eager to unite with the ocean; in spite of what may be mirrored in the pools along its course, it moves unstoppably onward and valleyward'.

Ein Strom entrauscht umwölktem Felsensaale,
Dem Ozean sich eilig zu verbinden;
Was auch sich spiegeln mag von Grund zu Gründen,
Er wandelt unaufhaltsam fort zu Tale (ll. 1–4).

Here, unlike in 'Song in Praise of Mohammed', the singer of the song, the speaker of the poems gives neither excessive nor inappropriate physical attributions to the natural object. In the sonnet, the metaphysical attributes of consciousness and self-will ascribed to the water by the speaker is neither inappropriate nor excessive, since they represent characteristics of both the subject (the speaker) and of the object (the spring/brook) and, above all, are stated in such an ambiguous way as not to violate the physical nature of the one with the attributes of what it symbolizes; thus, the symbolic relationship remains open.

There is yet another aspect of 'Overwhelming Astonishment' that needs discussing—one that is indeed more crucial to the larger subject of the hydrological-cycle: this sonnet of Goethe's is one of those examples where the larger archetypical scheme of the water-cycle, as mapped out by Northrop Frye, has been 'short-circuited' in some natural way (it, of course, has nothing to do with the supernatural spring passage of the High German Romantics or, for that matter with Mephistopheles' created sources). Either through personal empirical observation or a knowledge of the underlying theory—and in Goethe's case it was both—the poet finds and exploits these little deviations in the great complexity of the hydrological-cycle.

Written about the same time as Goethe's sonnet, Wordsworth's *The River Duddon: A Series of Sonnets* (*Poetical Works* 3: 244–61)[52] uses the stream's spring to represent human birth and infancy, and both streams have their source/birth among the cloudy high mountains—hence, a relation between the river and the cycle of life is established. The great difference between Goethe's sonnet as well as his 'Song in Praise of Mohammed' and the sonnet series of Wordsworth lies in the fact that, whereas Goethe uses an unspecified river as an allegorical symbol, Wordsworth's sonnet series describes a real river—hence, it traces its descent in part from the topographical tradition. Many of the sonnets, moreover, describe scenes and events along the river. This work taken as a whole embodies the essence of what my study is attempting: to capture a sense of the complexity involved, not just in the water-cycle, but in the way it touches human existence in the greater and smaller coursings of the quarters of this cycle. What is more: *The Duddon River*—as I will show throughout this and the coming chapters—duplicates to some degree in art what Stephen Lansing was attempting to do in his application of the complexity theory to anthropology. Lansing finds a stream in Bali to be a social, mythical, and agricultural nexus. Wordsworth's sonnet series has an almost identical subject matter—it traces a stream to find all the ways that it touches human lives on a mundane and even on a mythic level.

In the first of the *River Duddon* sonnets, Wordsworth, by way of setting forth his theme, outlines the basic categories of springs much like those already treated above in this chapter. Indeed, the poet offers a multicultural compendium of the source-types of world literature; moreover, it is a fairly complete survey of the sources I have just treated in this chapter. First, the famous Horace's ode to the Spring of Bandusia ('O fons Bandusiae' 3:13) is employed by Wordsworth to evoke Greco-Roman antiquity:

> NOT envying Latian shades—if yet they throw
> A grateful coolness round that crystal Spring,
> Blandusia, prattling as when long ago
> The Sabine Bard was moved her praise to sing [...].

In these lines, we find a civilization-cycle being appealed to; then, the circuit of the globe is traced to an Oriental fountain—perhaps, an artificial one similar to that found in *West-Easterly Divan*:

> Careless of flowers that in perennial blow
> Round the moist marge of Persian fountains cling [...].

It is not only the historical cycle of world literature that is being referred to but also a variety of topography from all over the globe. First, the cataclysmic torrents of the high mountains:

52 *Wordsworth's Poetical Works*, eds. Ernest de Selincourt and Helen Darbishire, 5 vols. (Oxford: Clarendon, 1940–49). Other than *The Prelude*, all poems by Wordsworth come from this edition, which will be referenced in the text as *Poetical Works*. For somewhat superficial treatments of water symbols in the *The River Duddon* sonnet series, see the following study: Stewart C. Wilcox, 'Wordsworth's River Duddon Sonnets', *PMLA Imagery: A Study in Poetic Vision* (New Haven: Yale UP, 1952) 93–96.

Heedless of Alpine torrents thundering
Through ice-built arches radiant as heaven's bow;

but, the latter, a more elevated source, is rejected as the poet seeks

the birthplace of a native stream
All hail, ye mountains! hail, thou morning light! (sonnet 1).

Notice in the last line of this sonnet that yet another cycle or pattern in space or time is introduced: that of the cycle of the day. Wordsworth narrates the poem as if he were in the act of following the course of the river for the duration of a day. Concerned as it is with a sense of a real place in England, Wordsworth elevates it as a worthy poetic subject, the River Duddon by historical and international references, even when they are mentioned to point a contrast. The other noteworthy aspects of the Duddon's headspring are: it is a natural spring; it is not of the sublime Chamonix sort; it is closely connected with human birth.

Unlike Goethe in 'Song in Praise of Mohammed' and 'Overwhelming Astonishment', Wordsworth traces the source beyond the spring—'the cradle'—to the sky quarter—'the clouds'; moreover, a typically Wordsworthian theme—(proto)environmentalism[53]—creeps in:

CHILD of the clouds! remote from every taint
Of sordid industry thy lot is cast;
There are the honours of the lofty waste [...] (2).

In this passage, 'the lofty waste' of the mountains is important, not for adding dynamic qualities to the stream, which would make it a potential analogue for supernatural processes, but for the fact that it keeps the stream a pristine example of unspoilt nature. What suggestions there are of the supernatural are personifications of various natural phenomena: frost, wind, and desolation preside over the stream's birth like Goethe's 'Good Spirits' / 'Gute Geister' ('Song in Praise of Mohammed', l. 6):

Not seldom, when with heat the valleys faint,
Thy handmaid Frost with spangled tissue quaint
Thy cradle decks;—to chant thy birth, thou hast
No meaner Poet than the whistling Blast,
And Desolation is thy Patron-saint! (sonnet 2).

Wordsworth, like Goethe, describes his mountain streams as breaking from their sources with a turbulent, self-will of youth:

Starts from a dizzy steep the undaunted Rill
Robed instantly in garb of snow-white foam (sonnet 4).

This element of self-will continues through the remainder of Wordsworth's sonnet series, as it does through that of Goethe's two river poems; but, since they all follow the water-cycle, I will have more to say on this in the following chapters.

---

53 Jonathan Bate gives *River Dutton* one of the few sensitive treatments it has received; appropriately this is in his *Romantic Ecology: Wordsworth and the Environmental Tradition* (London: Routledge, 1991) 19.

There will gradually appear important distinctions between the two poets' treatment of the stream/human: rather than tracing man's life with an allegorical river, as Goethe has done, Wordsworth traces the course of a river in anthropomorphic terms. As one example of the subtle relation in his work of the course of the river to the life of a man, let us examine sonnet nine, entitled 'The Stepping-Stones', where the 'growth' stream is discussed when the tiny 'Rill' that has just emerged from the source has increased into a 'Brook':

> The struggling Rill insensibly is grown
> Into a Brook of loud and stately march [...].

The poet juxtaposes the willfulness of the stream with that of a child, who is testing his courage by jumping, at a run, the very stepping-stones that try the current of the growing stream—perhaps, an autobiographical touch: 'the clear waters'

> pursue their race
> Without restraint. How swiftly have they flown,
> Succeeding—still succeeding! Here the Child
> Puts, when the high-swoln Flood runs fierce and wild,
> His budding courage to the proof (sonnet 9).

Here, then, both child and stream have a separate but analogous existence. On the natural, non-symbolic level of the poem, the child is one of the many introduced or alluded to in the poem whose lives are influenced by the River Duddon as it moves along its course.

It must also be pointed out that the concepts of the spring and of the river in the above poems of Goethe and Wordsworth serve only in the opening parts of the overall structures of the works. In these poems, which I will treat at greater length in subsequent chapters, the courses of rivers are fully developed by chronological narrative into either allegories or more loosely stated analogues for the lives of men: these streams flow from a source down to the sea—in a single temporal continuity—as a man's life passes from birth to death. This limited span is broadened by the implication of the water-cycle, which is more clearly implied, as we shall later see, in the ultimate sonnet of *The River Duddon* series.

The two series or cycles of paintings I am about to consider—Caspar David Friedrich's *Cycle of Life's Ages / Lebensalter-Zyklus* and Thomas Cole's *Voyage of Life*—lend themselves particularly well to comparison with Goethe's and Wordsworth's poems because the symbolic meaning of the spring, as well as the other quarters, can be clearly deduced from a context and takes on particular resonance in being part of a larger series which employs the entire cycle.

The fact that Friedrich's series forms a sort of narrative of life's stages locates this sepia series close to poetry. As was shown in examining *River Duddon* and 'Song in Praise of Mohammed', the spring and the small rivulet beneath offer metaphors, respectively, for birth and childhood; the same is true of the 'Spring' / 'Frühling' from the *Cycle of Life's Ages*. Friedrich juxtaposes the young children with a 'small stream which is still near its source'.[54] The herbage reveals that the

---

54 Börsch-Supan, *Friedrich*, Tate Gallery, 82. Also see Erika Platte, *Caspar David Friedrich: Die Jahreszeiten* (Stuttgart: Reclam, 1961) 8, 12–13.

season is that of early spring; the subtitle of the work is *The Seasons / Die Jahreszeiten.* The early points of three cycles—of seasons, water, and life—are all interwoven in one panel of the allegory.

A strikingly parallel series—one which employs the hydrological-cycle, among others, to convey the cycle of life—is by the American painter Thomas Cole. His use of the water-cycle is very symbolical, indeed allegorical. He opens with a rock-born spring. Shelley's and Stolberg's stream springing full-blown from a rock cave has parallels to what Cole has visually rendered and has verbalized in his exhibition notes, which are worth quoting *in extensio*:

> A stream is seen issuing from a deep cavern, in the side of a craggy and precipitous mountain, whose summit is hidden in clouds. From out the cave glides a boat, whose golden prow and sides are sculptured into figures of the hours: steered by an Angelic Form, and laden with buds and flowers, it bears a laughing Infant, the Voyager whose varied course the artist has attempted to delineate. On either hand the banks of the stream are clothed in luxuriant herbage and flowers. The rising sun bathes the mountains and flowery bands in rosy light.

Cole then sets out in considerable detail his allegory:

> The dark cavern is emblematic of our earthly origin and the mysterious Past. The Boat, composed Figures of the Hours, images the thought, that we are borne on the hours down the Stream of Life. The Boat identifies the subject in each picture. The rosy light of the morning, the luxuriant flowers and plants, are emblems of the joyousness of early life. The close banks, and the limited scope of the scene, indicate the narrow experience of Childhood, and the nature of its pleasures and desires. The Egyptian Lotus, in the foreground of the picture, is symbolical of human Life. Joyousness and wonder are the characteristic emotions of childhood (27).[55]

Cole's voyage motif gives more centrality to the water-cycle than does Friedrich in his childhood sepia, where the rivulet runs alongside the children and shares place with the springtime foliage. Cole uses an oily facture of the paint in rendering the placid and reflective surface of the newly emerging stream.

Not only do Goethe's 'Song in Praise of Mohammed', Wordsworth's *Duddon River*, Friedrich's *Cycle of Life's Ages*, and Cole's *Voyage of Life* represent complex interlinkings of the cycles (especially the addition of the water-cycle), they also represent a combination of what Stephen Jay Gould has termed 'Time's Arrow' and 'Time's Cycle'. In discussing a frontispiece from Thomas Burnet's *Sacred Theory of the Earth*, Gould shows how Burnet combines his cycle of the natural processes, which have changed the earth's surface, with the linear or 'arrow' conception of Christian history moving from creation to apocalypse. Gould further demonstrates how the combinations of the cycle with the linear concept allows the thinker to develop a much richer vision of things. Gould observes how Lyell combines the arrow with the cycle to create the first comprehensive conception of geology. In the realm of evolutionary biology, he demonstrates how the linear or historical notion of homology and the cyclic or synchronic idea of analogy combine: 'The arrow of homology and the cycle of analogy are not warring concepts, fighting for hegemony within an organism. They interact in tension to build the distinctions and likenesses of each creature'. Gould extends his analysis to fine arts, both modern and medieval:

---

55 The notes are reproduced in Earl A. Powell, *Thomas Cole* (New York: Abrams, 1990) 84–92. This edition will be referred to throughout this book.

he shows how artists added to the linear Christian narrative the cyclic conception of New Testament events reproducing, by analogy, the Old Testament ones.[56]

Cole, Friedrich, Wordsworth, and Goethe achieve a similar richness by superimposing a *linear* destiny of a human—whether secular or religious in conception—upon the water-*cycle* metaphor. This complex interlinking of the cycles—especially the addition of the water-cycle—offers a complex and holistic way of dealing with the complexities of life—especially, as we will see in later chapters, with the shock of growth and other sorts of change.

---

56 Stephen Jay Gould, *Time's Arrow/Time's Cycle: Myth and Metaphor* (Cambridge: Harvard, 1987) 20–22, 174–78, 198, 181–90.

CHAPTER TWO: FLOW

After treating the Baroque and eighteenth-century (Neo)Classical backgrounds, the Romantic river per se will be treated first; then, the intrusive waterfalls, which break the general smoothness of rivers' flow, will be discussed; and finally, those rivers and waterfalls clearly part of the water-cycle will be analyzed. In the following treatment of rivers and to a greater degree than in the previous source chapter, Lansing's complexity theory will be of even greater importance. Just as he treated the whole social, religious, economic, and technological complexity of life along the streams of Bali, so too will I discuss how Romantics seek to convey how the rivers were shaping the diversity of life along their banks. Because of water milling and navigation, the actual rivers—upon which the Romantic-culture artifacts are modeled—have shaped the human geography of Europe and America. Today we find it hard to conceive of the extent to which rivers at that time constituted such a physical and mental shaping force. Let me, by way of introduction, just suggest one of the more impressive indications of the mentally shaping force possessed by these streams. In writing of a great engineer-scientist pioneer of early energy physics, a modern science historian makes clear the conceptual metaphor that underpins the energy unit of the time—the caloric: 'Sadi Carnot still handled caloric as flowing from a real reservoir of heat down a continuous gradient'.[1] The shaping model or metaphor for Carnot's thinking, then, was that of a still essentially watermill world. Energy and power could only find expression in the stream's carrying its water from springs, glaciers, or lakes set at a great height, which gives the water its force of potential and actual energy. And natural science was only one of many aspects within Romantic culture inspired by the flow of streams in their cycle context: this chapter examines the rich artifacts treating streams and falls. Part of the reason for this creative and compelling nature of the Romantic model of the water-cycle can be explained by how it compares with those of previous European cultures and styles.

In the opening section of this chapter, considerable space will be spent on Baroque works—particularly a poem by Henry Vaughan. This is necessary to understand precisely how the Romantics made significant contributions to the rhetoric of water phenomena. Vaughan's lyric and other contemporaneous works are important for their lacunae of knowledge; however, the overwhelming importance of these lacunae will not be made absolutely clear until chapter four—this suspended mode of presentation, as I have said before, is early on demanded by the nature of the cycle subject. Indeed, the full significance of much that is broached in this and the other two opening chapters will not emerge until the final chapter—'The Link'.

Toward a Romantic Model

It is the waterfall that gives the most dramatic form to flow, which is why the Baroque artists favored it. The poet Henry Vaughan demonstrates a fascination for rushing, falling water comparable to that found in works of the painter Jacob van

1 Charles Coulston Gillispie, *The Edge of Objectivity: An Essay in the History of Scientific Ideas*, pb. ed. (Princeton: Princeton UP, 1973) 241.

Ruisdael. Vaughan's complex employment of a water image at the core of his lyric 'The Water-fall,' in his collection *Silex Scintillans,*[2] has much in common with the treatment of imagery in Ruisdael's *A Waterfall*. To understand both works, however, it is necessary to examine them in accordance with the contemporary science of hydrology. To do this, the scientific method must be considered on these levels: (1) the simple observation of things in nature; (2) a sufficient fascination generated by these phenomena for poets, painters, or scientists to make their various uses of them; (3) conceiving of the natural element as part of a larger whole, which is usually the point where scientists might formulate hypotheses on such processes and test them. At the third level, the poets may pick up one of the resulting scientific theories that place the natural element in a larger context; they may either learn of such a theory directly or indirectly from the general climate of opinion and may then either use or reject it; if they do decide to use the theory, this is because it offers them a metaphor for representing ideas cosmic in scope. Of course, these levels are only useful as tools to question which stage of the scientific method might be appealed to at any moment by the artist, rather than as *verba ultima*.

In the introduction to his translation of 'The Waterfall' by Derzhavin, Harold B. Segel makes a brief but very incisive and, for my purposes, very useful comparison of this poem with Vaughan's 'The Waterfall'. Segel's comparison not only is valuable in its succinct summary of Baroque usage of the falls but also represents—as far as it goes—a succinct model of traditional comparative methodology at its best:

> Elaborated throughout the poem as a dynamic metaphoric symbol of the philosophical theme [the transitoriness of all earthly things, the brevity of man's terrestrial sojourn, the omnipotence of death and the equality of all men before death] is the waterfall, the description of which Derzhavin based on his own observations of the Kivach waterfall on the Suna River in northwestern Russia. The brilliant turbulent course of the fall as it rushes to crash into the depths below depicts the course of man's life on earth in all its color and vitality—and in the inevitability of its descent, its fall into the abyss of death. But after tumbling into the depths the life of the waterfall is not over; it is merely transformed. The waterfall, as man's life itself, infers the poet continues to live, but in a different state. The waterfall emerges from the depths to flow on placidly, like a milky way. So too after death does the life of man continue eternally as the immortal life of the soul.[3]

Derzhavin was not the first poet, or the last, to find in the waterfall an appropriate symbol of man's life—terrestrial and eternal. The Baroque proclivity for themes to which Derzhavin himself was so strongly attracted led more than a single Baroque poet to write of the analogies between the course of a waterfall and the life of man. Segel compares Derzhavin's poem with Vaughan's 'The Water-fall'. Considerably shorter than Derzhavin's later work, and devoid of the Russian poet's rich use of color and sound as well as his evocative descriptions of various aspects of nature, Vaughan's 'The Water-fall' embraces essentially the same philosophic view and provides interesting formal material for comparison. Unfortunately, Segel's comparison

---

2 'The Water-fall', in *The Works of Henry Vaughan*, ed. L.C. Martin, 2nd ed. (Oxford: Clarendon, 1957) 537–38.

3 Harold B. Segel, *The Literature of Eighteenth-Century Russia: A History and Anthology* (New York: Dutton, 1967) 2: 286–303, 287.

ends at this point. Short as it is, it contains some perspicacious remarks. In addition to concisely setting forth the essential nature of the Baroque usage of water in poetry, he shows that this usage had enduring value lasting into the eighteenth century. Yet, even more importantly, Segel points out the aspect that most differentiates the eighteenth-century work from the seventeenth-century one: the former places great emphasis on descriptiveness and on 'color and sound'—or, rather, adjectives and nouns which convey these. Thus, whereas Derzhavin's waterfall has aspects which might be called either symbolic or descriptive, those of Vaughan's are relatively more limited to the symbolic.

We must consider precisely how painters or poets go about rendering—within the laws of their particular art forms—or exactly how they achieve the essence of the water image in the chosen language of their 'text'—by which I, of course, signify not only a poem but also a painting. The first step in this process is simple scientific observation of the natural phenomenon, seeking an understanding of how it sounds and appears.

Ruisdael's *A Waterfall*[4] demonstrates a fascination for observing water, as well as for the act of observing itself. He scrutinized running water tripping over canal gates, mill weirs, and rocks; sunlight shadowing the undersides of clouds, filtering through them, and falling on the tree leaves, which shimmer in the wind. Kenneth Clark has called Ruisdael 'the greatest master of the natural vision before Constable'.[5] Such a linking with the great Classical-Realist-Romantic painter John Constable would incline us then to give close attention to Ruisdael and see precisely to what extent his artistic vision did allow him to view and thereby paint the actual natural element or, on the other hand, how much of his rendering of water is—to use E.H. Gombrich's term—schemata;[6] that is to say, how much is imposed by the conventional approaches of his art and how much by reality? The schemata itself is a useful shorthand for the artist; it constitutes a starting place within the whole process of filling a canvas—the essential act of painting. When the artist has ostensibly sought to represent a natural element, the critic must pose the question: how far does the artist get beyond schemata and how close to nature?

In order to comprehend the extent to which Ruisdael gets beyond the schematic rendering of the waves, it would be useful to compare his manner of painting water to that found in Botticelli's *The Birth of Venus*.[7] In this Renaissance masterpiece, nature is secondary and waves can be implied symbolically with repeated caret-like strokes. Ruisdael's *A Waterfall* is dominated by spills, swirls, and waves. The canvas, indeed, contains two falls—one that seems natural (in the foreground) and another, a man-made one, a weir (in the background). Not only does the natural waterfall dominate the foreground, but the artist has also sought to render the

4 Jacob van Ruisdael, *A Waterfall*, National Gallery, London.

5 Kenneth Clark, *Landscape into Art*, rev. ed. (New York: Harper, 1976) 64; cf. Clark's views with those of Jakob Rosenberg and Seymour Slive who tend to emphasize the stylized and idealized aspects of Ruisdael's art, in *Dutch Art and Architecture, 1600 to 1800* (Baltimore: Penguin, 1966) 153–56.

6 E.H. Gombrich, *Art and Illusion: A Study in the Psychology of Pictorial Representation*, rev. ed. (Princeton: Princeton UP, 1961) passim.

7 Botticelli, *The Birth of Venus*, Ufizzi Gallery, Florence.

torrential flux of the water with meticulous care. The painter could very well have used one application for the water—a single mode of stroking or daubing with the brush stereotypically reproduced; but instead he used and carefully modulated three different brush applications: the first is rendered with oily, straight slashes to represent the water as it first slides over the rocks; the second, with swirling daubs; the third, with dapplings of the paint. Both the latter were rendered with white paint and with a dry-brush facture to represent foam.

It must not be forgotten, however, that one can find as much of art as of *the actual* in this and other Ruisdael's paintings containing waterfalls. Seymour Slive and H.R. Hoetink write the following:

> Paintings of northern landscapes with rugged mountains, giant firs, and powerful waterfalls cascading over huge boulders and dead trees form the largest category of Ruisdael's work, and he convinces us, more than any other artist, that these pictures are based upon his own intense, first-hand study of their elements. Strange as it may seem, Ruisdael did not know such scenes from his own direct observation. His vision of them is based on art, not nature. They are derived from works by the Alkmaar artist Allart van Everdingen (1621–75), who introduced and popularized northern landscapes in the Netherlands.[8]

These scholars' initial response to the 'intense, firsthand' quality may be the correct one; perhaps, they put a bit too much emphasis on documentation in drawing their conclusions. The fact is that Ruisdael might well have rendered his painting's components or elements—among them being falling water—from observation in his mundane environs—water falling through a sluice or over a broken weir. It is, then, perhaps safer to say that Ruisdael bases his vision as much on nature as on art. To be sure, Ruisdael's *Waterfall* is no isolated work; it forms one part of a larger manifestation; for as Hans Engelhard proves with his compendious study on the subject, the waterfall became an important part of the thematic repertory of European painting in the seventeenth century.[9] Ruisdael's painting is part of a cultural phenomenon that cuts across the boundaries between arts. Similarly, Vaughan's 'The Waterfall' offers but one example of a larger phenomenon.

Keeping in mind the distinctive techniques available to each art form, let us consider the opening lines of 'The Water-fall'. Vaughan's lyric-voice expresses a fascination with the various properties of water:

> With what deep murmurs through times silent stealth
> Doth thy transparent, cool and watry wealth
>   Here flowing fall,
>   And chide, and call,
> As if his liquid, loose Retinue staid
> Lingring, and were of this steep place afraid [...] (ll. 1–6).

It is certainly true that these properties are conveyed by means of somewhat abstract adjectives ('transparent' and 'cool') and nouns ('stealth' and 'wealth'). The imagery is customary for an idyllic scene; 'deep murmurs' are the sort of sounds that the streams in literature have made at least since the time of Theophrastus. Images are

---

8 Seymour Slive and H.R. Hoetink, *Jacob van Ruisdael* (New York: Abeville, 1981) 103.

9 Engelhard's dissertation is a helpful, factual catalogue of paintings, seeking to show how close each painting is to the geology of the topographical locus it seeks to portray.

important, but so are sounds. The fall of the water seems to find a correspondence in the short lines three and four. F.E. Hutchinson has observed exactly how replete this passage is with alliteration of sibilants and other types of consonant sounds.[10] Vaughan is employing *s, c, f,* and *l* sounds to ends similar to those of Ruisdael, who, as we have seen, manipulated brush-strokes of differing colors and tones to reproduce the experience of perceiving water. The only essential difference is that in one the experience is visual; in the other, aural.

Qualifications, however, must be made: Vaughan, as Hutchinson observes, uses alliteration 'more extensively than most of his contemporaries writing English verse'.[11] Further, it must not be forgotten that alliteration is used throughout Vaughan's *Silex scintillans* and not necessarily in conjunction with specific imagery, water or any other sort. The following lines from 'The Constellation' [12] can be taken as an example:

> No sleep, nor sloth assailes, but poor man
> Still either sleeps, or slips his span (ll. 15–16).

In reading these lines, we should nevertheless remember that just as Ruisdael could use a certain brush-stroke to render both water and say, silk, or other smooth fabrics, so too can Vaughan use alliteration for describing either water or sleep. It still remains to be asked, though: how much this alliteration is *nature,* and how much is *convention*—how much is *schematic,* and how much is *actual*? Well, when sound and images match, I would vote for the natural and the actual. It is also possible that Vaughan uses shorter (diameter) lines and enjambment in lines 7 through 10 in order to suggest the breathless plunge of the water:

> The common pass
> Where, clear as glass,
> All must descend
> Not to an end [...] (ll. 7–10).

Notice that the poet only breaks the flow with a single parenthetical phrase. It is also significant how the normative line length (pentameter) is suspended at the points where a waterfall is described. I think the following conclusions might safely be made concerning Vaughan's 'The Water-fall': the abstract sign representing the object in nature is no longer sufficient; the poet demonstrates an interest in the portrayal or re-creation of the object based on the close observation of its qualities. It may be said with some certainty that the alliteration, if perhaps not the imagery, of Vaughan's lyric demonstrates a simple empiricism.

---

10 F.E. Hutchinson, *Henry Vaughan: A Life and Interpretation* (Oxford: Clarendon, 1947) 161. Other secondary works concerning the poem: W. Nelson Francis, 'Vaughan's "The Waterfall"', *The Explicator* 14 (1956): item 57; Wulf Datow, '"The Water-fall" von Henry Vaughan (1621–1695)', *Die neueren Sprachen* 15 (1966): 410–20; Ted-Larry Pebworth, 'The Problem of *Restagnates* in Henry Vaughan's "The Water-fall"', *Papers on Languages and Literature* 3 (1967): 258–59; Alan Rudrum, '*OED* Defended: A Crux in Henry Vaughan's *Waterfall*', *The Anglo Welsh Review* 25 (November 1975): 83–86.

11 Hutchinson 161.

12 'The Constellation', *Works* 469–70.

What then does this new empiricism offer poets, artists, and composers? It offers a unilluded perception of water phenomena and a freshness of vision backed but not dominated by the aesthetic conventions for rendering the sound or sight of water. We will see the growth of this empiricism over the century to follow Vaughan's beginnings; we will see this empiricism culminate among the Romantics. Before this, however, we must again look further into the seventeenth and the eighteenth century to find the growth of the new sensibility.

In addition to his comments on the water-cycle, Frye offers the following paradigms for river images and symbols in sixteenth- and seventeenth-century poetry: 'Water-symbolism centers on the disciplined river, in England the Thames, which runs softly in Spenser and in neo-Classical rhythms in Denham, a river whose most appropriate ornament is the royal barge'.[13] Frye first refers to the famous refrain in Edmund Spenser's *Prothalamion*: 'Sweete *Themmes* runne softly, till I end my song'.[14] Closer, however, to my interest is Frye's second reference: to John Denham's *Cooper's Hill*[15] and, more specifically, to the famous passage in which the poet seeks a model for his 'neo-Classical rhythms' in the stately, 'disciplined' flow of the Thames:

> Oh could I flow like thee, and make thy stream
> My great example, as it is my theme!
> Though deep, yet clear, though gentle, yet not dull,
> Strong without rage, without ore-flowing full (ll. 189–92).

Denham singles the Thames out from the rivers of the world for especial praise precisely because it is 'controlled'—that is, it had (at least, in Denham's time) the peculiar property among rivers of not flooding destructively:

> No unexpected inundations spoyl
> The movers hopes, nor mock the plowmans toyl:
> But God-like his unwearied Bounty flows;
> First loves to do, then loves the Good he does (ll. 175–78).

In short, the poet's river is an orderly and utilitarian stream. The Thames here is used by Denham as a symbol—the river's flowing into the sea represents death:

> *Thames*, the most lov'd of all Oceans sons,
> By his old Sire to his embraces runs,
> Hasting to pay his tribute to the Sea,
> Like mortal life to meet Eternity (ll. 161–64).

This symbolic use of a river closely resembles that of the Baroque poets as defined by Segel.

To further complicate matters, however, its descriptive passages and use of an actual locus made Denham's *Cooper's Hill* one of the most important models for English topographical poets. Concerning Denham's place in this poetic tradition,

---

13 Northrope Frye, *Anatomy of Criticism: Four Essays* (Princeton: Princeton UP, 1959) 154.

14 Edmund Spenser, *Prothalamion, Poetical Works*, eds. J.C. Smith and E. de Selincourt (Oxford: Oxford UP, 1970) 601–02.

15 John Denham, *Cooper's Hill, The Poetical Works of Sir John Denham*, ed. Theodore Howard Banks, 2nd ed. (Hamden: Archon, 1969) 59–87.

Robert Arnold Aubin offers the following appraisal: 'Denham may [...] justly be regarded as the father of at least the first generation of those prospect-poets who strew so thickly the stream of English poetry of the next two centuries'.[16] There is a strongly pastoral aspect to neoclassical streams, as Frederick Green points out. Unfortunately, in his important comparative study, *Minuet,* he makes only passing reference to the use of water; but, even these are helpful. In discussing eighteenth-century neoclassical imitations of Virgil's pastoral poetry, he observes: 'The action of the idyll or eclogue [...] must always take place in a rustic or solitary spot: the bank of a stream, a field enameled with flowers, a grotto, and, very occasionally, at the seaside'.[17] Green might have added that the field inevitably has rivulets flowing through it and the grotto features a flowing spring.

Another helpful tool can be derived from Frye's description of neoclassical water images and symbols as tame, placid, and controlled. If one accepts the binary generalization that neoclassical usage and romantic usage are antithetical, one might infer from Frye's remarks that romantic waters should be chaotic, turbulent, and uncontrolled. If one treats this as a hypothesis and as a tool of limited application, one might—as I shall attempt to do in the chapters that follow—test it and use it with relative safety. The concentration on the essence of the flow of water that we saw in Ruisdael is similarly absent in Wilson's paintings of the Dee River. This absence of freshness and vitality in rendering water of the river and the concentration on the environs of the loci (not to mention the obligatory human figures in the foreground) comes as no surprise; for, by the end of the eighteenth century, in the words of Kenneth Clark, 'landscape painting became mere picture-making according to certain formulas' and 'fact was confined to topography'.[18] In *Holt Bridge on the River Dee,* there is no attempt to render the river *qua* river. The same could be said of the many paintings Wilson rendered of the placid flow—actually the seeming stasis—of the Dee, the Thames, and other rivers.[19] The artist has given little attention to rendering the stream—by suggesting its liquid properties or aspects of flow. Although eighteenth-century landscape painting relies heavily on formulas, we should be careful not to view it as something which Turner, Constable, and Friedrich reacted against and as a stilted series of outmoded conventions from which these painters made a radical departure.

Neoclassical landscape and the related topographical painting must be understood before the Romantic revolution in this genre can be comprehended, and it must be viewed as, in part, a return (revolution) to the Baroque. Like Ruisdael and his contemporaries, J.-P. de Loutherbourg and his coevals brought about changes which, in the context of art history, can be viewed as radical in themselves. Loutherbourg,

---

16 Robert Arnold Aubin, *Topographical Poetry in Eighteenth Century England* (New York: Modern Language Association, 1936) 223–41, qtd. from 34.

17 Frederick Green, *Minuet: A Critical Survey of French and English Literary Ideas in the Eighteenth Century* (London: Dent, 1935) 235.

18 Clark, *Landscape into Art* 64; William Vaughan, *Romantic Art* (New York: Oxford UP, 1978) 178–79.

19 This reduction of rivers to reflective surfaces may be part of the backward-looking aspect of Wilson's art described by David H. Solkin, *Richard Wilson: The Landscape of Reaction,* London: Tate, 1982. Waterfalls, as will be discussed later, are a different matter.

while working within the topographical tradition, showed an interest and a facility for treating cataclysmic water phenomena, only to be surpassed by Turner. Rivers, of course, are not the exclusive property of the poets to be treated later in this chapter; for the topographical and descriptive poets before 1770 also include the images of brooks and rivers in their landscape painting and poetry. Perhaps the most striking comparison between the topographical and the later Turner is found in Clark's *Romantic Rebellion,* where he contrasts the two versions of *The Falls of Clyde.*[20] While Clark concentrates his discussion only to the light reflected in the flow of the streams of water and the mist thrown off from them, I intend to show that what is true of the light is true of the mist and water as well. Turner's 1802 version is a topographical work, like Wilson's *River Dee,* with all the environs of rocks and trees meticulously rendered. In the later version (circa 1832), the environs of the falls have dissolved, leaving a painting of pure color, light, and, above all, the motion of the water and mist. That is to say, the later version becomes a rendering of the quintessence of the waterfall itself, of its ceaseless flux and flow. In literature, we have seen this concentration on the flux of the water in prejudice to the surrounding details: the extreme concentration on the changeful element seems to imply its transformation into a symbol.

Certainly, it would be hard to argue for the existence of a topographical approach in music—beyond references in the titles of musical works or some allusion to local folk-music—because there is a real question of how descriptive it can be. Indeed, water music, as we have already seen, is even more fraught with dangers than water poetry or painting; Classicism—the Neoclassical equivalent in music—has its great water emblems, but I treat them later because they fall within our focusing period.

The following lines from Denham's *Cooper's Hill* may serve as an example of topographical usage in the realm of poetry; here a specific river is described as part of the scene viewed by the persona:

> My eye descending from the Hill, surveys
> Where *Thames* amongst the wanton vallies strays (ll. 159–60).

The following passage from Thomson's *Summer* offers a paradigm for a stream described without symbolic overtones:

> [The stream] purls along
> The vocal grove, now fretting o'er a rock,
> Now scarcely moving through a reedy pool,
> Now starting to a sudden stream, and now
> Gently diffused into a limpid plain [...] (ll. 480–84).

Denham intends a specific locus; Thomson evokes a generalized English landscape (stylized, in part, under the influence of Italian and French Painters[21]). On the

---

20 Kenneth Clark, *The Romantic Rebellion: Romantic versus Classic Art* (New York: Harper, 1973) 246–47. There has in the last decade been an upsurge of interest in Turner and the topographical treatment of specific rivers: David Hill, *Turner on the Thames* (New Haven: Yale UP, 1993); Ian Warrell, *Turner on the Loire* (London: Tate Gallery, 1997).

21 Elizabeth Wheeler Manwaring, *Italian Landscape in Eighteenth Century England* (New York: Oxford UP, 1925) passim.

other hand and clearly in the Early Romantic camp, Robert Burns' use of a specific river in 'Sweet Afton' portrays the gentle flow of the stream metaphorically as a correspondent to the speaker's love song. Goethe's treatment of the river images and symbols in *Werther* and 'To the Moon' / 'An den Mond', evidences more strongly this subjective stamp of the narrator upon the imagery.[22]

Flow and Permanence

In the first chapter, we examined the Homeric elements in the fountain images, which are used to symbolize a relative state of equilibrium in the protagonist's mind. It is not surprising that when, as Werther himself puts it, 'Ossian has driven out Homer from my heart',

> Ossian hat in meinem Herzen den Homer verdrängt (464),

there is a corresponding change in the water imagery from the Homeric fountain to the Ossianic torrent. This passage is crucial enough to be analyzed as a whole. Although primarily descriptive in intent, the relation between object and subject lacks the detachment Werther seeks and finds in external nature-states corresponding to his mental ones. Werther's pent-up passion for the now-married Charlotte finds an analogue in the snow and ice; his imminent attempted seduction of Charlotte, in a sudden 'thaw' which had set in:

> Es war plötzlich Tauwetter eingefallen;

ironically, this thaw causes a flood in the very place where, two springs before, Werther had experienced happiness—the valley around the village of Wahlheim: 'I had heard that the river had overflowed its banks, all the brooks were swollen, and from Wahlheim on down, my beloved valley was flooded!'

> ich hatte gehört, der Fluß sei übergetreten, alle Bäche geschwollen und von Wahlheim herunter mein liebes Tal überschwemmt! (482).

This is the same river which, in the happier times of two summers before, Werther had described in vastly different terms, during the first book of the novel: 'with the gentle stream—gliding along between the whispering reeds—mirroring the charming clouds, which the gentle evening breeze wafted across the sky'.

> der sanfte Fluß [, der] zwischen den lispelnden Rohren dahin gleitete und die lieben Wolken abspiegelte, die der sanfte Abendwind am Himmel herüberwiegte (430–31).

---

22 Daniel V. Hegeman handles the very difficult subject of river images and symbols in Goethe's works in a superficial and often inaccurate manner in his 'The River in Goethe's Writings', *Monatshefte* 44 (1952): 135–44. Much more interesting is A. Fuckel's brief but incisive essay which treats Goethe's use of the river as an allegorical symbol for the course of human life, which touches on Goethe's contemporaries including Stolberg, and which indicates a seventeenth-century source for such an allegorical use of the river ('Das Bächlein' by Procopius von Templin); 'Der Fluß als Sinnbild des Lebenlaufes in deutschen Dichtungen', *Zeitschrift für den deutschen Unterricht* 27 (1913): 835–58.

Another tranquil scene with this valley and its river comes at the end of the first book of the novel, introducing an instance when Werther is about to share a quiet moment with Lotte and Albert before he attempts to leave and find a new life away from them: 'I stood on the terrace under the tall chestnut trees and, for the last time, watched the sun go down over that dear, lovely valley—over the softly-flowing river'.

> Ich stand auf der Terrasse unter den hohen Kastanienbäumen und sah der Sonne nach, die mir nun zum letztenmal über dem lieblichen Tale, über dem sanften Fluß unterging (436).

The hopes of this new life that close the first book are dashed at the opening of the second. The river of the second Ossianic half, then, will ultimately be described in a completely different mood.

The winter torrent has attained such force that the speaker of the poem—the actual poet's persona—compares it to a stormy sea: 'A terrible spectacle—to see in the moonlight the turbulent waters swirling down from the crags, out over crops and meadow and hedges and all, and up and down the wide-reaching valley: a stormy sea amid the howling of the wind'.

> Ein fürchterliches Schauspiel, vom Fels herunter die wühlenden Fluten in dem Mondlichte wirbeln zu sehen, über Äcker und Wiesen und Hecken und alles, und das weite Tal hinauf und hinab eine stürmende See im Sausen des Windes! (482–83).

While the fragmented syntax of the opening and closing phrases conveys the shock and horror, the expansive Ciceronian rhetoric of the center with the rolling repetition of *and/und* underscore the vision of flooding.

Since the above use of moon, valley, field, and bush images in conjunction with that of a winter torrent is paralleled in several passages of the other work by Goethe being examined in this section, the lyric 'To the Moon' / 'An den Mond' (1: 71–72), it seems fitting to integrate the discussion about this lyric into that about the Werther passage: 'Quietly filling again bush and dale with incandescent mist—finally liberating for once my entire being—casting over my domain your soothing gaze, like a friend's gentle consideration of my fate'.

> Füllest wieder Busch und Tal
> Still mit Nebelglanz,
> Lösest endlich auch einmal
> Meine Seele ganz;
>
> Breitest über mein Gefild
> Lindernd deinen Blick,
> Wie des Freundes Auge Mild
> Über mein Geschick (ll. 1–8).

As is Goethe's want, each image is packed with multifaceted significance. 'Nebelglanz', literally Englished as 'mist-gleam', can mean that the moonlight is diffuse and outline-blurring like mist, or it can describe the effect that moonlight has when shining through the mist. The phrase 'mein Gefild' might be simply translated as 'my field' or, as I have done, 'my domain'—well, of course, neither satisfies: one makes Goethe sound like the complacent landowner; the other, like the proud monarch. The 'mein' is at once the catching point for the translator and the indication to the critic that we have in 'To the Moon' one of those 'sympathetic linkings of the mind and the moon-landscape', which Ronald Peacock discovered in an earlier lyric, 'To Luna' / 'An Luna'.[23] And Barker Fairley finds, in poetry of the early Weimar period, a clear link between the great calming influence upon the erstwhile Storm and Stress poet—his beloved Charlotte von Stein—and the metaphor of a clear moon in a peaceful night sky.[24]

This moonlit landscape as mindscape links the sense of hope and permanence with that of change and despair. In the stanzas that follow, the steady light of the moon—a metaphor for the soothing presence of a friend, Charlotte—finds its contrast in the stream—a correspondent for the more restless aspects of the persona/poet's soul, especially when the latter addresses the stream: 'When on a winter's night in your fury you overflow':

> Wenn du in der Winternacht
> Wütend überschwillst [...] (ll. 25–26).

This juxtaposing of the permanence of the moon with the flux of the stream offers us yet another example of Goethe's forcing a signifier to represent a quality not usually associated with it: the moon, after all, is the age-old symbol of change and inconstancy. When, however, seen shining in a clear atmosphere, the moon can be used in precisely this way, at one single instance in time and perched precariously on the edge of falling into another, contradictory significance. According to my reading of the poem this tectonic stress among both the signifier and signified as well as the other symbol complex is intended by Goethe.

There is, further, a significant difference between the passage in the novel and that of the later lyric: whereas in the lyric, the moonlight lends permanence to the changing of the stream; crucially, the moonlight of Werther's landscape is being rendered seemingly inconstant by the passage of a dark cloud across the winter sky; important, too, is the manner in which the rhetorical vector of the sentence spills out into the last clause: 'And then when the moon came out again and rested over the black clouds and, in the frightfully splendid reflection, the flood rolled and roared before me: I was overcome by a shudder and again a longing!'

> Und wenn dann der Mond wieder hervortrat und über der schwarzen Wolke ruhte und vor mir hinaus die Flut in fürchterlich herrlichem Widerschein rollte und klang: da überfiel mich ein Schauer und wieder ein Sehnen! (4: 483).

In the lyric, then, the steady shine of the moon offers a vision of permanence as a counterweight to that of change; in the Storm and Stress novel, the scudding clouds makes the light from the moon flicker, and when the clouds break to allow the moon to shine with a steady light and to offer an emblem of stasis before the scene of mutation, the hopelessly overwrought persona focuses on its shattered reflections cast from the waves of the mäelström before him.

The ultimate vision of *Werther*, then, is one of change. Permanence appears only as a passing foil, within a complex layering of images conveying change; the moon, which is more often the emblem of mutability, suggests a permanence which the

---

23 Ronald Peacock, 'Goethe as Lyric Poet', *Essays on Goethe*, ed. William Rose (London: Cassell, 1949) 21–44, qtd. from 26.

24 Barker Fairley, *A Study of Goethe* (Oxford: Clarendon, 1947) 105–06.

passing veil of clouds quickly dispels. The moon only lights the actual scene of change—the tragic vision below—which, supplemented by the stream's sound, brings forth a psychological leitmotif in what is heretofore its strongest manifestation: Werther's desire for the ultimate human metamorphosis—death. And at this point the scene of the winter-night flood becomes not only an analogue for the personal emotions but an active and effective agent: 'Alas, I stood with open arms facing the abyss and gasping forth a whisper, "down!" "down!" lost myself in the joy of dashing down my torments, my suffering! to storm off like the waves!. [...] Oh Wilhelm! how gladly would I have yielded up my human existence to rend the clouds and grasp the waves with that tempestuous wind!'

> Ach, mit offenen Armen stand ich gegen den Abgrund und atmete hinab! hinab! und verlor mich in der Wonne, meine Qualen, mein Leiden da hinabzustürmen! dahinzubrausen wie die Wellen! [...] O Wilhelm, wie gern hätte ich mein Menschsein drum gegeben, mit jenem Sturmwinde die Wolken zu zerreißen, die Fluten zu fassen! (483).

This objective destruction of a landscape in which Werther had once experienced something like happiness is a correlative for a subjective destruction which has long since occurred in Werther's mind; that is, he already has the state of mind ready to fully appreciate the dismal scene. In addition, the shock of seeing the place destroyed where he had experienced so many happy moments contributes to the depression leading to his eventual suicide. The images of calm and seeming stasis which Werther earlier experienced on the banks of the stream have been reduced by the violently changeful weather to destruction and disorder—all making clear to him how order is precariously perched upon the knife edge between modes of chaos.

All the imagery of this passage from Goethe's novel, then, conveys the notion of change—indeed, of cataclysmic change and the inevitable destruction of all that is beautiful; in 'To the Moon', from the early Weimar period, the stream has the aspect of change, but the moonlight, that of permanence; in two of Goethe's later poems to be examined below ('Dauer im Wechsel' and 'Lied und Gebilde') the stream symbolizes the paradox of permanence in change. In addition, both of the torrential riverscapes from the pre-Classical works of Goethe bear the subjective stamps of their personae. While each riverscape is partly descriptive, it is also symbolic in that a relationship, if open and unspecified, exists between the persona's state of mind and the externals he describes. The two lyrics to be considered next use the stream image less for its own sake and more as a symbol to express an abstract concept—the permanence-in-change paradox inherent in the relation of the artist to his art.

A broadly comparable juxtapositioning of the symbol for mutability with a symbol for permanence of the stream and moon in Goethe's 'To the Moon' is to be found in a passage from 'Mont Blanc', where Percy Shelley sets a stream, representing the process of change inherent in the universe, against the perennial snows of the mountain peaks. Shelley, like Goethe, implies a larger permanence lying beneath momentary changes. A significant difference lies in the cataclysmic aspect of the change: the seething tumultuous quality of the torrent in 'Mont Blanc' (1: 229–33) is more akin to those of Goethe's Storm and Stress works. Moreover, there is a philosophically partisan appeal—here, it is made to Neoplatonism—that is absent in the German work. Percy Shelley has a symbol (the mountain) that seems to broadly represent the Neoplatonic One, place of permanence, and another (the torrential

Arve) to represent the Ideals—that is, the powerful emanations from this Divine Source down into the realm of the transient. Early in the poem, Shelley paints the

> awful scene,
> Where Power in likeness of the Arve comes down
> From the ice gulfs that gird his secret throne,
> Bursting through these dark mountains like the flame
> Of lightning thro' the tempest (ll. 15–19).

The perennial ice here offers a perfect analogy for power—as yet potential, arrested in a solid state. Shelley's use of a mountain and its 'eternal' ice—which he had enough science to know to be subject to constant stress from wind and the pressures caused by the piling up of snow—bears a striking parallel with Goethe's frequent opposition of signifier with signified. The superficial permanence of the ice, known to be only apparent, has been taken from among terrestrial flux to represent the Eternal—much as Plato took the earthly beauty of the male figure to construct a sense of Real Beauty in the *Symposium.*

The mountain and the Neoplatonic One it represents dominate the poem in recurring, if varied, leitmotifs:

> Far, far above, piercing the infinite sky,
> Mont Blanc appears,—still, snowy, and serene—
> Its subject mountains their unearthly forms
> Pile around it, ice and rock; broad vales between
> Of frozen floods, unfathomable deeps,
> Blue as the overhanging heaven, that spread
> And wind among the accumulated steeps (60–66).

Percy Shelley is offering an image of true substance here: we must remember that this vast sea of ice is in essence unlocked energy in the form of frozen water located on a very high place, and it was by the still predominantly water milling technology of the poet's time (as the hydroelectric equivalent of earlier decades of this century) this potential energy could eventually, after a thaw, be exploited downstream. The thaw, indeed, was ongoing and generating a water energy, the power of which was and is made manifest in the falls and cataracts of the French Alps, Europe's highest mountains.

The high separateness of the snow-covered peaks is taken as a pure emblem for the fact that

> Power dwells apart in its tranquility,
> Remote, serene, and inaccessible:
> And *this,* the naked countenance of earth,
> On which I gaze, even those primæval mountains
> Teach the adverting mind (ll. 96–100).

In one of the greatest readings of this or any other lyric of Shelley, Earl R. Wasserman points out how the poet has transformed the topographical actuality into a symbol for the most abstract and complex vision of permanence and change:

> just as in the topographical structure the Ravine at the base of the mountain symbolizes the world of ordinary experience, so the peak of Europe's highest mountain symbolizes the transcendent:

and since Mont Blanc gives its name to the title of the poem, we might well expect that the essence of the poem will be the revelation handed down from those heights.

Wasserman sees the peak as a symbol for the distant Neoplatonic realm of the Ideas, which act as a Form-giving Power upon the otherwise mutable world below: 'The gleams of the remoter world are, of course, symbolically imaged by the snow at the summit of the mountain'.[25] The Neoplatonic power is a permanent singularity; however, its emanations are changeful variations:

Mont Blanc yet gleams on high:—the power is there,
The still and solemn power of many sights,
And many sounds, and much of life and death (ll. 127–29).

Paradoxically, the processes of earthly change find their sources in the 'far fountains' (ll. 101–16)—which are the eternal snows of the mountaintops.

The same juxtaposing of a symbol of the changing rush of a stream with the permanence of a mountain peak glowing in the sun or the moon set in a clear winter's sky that we found in Shelley's 'Mont Blanc' and Goethe's 'To the Moon' is also to be found in a landscape without human figures, Goya's etching with aquatint, *Landscape with Waterfall*, which dates from sometime before 1810. This diminutive work, measuring only 16.6 x 28.5 centimeters, deserves considerable attention, first, because of its historical context and, second, because of the work itself. This work is a rare example of Goya's foray into the landscape genre.[26] Three images dominate the three main grounds of the etching. A castellated city in the rear, filling the background, is rendered with the finer-pointed lines we associate with the classical style of engraving. By contrast, the huge, dark mass of rock, looming in the foreground, is richly rendered with thickly engraved lines, and its volume is further suggested by a gray wash. By means of these contrasting techniques, Goya has created a sense of solidity and permanence in his engraving of the rock; whereas his fine-lined technique has made the castle seem insubstantial, even transient. The dominant image of the work, however, is the waterfall of the middle ground. The flowing water of this falls is suggested by parallel lines and by highlights of gray tinting; the effect of the spray is achieved by Goya's allowing the white surface of the paper showing through.

Both Goya's techniques for rendering these images and his juxtaposing of them combine to suggest something beyond the things themselves—that is, they seem to function as symbols. The insubstantial quality of the castle image may offer an emblem for the transience of things human. This, as I have already suggested, is particularly true when the building is seen in relation to the extremely substantial mass of rock in the foreground. It is, however, the waterfall that lends to the combination its full resonance: water is a fluid and unstable element; it is always in a state of pure flux, and that is the essential point: the flux is *ceaseless*; the process

---

25 Earl Wasserman, 'Mont Blanc' in George M. Ridenour, *Shelley* (Englewood Cliffs: Prentice, 1956) 69–102, qtd. from 90, 97, respectively.

26 In another rare example of an extensive presentation of landscape *The Meadow of San Isidro* (1788; Oil on canvas, Museo del Prado) is rendered with the Manzarnares River cutting across the horizontal axis of the painting; therefore, in his rare departure from predominantly figure paintings, Goya has a tendency to give predominance to water phenomena.

of the water's flow is unending. It seems no accident that one of the very rare times Goya—master as he was of the human figure in motion—executed a pure landscape, he chose the rush and turbulence of a waterfall.

How far will this quest for permanence in the face of change go? How can permanence be gained without creative change? And at what cost? Before the French Revolution, Goethe could have his Werther write, 'Alas, I stood with open arms facing the abyss and gasping forth a whisper, "down!" "down!". [...] Oh Wilhelm! how gladly would I have yielded up my human existence to rend the clouds and grasp the waves like that tempestuous wind!' After (and, of course, after a great deal of other personal and poetic transformations), he rejected such surface display of emotional chaos completely and created the strangely inhibited, hot-house emotional world of *Elective Affinities* / *Die Wahlverwandtschaften*; people suffer, languish, and die with hardly more than a whimper; order, through self-control, as well as control of others and things, is all—or seems to be all. The supposed permanence of conventions, however, is subjected to a subtle change that transforms everything into chaotic desolation. The seemingly hopeless attempt of achieving order on the very brink of chaos that is found in the earlier novel is repeated in the later; the Neoclassical surface only tends to heighten the contrast of the two conflicting forces. Four respectable characters—Edward, Lotte, her niece, and the Captain—are shaken out of the order and morality of their society by the growth of adulterous desires. Underlying the psychological drama of the four principal characters is a similar attempt to create a sort of micro-Utopia on Edward's estate; I will discuss this in the next section, where I treat land development and the stream.

'Humanized Landscapes' on the Banks of Streams

Jonathan Bate—in his 1991 study, *Romantic Ecology*—has offered us a new direction after what some (but not I) would call the Marxist Romantic orgy of the 1980s. What I do not like is his fierce polemic stance against the endeavor of these very devoted scholars who follow as well as revise Marx. Bate proposes an opposition that may never have existed in the West: between the Marxian 'red' and the environmentalist 'green'. We may not all take cultural materialism whole cloth, but there is a great deal that we have learned during the previous decade that should not be lost—and as I shall attempt to show—we still have a lot to learn. P.M.S. Dawson read Bate's intent very well when he wrote: 'Bate's neo-Stendhalian counterpoising of the red and the green, of Marxism and environmentalism, reveals clearly enough what he hopes to be saved from by his proposed turn to ecocriticism'.[27] Bate's book—for all that it is a pioneering work—is often more anti-Marxian than it is pro-environmentalist.

27 P.M.S. Dawson, '"The Empire of Man": Shelley and Ecology', Betty T. Bennett and Stuart Curran, *Shelley: Poet and Legislator of the World* (Baltimore: Johns Hopkins UP, 1996) 232–39.

As I have earlier observed, Bate would have it that, for Karl Marx, nature is merely a source of raw materials for human industry.[28] A different reading of what Marx said about nature is to be found among the anthropologists. As a sidelight to his field application of complexity theory to the irrigated rice fields of Bali and its hydrologic system, J. Stephen Lansing pieces together and elaborates insights of Marx concerning 'humanized nature'. Lansing finds this nature shaped by human agency, yet remaining still natural, in the irrigated fields of Bali, and I feel it is represented by the Constable's renderings of Hampstead Heath or—better yet—of the canalized River Stour flowing through Suffolk's grain-rich Dedham Vale. Admittedly, Lansing might seem to be indulging in a bit of Revisionism—which as a purist, I tend to reject; however, unlike many modern instances, it seems for elaboration of the implicit more than outright revising.

What I propose, in light of Marx and Lansing, is a comparative reading of these texts: landscapes painted by Constable, '*Scene on a Navigable River' (Flatford Mill)*; a landscape constructed by Bali rice farmers; two landscape passages by Austen; one by Goethe; and one by J. Hector St. John de Crèvecoeur. While allowing for an essentially positive reading of these 'humanized landscapes', I will indicate the potential downsides existing in them—both ideologically and environmentally speaking. Each landscape, after all, has its own economic and political context.

The Bali landscape, as described by Lansing, provides a useful referent because I am afraid that there are some 'Greens' who will not accept a defense, on its own terms, of the humanizing of European and American landscape, even when it comes from the early nineteenth century. Human modifications of the landscape have commonly been seen as an indication of Western aggressiveness toward and destruction of nature. A widespread bias against a Western development of land has been the result. On the other hand, Lansing found in the landscape of Bali a perfect balance between human effort and nature—a marvelous system of management as religious ritual—agriculture that works with the ecosystem and not against it. Since I have discussed all this in the introduction, I will now only reiterate that the Bali landscape offers an example of a *humanized* landscape, which is not *denaturalized.* Lansing discovered in this Balinese landscape a concept of hydrology and a system of hydraulics that successfully negotiates between the chaos of disease on the one hand and another chaos of drought on the other. Nature is altered but not forced.

Lansing's complexity theory allows for the emergence of an interesting aspect of the works treated below by Austen, Constable, and Goethe. This has to do with the tendency to force these works into dichotomously arranged periods—(Neo)Classicism and Romanticism. Two of the finest of recent treatments of Goethe's *Elective Affinities* demonstrate this: one seeks to show how it is a product of Weimar Classicism and the other article seeks to show how it is a product of Romanticism—they offer convincing arguments based on elements that are, indeed, contained in the work.[29] This balanced debate on the novel is possible simply because its

---

28 Jonathan Bate, *Romantic Ecology: Wordsworth and the Environmental Tradition* (London: Routledge, 1991) 57.

29 I have already discussed in the introduction the important attack on dichtomizations in Michel Foucault's *The Order of Things: An Archaeology of Human Sciences*, where he shows that eighteenth-century biology should not be divided in any strict sense between the believers in

(Neo)Classical elements and Romantic elements exist in dialectical tensions that create the greatest possible resonances. While period terms have a certain merit showing a particular consensus of interest at a particular time, they, as said before, have their reductive dangers; this is especially true when one is dealing with period terms that can also be employed as style terms—such as (Neo)Classical and Romantic.[30] There is—as has been discussed again and again throughout these pages—a constant tension and interplay of (Neo)Classical and Romantic qualities between 1760 and 1870, and nowhere is this more true than in the complex works of Constable, Austen, and Goethe.

Jane Austen, like Goethe, has never fit conveniently with her Romantic contemporaries. She, too, shows an interest in landscape per se—not just the somewhat abstract estate that Alistair M. Duckworth describes. His study crystallized much of my thinking about the socioeconomic, as well as ethical aspects of the estate, like ownership, inheritance, and caretaking; however, he omits from his purview the palpable sense of the landscape, of its soil, and of its waterways, as he does the profound correspondence that these ecological aspects of the land would have with the psychology and the culture of its owners.[31] It is precisely the greatness of Austen that she is able to make us feel the reciprocity between humans and their land—each making profound impressions upon the other, each shaping the other. It is, of course, hard to know to what degree she thought this interchange of influence happened in actuality; however, it is certain that she has aesthetically contrived her rare but telling employment of landscape to reveal the psychology of her characters: deep ecology, in Austen's novel-world, is the road to deep psychology. Her heroines—caught as they are in a patriarchally fabricated dependency upon males—desperately need the ability to read the character of men whom they will be dependent upon to a degree that is inconceivable in our modern Western culture. In her works, too, the land is both an aesthetic, as well as a utilitarian object of appreciation.

The aesthetic humanization of landscape is an important theme of Jane Austen, who has her protagonist in *Pride and Prejudice* finally achieve a correct appraisal of her future husband, Darcy, by reading the landscape of his estate at Pemberley—particularly the treatment of 'a stream of some natural importance' which 'was swelled into greater, but without any artificial appearance'. Human artifice has been applied but to give a natural feel. The protagonist's first impression is that nature and man are in harmony: 'Elizabeth was delighted. She had never seen a place for which nature had done more, or where natural beauty had been so little counteracted by an

---

rigid classification by stasis, like Linnaeus, and by dynamism, like Buffon; *Les Mots and les choses: Une Archéologie des sciences humaines* (Paris: Gallimard, 1966) 138–39, 163.

30 On the (Neo)Classical side is Stuart Atkins, '*Die Wahlverwandtschaften*: Novel of German Classicism,' *The German Quarterly* 53 (1980): 1–45; rpt., *Essays on Goethe* (Columbia: Camden, 1995) 137–81. On the Romantic side is Ellis Dye, 'Goethe's *Die Wahlverwandtschaften*: Romantic Metafiction', *Goethe Yearbook* 8 (1996): 66–92.

31 Alistair M. Duckworth, *The Improvement of the Estate: A Study of Jane Austen's Novels* (1971; Baltimore: Johns Hopkins UP, 1994).

awkward taste' (245).[32] From the windows of the great-house, she again surveys the grounds: 'Every disposition of the ground was good; and she looked on the whole scene, the river, the trees scattered on its banks, and the winding of the valley, as far as she could trace it, with delight' (246). Elizabeth is, of course, thinking more of landscape gardening than utilitarian applications of the landscape, but her feelings are only transformed by the housekeeper's information that Darcy 'is the best landlord' (249).

After seeing the house, Elizabeth in company with her all-important aunt and uncle, the Gardiners, goes for a walk around the grounds, and it is no accident, but very much a part of the artistry and significance of the novel, that their wanderings are always relative to and often following alongside the stream. 'As they walked across the lawn toward the river', Elizabeth has her famous encounter with Mr. Darcy, who has returned unexpectedly. After Mr. Darcy leaves, Elizabeth and the Gardiners continue their amble: 'They had now entered a beautiful walk by the side of the water, and every step was bringing forward a nobler fall of ground, or a finer reach of the woods to which they were approaching; but it was some time before Elizabeth was sensible of any of it' because her thoughts were still on the encounter with a seemingly changed Darcy. It is the great-house and not the surroundings that give *body* to Darcy's presence and situate the body of Darcy in her thoughts:

> Her thoughts were all fixed on that one spot of Pemberley House, whichever it might be, where Mr. Darcy then was. She longed to know what at that moment was passing in his mind; in what manner he thought of her, and whether, in defiance of every thing, she was still dear to him (253).

The Gardiners' comments and the scenery of the river, as well as the surrounding woods and hills, finally draw her out of her lonely self-obsession—objectifying her perception, as Goethe, Austen, and other Neoclassical-Realist-Romantics conceive it.

I cannot help seeing this as sort of 'hero journey' of a departure and return—where the journey is through a landscape, that is also a mindscape, of self-discovery:

> they pursued the accustomed circuit; which brought them again, after some time, in a descent among hanging woods, to the edge of the water, in one of its narrowest parts. They crossed it by a simple bridge, in character with the general air of the scene; it was a spot less adorned than any they had yet visited; and the valley, here contracted into a glen, allowed room only for the stream, and a narrow walk amidst the rough coppice-wood which bordered it. Elizabeth longed to explore its windings [...] (253–54).

Mrs. Gardiner 'is not a great walker', and the group soon must make the return part of the circuit. As before, there is nothing natural or human in Austen's taut narrative that is not suffused with a significance that is only later to prove crucial. The journey also ends in a revelation:

> they took their way towards the house on the opposite side of the river, in the nearest direction; but their progress was slow, for Mr. Gardiner, though seldom able to indulge the taste, was very

---

32 *The Works of Jane Austen*, ed. R.W. Chapman, rev. by Mary Lascelles, 3rd ed. 6 vols. (Oxford: Oxford UP, 1969) 245. Throughout the text of this entire book, all quotations from the works of Austen will be taken from the Chapman-Lascelles edition.

fond of fishing, and was so much engaged in watching the occasional appearance of some trout in the water, and talking to the man about them, that they advanced but little. Whilst wandering on in this slow manner, they were again surprised, and Elizabeth's astonishment was quite equal to what it had been at first, by the sight of Mr. Darcy approaching them, and at no great distance (254).

The path reveals a Darcy, not merely changed, but totally transformed in demeanor from the one Elizabeth rejected at Rosings.

It would be wrong to underestimate the influence upon Elizabeth of Darcy's natural unaffected treatment of the Gardiners, to the point of his offering—'with the greatest civility'—Mr. Gardiner the freedom of his watercourse and the loan of fishing tackle (255). Since the thought that had but an hour before reconciled her to having missed the chance of being 'mistress' of Pemberley was that the Gardiners would not have been welcomed 'as visitors' given the snobbery that Darcy had displayed throughout the first part of the novel (246), since her uncle and aunt are the London nexus that ultimately bring the couple together, and since Austen strategically focuses the very last paragraph of the novel upon Darcy and her relation with the Gardiners, there can be no doubt about the importance of such a humble item as fishing tackle. It may be, indeed the very insignificance and commonness of this cultural artifact that gives it charm when mentioned by the new Darcy. This new Darcy is only partly configured by being in the relaxed surroundings of his estate and its stream; however, the 'circuit' along the river certainly re-contextualized and re-presented Darcy to Elizabeth. On all sort of levels, then, the stream plays a great part in creating the softening natural ease that humanizes/naturalizes the interaction of Darcy and the others. Significantly, Elizabeth describes Lady Catherine's estate, Rosings, as having many fine features, but there is no mention of any undulating stream running through it to soften the landscape and the people by it. What we have at Darcy's Pemberley, on the other hand, is a humanized landscape that in its turn naturalizes the human.

In *Emma*, the dynamic of land and people is different from that in *Pride and Prejudice*. The protagonist already knows Mr. Knightley; she has no earthshaking need to read his character in the landscape of his estate; yet, Mr. Knightley's changes to his estate are even more interesting because they all have a utilitarian side. Jane Austen takes considerable care in giving us a sense of how the estate is always being transformed and improved quite directly and materially by Mr. Knightley. In part one of the novel, Mr. Knightley is described in conversation with his brother, Mr. John Knightley, about the 'plan of a drain [i.e., hydraulics], the change of a fence, the felling of a tree' (100) or the moving of a path by 'turning it more to the right that it may not cut through the home meadows' (106). I think it is by design that Austen rather neglects the Donwell Abbey building and expends more words on 'its ample gardens stretching down to meadows washed by a stream' and of 'its abundance of timber in rows and avenues, which neither fashion nor extravagance had rooted up' (358). This latter is a satirical swipe at the dogmatic and faddish application of the Picturesque strictures to estate planning, especially of the park portion. Jane Austen places emphasis on the productive side of the estate: trees are for timber or for apples in sufficient supply not only to meet the needs of the people of the Abbey, but also the direr need of the Mrs. and Miss. Bates of neighboring Highbury; fields with 'clover, which was to be begun cutting on the morrow' (361).

Indeed, the gardens, orchards, fields, and fishponds of Donwell come in for some of the most extensive and elaborate bit of landscape word-painting to be found in the works of Jane Austen: it might be further observed that the picturesque charm of Box Hill is sorely neglected by comparison. And it is along Donwell stream that it finally dawns upon Emma that she loves and is jealous of Knightley; it is the obvious care with which he shows Miss Smith, Emma's protégée, around that causes Emma's jealousy; she fails to understand that the tour is not really to show the Abbey, but the Abbey-Mill tenacy. The verbal survey of the landscape culminates on the stream:

> The considerable slope, at nearly the foot of which the Abbey stood, gradually acquired a steeper form beyond its grounds; and at half a mile distant was a bank of considerable abruptness and grandeur, well clothed with wood;—and at the bottom of this bank, favorably placed and sheltered, rose the Abbey-Mill Farm, with meadows in front, and the river making a close and handsome curve around it (360).

Abbey-Mill Farm is cultivated by the efficient and highly regarded tenant of Mr. Knightley, Robert Martin—future husband of Emma's protegée. We are treated to a view of 'all its appendages of prosperity and beauty, its rich pastures, spreading flocks, orchard in blossom, and light column of smoke ascending' (360). Jane Austen's phrase—'prosperity and beauty'—is pithy and freighted with the deepest significance; the financial and the aesthetic sides of the estate are thereby joined in a single phrase: it is beautiful, not in spite of its utility but, in part, because of it.

I have never understood why Constable and Wordsworth's names are inevitably linked in comparisons: to me it is Austen and Constable who are the real soul mates of English Romanticism (and they in turn link beyond with Goethe). Constable rejects the sublime topography of the Lake Country with its wild uncultivated landscape for the human-shaped wheatfields, cloverfields, and pastures that expand on either side of the Stour River. Constable further lovingly represents both the hydrology and hydraulics of the Stour-Valley Canal running past the family watermills. We see a stream which drives the mills and is canalized to float boats to carry the grain, and yet—thanks in part to old-fashioned organic farming on the shore—offers a safe place to fish. The Tory Constable offers a perfect Western example of Marx's humanized nature, yet still not denaturalized landscape. Constable's *Flatford Mill. 'Scene on a Navigable River'* can be read—if one disregards the second and actual title of the work—as a deeply personal glorification of family vested interests. Everything that we see is part of the Constable family holdings: the fields, where presumably grain is being harvested (it might of course be hay), are owned by the family—above all, so is the mill, and, in an intricate corporate way, so is the canal and, perhaps, the barges. Indeed, John Walker states categorically that these are barges of the Constable family.[33] Ian Fleming-Williams writes:

> The canalised Stour and its traffic had been the concern of his family for three generations. His father (who had inherited Flatford Mill) and four of his uncles were Commissioners, canal-users and owners of river rights, who managed the river for its proprietors—collectively, The Stour Navigation Company. As a leading Commissioner, his father had attended the meetings at

33 John Walker, *John Constable* (New York: Abrams, 1978) 94.

Nayland that dealt with the Company's business for thirty years or more. His younger brother, Abram, succeeded, and as a Commissioner also attended the meetings for many years.[34]

The workers are most probably in the employ of Constable's family. They are—in nineteenth-century capitalistic parlance—'the hired *hands*' of the Constable family, who as managers articulate those *hands*.

This painting celebrates a canal technology that was very modern—even if around a hundred years old.[35] A modern equivalent to Constable's decision to render canal boats, lock mechanisms, and hay wains would be a contemporary painter's rendering an operational farm truck or tractor with its hundred-year old technology of the internal combustion engine. In *Boat Passing the Lock*, the mechanics of contemporaneous technology is being celebrated.

Of course, one can take the Marxist reading onto a more obviously political level, as did John Barrell when he wrote that the figures of human labor 'are absorbed by the landscapes they cultivate'. Barrell claims that it is not only 'that Constable reduces all labourers to serfs, but that in the very same act he presents them as involved in an enviable, an almost relaxed relationship with the natural world, which allowed his nineteenth-century admirers [...] to ignore the fact that the basis of his social harmony is social divisions'.[36] Barrell is right to use the pre-Capitalistic term 'serfs'; indeed, his application of the Marxist critique of Capitalism to Constable seems rather like trying to apply it to the Brothers Limbourg. (This sort of confusion inevitably enters into all attempts to see an Agrarian *Capitalism* at such an early date, when shreds of Feudalism overlap with protean forms of Marx's true Capitalism.) To use Jonathan Bate's labels, Barrell's is the Revisionist 'Red' viewpoint, and a limited one at that; however, add to this a 'Green' viewpoint, and one might have something like a comprehensive view of this enigmatic painter. Fragmented, these laborers, these hired *hands* may certainly be separated from the controlling class of *minds*; however, they are unified with the natural environment, as Barrell admits. There is in Constable—and especially this work—a process of generalizing the ecological interaction of the human and the natural, which entails taking it beyond the limited family and economic context I have already mentioned.

The actual title, *Scene on a Navigable River*, not the commonly used title, *Flatford Mill*, demonstrates that Constable is seeking a generalized vision of an environment filled with interrelated human activity—the boats probably carry the grain that has been harvested from the fields on the left and ground in the mills. If this is the case, then, rendering the facial features of the workers to look like Tom, Dick, or Harry is irrelevant. Constable is just not essentially a portrait painter; he in fact disliked working in the genre; you cannot approach his landscapes in the same way that you can approach the essentially portraitist Gainsborough; and I am afraid that seemingly formal concerns are essential to achieving a full ideological reading

---

34 Ian Fleming-Williams, 'John Constable at Flatford', *Connoisseur* 178 (1980) 216–19.

35 I found Paul Mantoux's work useful for orienting the canal technology in the culture of England: *The Industrial Revolution in the Eighteenth Century: An Outline of the Beginning of the Modern Factory System in England* (New York: Harper, 1961) 102–35.

36 John Barrell, *The Dark Side of the Landscape: The Rural Poor in English Painting 1730–1840* (Cambridge: Cambridge UP, 1980) 133.

in painting or literature.[37] The lock worker in *Boat Passing a Lock* is placed on the vertical axis and is thereby not 'absorbed in the landscape' but is rather in balance with it. Constable probably made the Stour as famous a river as the Thames or any other English river by giving it a prominent place in the composition of many of his landscapes. His choice of the Stour says a great deal about his art. In order to contrast Constable's landscapes with those of his great compatriot, John Canaday observes that Turner 'discovered that the Rhine was his river, with its turbulent falls, its forested banks, and its castled crags, just as the peaceful Stour was Constable's'.[38] In the two representative river paintings, *Flatford Mill from a Lock on the Stour* and *The Mill over the Stream,* the Stour dominates the lower center of the canvas—indeed the definite pyramidal structure of the river (flowing toward the top of the pyramid) gives each painting its compositional strength. In the immediate foreground of *The Mill over the Stream,* froth from the water's flow is portrayed by chiaroscuro achieved by white brush-strokes.

Not only did Constable lavish loving care on the details of the flow of the River Stour, he sought to demonstrate the complexity of the various roles it played in the lives of the people along the river: he not only showed it flowing freely through Suffolk's grain-rich Dedham Vale, but showed how human agency had carefully controlled hydrology in what we might term its technological aspect—by hydraulics. Constable seemed never to tire of rendering the sluice and locks mechanisms, as well as watermills, which were, if not the high technology of the times, at least the standard and the characteristic. What they were not, circa 1820, was picturesquely quaint. The Tory painter represents in his landscapes not a rape of nature by humans, but rather a pact of two equals: human invention and natural creation. *Flatford Mill. 'Scene on a Navigable River'*—not at all picturesque—demonstrates this focus on the mechanisms and human energy involved in the passing of a boat through the lock; as one art historian put it: 'the operation of the locks became central features'.[39] What is more, there is in the center left of the canvas the reason of being for the lock and the canal: the grain mill to which it allows the bringing of the grain and the transportation to other places of the milled grain. *Flatford Mill* can be read—if one disregards the second and actual title of the work—as a topographical work; however, this is not true of the next canvas to be considered.

The *A Boat Passing a Lock* focuses upon a young worker in the process of opening the sluice—he and the level mechanism are at the center. This is humanized nature—to use the Marxist term—similar to the landscape in Goethe's *Elective Affinities / Die Wahlverwandtschaften* which reflects, as we have already considered in this chapter, the owner's care to humanize but not denaturalize nature by his hegemonic prevention of a stream flooding his tenant's village. We, however, must

---

37 See Thomas O. Beebee, *The Ideology of Genre: A Comparative Study of Generic Instablity* (University Park: Pennsylvania State UP, 1994); I have merely replaced literary genre with visual-arts one.

38 John Canaday, *The Lives of the Painters* (New York: Norton, 1969) 7440.

39 Leslie Parris, Ian Fleming-Williams, and Conal Shields, *Constable: Paintings, Watercolours, & Drawings* (London: Tate Gallery, 1976) 101; concerning the picturesque in Constable, see: Matthew Brennan, *Wordsworth, Turner, and Romantic Landscape: A Study of the Traditions of the Picturesque and the Sublime,* (Columbia: Camden, 1987).

remember that this improvement of the estate was one part of a program that was in the main an aesthetic one involving not just streams but lakes. Constable, then, is offering us an aesthetic vision of landscapes, put by the latest (for him) human technology to utilitarian purposes.

Similarly combining use and beauty, the landscape in Goethe's *Elective Affinities / Die Wahlverwandtschaften* (9) reflects: the owner's care to humanize but not denaturalize nature; his own, his wife's, and friends' psychology; his hegemonic prevention of a stream flooding his tenants' village.[40] The irony will be that not only can the central characters of the novel not control the outer environment, they are even worse at being able to control their own inner fluid nature with its 'elective affinities'—Goethe means quite literally, their biochemical reactions, in today's wording.

In discussing one of the details of his landscapes—overall, Edward focuses on the problem of flood control for the village below his manor house and says to his friend and architect, the Captain: 'The hill where the manor is situated forms a salient; the village is built opposite, in a fairly regular semicircle; in between, flows the stream, and to guard against its flooding, one villager erects stones, another stakes, yet another beams, and his neighbour planks—none coordinating their efforts with another, but rather causing damage and detriment to himself and others'.

> Der Schloßberg verläuft sich in einen vorspringenden Winkel herunter; das Dorf ist ziemlich regelmäßig im Halbzirkel gegenüber gebaut; dazwischen fließt der Bach, gegen dessen Anschwellen sich der eine mit Steinen, der andre mit Pfählen, wieder einer mit Balken, und der Nachbar sodann mit Planken verwahren will, keiner aber den andern fördert, vielmehr sich und den übrigen Schaden und Nachteil bringt.

Our masterplanner for social order has a corporate prescription: 'If the people were each willing to lend a hand, it would be possible without too great a subsidy to erect here a semi-circular wall [...] to create the most beautiful space, to encourage neatness, and by operating on a large scale, eliminating for ever all this inadequate patch-work'.

> Wollten die Leute mit Hand anlegen, so würde kein großer Zuschuß nötig sein, um hier eine Mauer im Halbkreis aufzuführen [...] den schönsten Raum herzustellen, der Reinlichkeit Platz zu geben und durch eine ins Große gehende Anstalt alle kleine unzulängliche Sorge auf einmal zu verbannen (9: 55).

This collective effort must be transformed into corporate effort—but let it be understood that it is still the feudal sort, rather than the capitalistic sort.

Edward likes the idea but is unwilling to undertake it unless he has absolute control over the effort: '"I prefer having nothing to do with townspeople and peasants unless I can have absolute control over them", said Edward'.

> Ich mag mit Bürgern und Bauern nicht zu tun haben, wenn ich ihnen nicht geradezu befehlen kann, versetzte Eduard (55).

The Captain readily agrees that it is 'so hard to consult on such things—especially with the lower classes, who understand everyday affairs quite well but seldom look

---

40 It must be understood that I can only deal with a small part of so central a theme as estate improvement; see Claudia Brosé, 'Park und Garten in Goethes *Wahlverwandtschaften*', *Park und Garten im 18. Jahrhundert* (Heidelberg: Winter, 1978) 125–29.

beyond tomorrow. [...] Everything of benefit to the general welfare should be supported by the absolute sovereign power'.

> [...] so schwer Rat zu pflegen, besonders mit der Menge, die im Täglichen ganz verständig ist, aber selten weiter sieht als auf morgen. [...] Alles eigentlich gemeinsame Gute muß durch das unumschränkte Majestätsrecht gefördert werden (9: 56).

Such sentiments anticipate the Congress of Vienna and the conservative Metternich settlement of Europe by less than a decade. This high-handed attitude toward the inability of the common people to conceive higher, coordinating systems offers an interesting parallel to the 'experts' and the Balinese farmers as described in the chaos-theory research of Lansing. In light of this research, can one help wondering if these German townspeople and peasants were not a great deal more efficient than this account would have us believe?

It would also be wrong to think that Goethe fails to infuse some irony in the sociopolitical sentiments spoken by his characters. He brings the retaining wall back into his narrative—as with the fountain and the streams in *Werther*—at a strategic moment as objective correlatives for the psychological transformations in his characters. It is the sight of the various building projects of which this newly completed retaining wall is a part—as well as the transforming effect all this construction has had on the landscape—that betrays Charlotte into a small but significant action that, as it were, opens the floodgate to her adulterous love for the Captain: 'It made a glorious sight in the beautiful weather. Charlotte was surprised and moved by the spectacle, and she warmly pressed the Captain's hand'.

> Es war bei dem herrlichen Wetter ein wunderschöner Anblick. Charlotte fühlte sich überrascht, gerührt und drückte dem Hauptmann herzlich die Hand (9: 70).

Whether it is human affection or the water in which Goethe finds an objective correlative, fluidity betrays all human strivings toward stasis. This fluidity is most perfectly realized as a river; this stasis, as the institution of marriage. Any attempts to control its flow, like the passions of an individual, are always posed on the precipitous edge of chaos. Viewing Metternich through this symbol lends his political actions a sort of aura of futility: in lieu of a more elegant way of saying it precisely, it is the proverbial boy with his finger in the Netherlandish dike. Post-Revolutionary Goethe seems, at once, to have admired and despaired of control.

If we return again to *Werther,* there is a passage in this pre-Revolutionary work that makes the social and even the political connection of the flooding stream very clear. In a letter dated 26 May, Werther—by interweaving the personal and the public—describes a hypothetical young man who is told to control his amorous passions. In my translation, I have italicized the crucial phrases that connect the sociopolitical with the hydrological:

> And then a Philistine, *some man of public office,* comes and says to him: 'My dear young man! To love is human, but you must love within human limits! Divide your hours—these for your work and those for recreation with your girl. Assess your income and, if once your necessities are seen to and there is some left over, I shall be the last to urge against giving her a present, though not too frequently: on her birthday, say, or her nameday [...] and so forth'.— If the person addressed obeys, he will turn out a useful young fellow, *and I should personally advise any prince to appoint him to his cabinet*; but his love will be finished, and, if he is an artist, so will his art. Oh, my friends! You ask why *the stream of genius* so rarely spills forth, so rarely

floods and thunders and shakes your astonished soul?—Dear friends, on either bank dwell *the self-composed gentlemen, whose summerhouses, tulip beds, and cabbage patches would all be devastated,* and who therefore know how to avert future danger *by damming and diverting* at the right moment'.

Und da käme ein Philister, ein Mann, der in einem öffentlichen Amte steht, und sagte zu ihm: Feiner junger Herr! Lieben ist menschlich, nur müsst ihr menschlich lieben! Teilet eure Stunden ein, die einen zur Arbeit, und die Erholungsstunden widmet eurem Mädchen. Berechnet euer Vermögen, und was euch von eurer Notdurft übrig bleibt, davon verwehr' ich euch nicht, ihr ein Geschenk, nur nicht zu oft, zu machen, etwa zu ihrem Geburts- und Namenstage usw.—Folgt der Mensch, so gibt's einem brauchbaren jungen Menschen, und ich will selbst jedem Fürsten raten, ihn in ein Kollegium zu setzen; nur mit seiner Liebe ist's am Ende, und wenn er ein Künstler ist, mit seiner Kunst. O meine Freunde! Warum der Strom des Genies so selten ausbricht, so selten in hohen Fluten hereinbraust und eure staunende Seele erschüttert?—Liebe Freunde, da wohnen die gelassenen Herren auf beiden Seiten des Ufers, denen ihre Gartenhäuschen, Tulpenbeete und Krautfelder zugrunde gehen würden, die daher in Zeiten mit Dämmen und Ableiten der künftig drohenden Gefahr abzuwehren wissen (4: 391–92).

Of course, this is Werther speaking, not Goethe—not all of Goethe anyway. Certainly, a part of Goethe must have been sympathetic with Werther's 'Philistine', since he, within just a few years, was going to have a prince 'appoint him to his council' and was quite wholeheartedly, for a while at least, to become a 'man of public office'. By the time Goethe writes *Elective Affinities,* he is clearly on the side of those who sought order; yet, Goethe is not of the same mind about the long-term efficacy of control as the Captain, Edward, or Metternich.

By comparison with a Metternich, one must admit that a Pitt, a Burke, or a Castlereagh comes off very well indeed. Even at its most repressive during the 1790s and the 1810s, England compares well with a Hapsburg-dominated Central Europe. While I am not fooled by thinking that Constable was motivated by any sense of democracy, the workers on the canal boats and locks do have a central place in his vision of his ancestral holdings. Michael Rosenthal has observed the negative influence of the 1822 Corn Law upon the agriculture interests of the Constable family and the effect all this had on the gloomy visions conveyed by the later works of Constable—the late *Salisbury Cathedral, Stonehenge,* and *Hadleigh Castle.*[41] Around the same year Rosenthal also points out another event that would contribute to the painter's gloomier turn: the frequency of arson due to social unrest among agricultural laborers.[42] In the sunny, early pre-1822 visions, the landscape has been transformed into a play where the human community labors hopefully. Although we do not need to be duped by this, the treatment of tenants, as well as landscapes, by Darcy and Knightley mediates the landowner orientation of Austen's novels and most post-Communist-Manifesto sensitivities. Landowner and tenant, and a whole host of ancestors, are shown to be part of humanizing of the landscape with taste and economy.

The concept of a 'humanized landscape' has a larger use. The usual ecological model that is too often appealed to by many modern environmentalists consists of an innocently wild ecosystem that is invaded by hordes of humans—resembling a virus

41 Michael Rosenthal, *Constable* (London: Thames, 1987) 159.

42 Michael Rosenthal, *Constable: The Painter and his Landscape* (New Haven: Yale, 1983) 207.

invading a healthy body. The setting aside of wild areas of nature can be viewed with some justice as an attempt to preserve them for an economically privileged few, who have the leisure and affluence to backpack into them. This model has mistakenly passed into the generalized public mind for the only end of environmentalism/ecology: setting up Federal Wilderness Areas in Alaska, Utah, or Idaho. And when practiced on a limited scale, few would question this as a useful policy. When, however, carried to extremes and given emphasis beyond reason, such a model also limits the vast political and intellectual endeavor that should be ecology in the fullest sense. The disaster caused by the Green Revolution in Bali demonstrates the danger of one-model-only thinking.

The current wilderness model of ecology derives, in part, from Wordsworth and, perhaps, from Turner—whose works, which render these wild mountain areas with little or no human presence, are especially favored. Ecology, however, should not consist of the exclusion of humans from nature; from a Spinozean point of view, humans and all they make are part of Nature/God; nor should it be the setting up of exclusivity barriers that limit humans intruding into nature to a few elite; it is rather the study and politics of how humans and nature can optimally interact. Above all, nature is not just a wilderness or a rain forest, and ecology is as applicable to Ohio and New Jersey as it is to Alaska or Costa Rica. The humanized landscape provides a vision that mediates between the needs of humans and the needs of nature beyond man—that mediates between the chaos of environmental destruction by humans or an order contingent on the relative banishment of a human presence. Humanized landscape is a balancing act; one possible model lies in Bali; another, in the Europe of Austen, Goethe, and Constable. These will provide a counterweight to the Turner-Wordsworth bias.

But before I close this discussion of developed landscape, I would like to treat two authors' visions that, perhaps unknown to them, would have offered an early warning as to what can happen if wise limits are not placed on the 'humanizing' process; moreover, this is a landscape whose political implications are worlds apart from the England of Castlereagh, let alone the Central Europe of Metternich.

In 1821, Goethe published a prophetic vision of the new Capitalistic world of fragmenting specializations and modern corporations. He pictures a league that combines aspects of the Jesuits and the Masonic illuminati and that seeks to transform society. The league also resembles a modern joint-stockholding company with hard-working executive-owners, who have renounced the usual emotional comforts of life for sacrifice to the needs of the organization. A new member, Lenardo, has a chance to make a contribution: 'through a region not blessed by nature and, among other properties, turned over to him by his uncle, a canal of late has been proposed, which would pass through our common holdings and which—should this happen through our united efforts—will increase in untold value'.

> [...] durch eine von der Natur weniger begünstigte Gegend, wo ein Teil der Güter gelegen ist, die ihm der Oheim abtritt, ward in der neuern Zeit ein Kanal projektiert, der auch durch unsere Besitzungen sich ziehen wird und wodurch, wenn wir uns aneinander schließen, sich der Wert derselben ins Unberechenbare erhöht.

The vision of a humanized landscape with thriving early nineteenth-century cottage industries is not an unattractive one: 'Here, he can very easily pursue his overpowering desire to start completely afresh. On both sides of this waterway, one can find much uncultivated and uninhabited land; here might spinners and weavers settle; masons, carpenters, and blacksmiths build modest houses for the spinners and weavers, as well as for themselves'.

> Hierbei kann er seine Hauptneigung, ganz von vorne anzufangen, sehr bequem enwickeln. Zu beiden Seiten jener Wasserstraße wird unbebautes und unbewohntes Land genugsam zu finden sein; dort mögen Spinnerinnen und Weberinnen sich ansiedeln, Maurer, Zimmerleute und Schmiede sich und ihnen mäßige Werkstätten bestellen [...] (6: 263).

One cannot help wondering if the area that the transportation canal is passing through is wetlands; modern hydrology in unity with ecology has proven the value of such lands. The idea that land is 'waste' land is an idea we have learned to reconsider.

St. John de Crèvecoeur, one of the great (proto)ecologists of Early Romantic culture, puts into the mouth of a key naturalist of the day an attitude toward nature that would dangerously dominate the next two centuries of part-progress and part-disaster in Western civilization. Crèvecoeur sets the scene by describing how the Schuylkill river winds 'through delightful meadows' and human-engineered banks that 'confine its stream'. The author has the naturalist John Bertram state that

> the Schuykill in its many windings once covered a great extent of ground, though its waters were but shallow even in our highest tides; and though some parts were always dry, yet the whole of this great track presented to the eye nothing but a putrid swampy soil, useless either for the plough or the scythe.

The corporate instincts of American capitalism, however, intervened, as Bertram's narrative continues:

> The proprietors of these grounds are now incorporated; we yearly pay to the treasurer of the company a certain sum, which makes an aggregate, superior to the casualties that generally happen either by inundations or the musk squash. It is owing to this happy contrivance that so many thousand acres of meadows have been rescued from the Schuylkill, which now both enricheth and embellisheth so much of the neighborhood of our city (184).

This corporate group of equal individuals organizing to take control of the land alongside a river contrasts favorably with the autocratic approach outlined in *Elective Affinities*.

The author describes his tour guided by Bertram along these humanized banks:

> We went to view his favorite bank; he showed me the principles and method on which it was erected, and we walked over the grounds which had been already drained. The whole store of Nature's kind luxuriance seemed to have been exhausted on these beautiful meadows; he made me count the amazing number of cattle and horses now feeding on solid bottoms, which but a few years before had been covered with water. Thence we rambled through his fields, where the right-angular fences, the heaps of pitched stones, the flourishing clover, announced the best husbandry, as well as the most assiduous attention. His cows were then returning home, deep-bellied, short-legged, having udders ready to burst, seeking with seeming toil to be delivered from the great exuberance they contained; he next showed me his orchard, formerly planted on a barren sandy soil, but long since converted into one of the richest spots in that vicinage (186).

When in the course of American history there yet remained plenty of wetlands, this would not be such a environmentally disastrous attitude; however, in that these passages point toward a future trend that would eventually go out of control, they might suggest to us today the dangers of carrying humanized nature too far. Like everything viewed in the light of complexity theory, it is a matter of balance.

The River Itself

Between 1760 and 1870, it was far less a question of how humans and their landscape shaped the rivers, but of how rivers had and were shaping still the human landscape. Even more than mountains and certainly to the same degree as seas, rivers shaped how humans conceived of the land and how they traced their aspirations across that land. Query II of Jefferson's *Notes on the State of Virginia* treats rivers, yet not just the rivers of Virginia, but also those of the known parts of the continent all the way southwest to Santa Fe. It seems on the surface one of those catalogues that he loved to make: river after river is treated. However, reading on, one soon sees that he is tracing out a great network of navigation, trade, and—yes, perhaps—an imperialist agenda, that would eventually become Manifest Destiny. He puzzles over the navigability and the interconnection of the rivers. And above all there is that great Mississippi, Ohio, and Missouri complex, which he would later secure for the United States with the Louisiana Purchase (32–45). Perhaps even more eloquent than the *Notes* is the map that they inspired: the one he instructed Lewis and Clark to draw up during their expedition, which in its turn relied on the rivers for navigation. The map almost entirely consists of a network of rivers, which in their turn resemble the arteries and veins of a vast anatomical drawing; the rivers run through this apperception of the American landscape like a chain of obsessive thoughts. Rivers, similarly, rush through Romantic poetry, painting, and even music.

Rivers—simply because they embody those particular qualities of flow existing in nature which the poet chooses to perceive and to utilize—demand a relatively extensive treatment. Therefore, it is significant that, unlike his 'Overwhelming Astonishment' and 'Song in Praise of Mohammed', Goethe's 'Permanence within Change' / 'Dauer im Wechsel' (1: 85–86) makes use of the true essence of the river image. In the former lyrics, there is an inconsistency between the symbol of the river and the thing it symbolizes: for, as was discussed in chapter one, in the former three lyrics, the qualities of the river symbol bear remote resemblance to aspects of the phenomenon as it exists in nature: the latter being a far more complex and multifaceted blend of permanence and change. This blend means that the essence of the river can only be conveyed through paradoxes. Although the water moves within the body of the river from the river's beginning to its end (the aspect of change), the river does not move from its topographical *locus* nor does it change in function (the aspect of continuity). This duality is the paradox of permanence in change which is inherent in the nature of a river and is found in 'Permanence within Change'. The river in this lyric of Goethe's later Weimar period is one of a series of symbols of change taken from the actual world. To attempt to simplify my way through this very difficult poem, I am going to take Goethe himself to be the speaker of this poem. The opening two stanzas offer symbols for both growth and the inevitable decay that follows. The series ends with the symbol which is most important for us—that of a stream, whose function never changes but whose contents do: 'With every rainfall, your fair valley will transform itself at once, alas, and in the same stream, you will never swim for the second time'.

> Gleich mit jedem Regengusse,
> Ändert sich dein holdes Tal,
> Ach, und in demselben Flusse
> Schwimmst du nicht zum zweitenmal (ll. 13–16).

This ever-changing substance within a more lasting essence is, after all, the real nature of rivers. That body of water we might arbitrarily call either 'the Mississippi' or 'the Rhine' is never the same—except, in name only.

This conception of change underlying seeming endurance—indeed the exact image employed to symbolize it—traces directly or indirectly back to the early Greek scientist/philosopher, Heraclitos, and his notion is preserved in the following fragment: 'Upon those that step into the same rivers different waters flow. [...] It scatters and [...] gathers [...] it comes together and flows away [...] approaches and departs'.[43] The editors and translators of this quotation from Heraclitos offer the larger philosophical context for the Presocratic philosopher's fascination with the image; similarly, I think it applies to Goethe's lifelong fascination with the river image because it offers an equilibrium between permanence and change. What image better offers itself as symbolic usage to the great conservative political evolutionist, who distrusted the great revolution of his day precisely because it lacked measure and balance in change? There are differences in the way that Heraclitos, the pessimistic prophet of change, and Goethe, on the other hand, approached the image of the river. Whereas the Greek philosopher used the symbol to support a metaphysics of change *without* permanence, we shall see that Goethe has used it for different purposes: in the stanzas that follow, it symbolizes the paradoxical unity of permanence *within* change.

In stanzas three and four, the Heraclitean usage of the ever-changing stream is developed into an extended metaphor for the human body of the poet—of Goethe himself. For like the river's body, his body changes in substance, in tissue: 'And you yourself! You will always look at walls and palaces—whatever has emerged before you rock-solid—through different eyes'.

> Du nun selbst! Was felsenfeste
> Sich vor dir hervorgetan,
> Mauern siehst du, siehst Paläste
> Stets mit andern Augen an (ll. 17–20).

His lips have changed, he goes on to say, and his legs; thus, Goethe-the-man is a name applied to an ever-changing body. While the structures that stood and now stand before his changed vision are only proportional foils for his changing perceptions—they only 'seem' to be 'rock solid'. Any sort of foil that these solid forms would seem to present for the fluidity of the stream is deceptive. The atoms that make up the river water, the human structures, or indeed the poet's cornea—these are all fluid when viewed from the *sub species eternitatis* of both

43 G.S. Kirk and J.E. Raven, *The Presocratic Philosophers: A Critical History with a Selection of Texts* (Cambridge: Cambridge UP, 1964) 195–96.

philosophy and science. The complexity with which the images are layered suggests a gamut of changeful possibilities.

The body of the speaker (whom I have taken to be Goethe) in 'Permanence within Change' has one sort of organ or appendage which is able to yield permanence in spite of its own change—the human hand is distinguished from all the other body parts because it is able to create forms (that is, create works of art): 'That hand—which gladly and gently extended to do good—which articulated structure—now, all these have changed. And what now in its place calls itself by your name came hither like a wave and, like that, hastens to [mingle with] the elements'.

> Jene Hand, die gern und milde
> Sich bewegte, wohlzutun,
> Das gegliederte Gebilde,
> Alles ist ein andres nun.
> Und was sich an jener Stelle
> Nun mit deinem Namen nennt,
> Kam herbei wie eine Welle,
> Und so eilts zum Element (ll. 25–32).

Through his place in this eternal cycle of poetic succession (expressed appropriately by an image of the One derived from a unity of the cycles' beginning with its end) and through his having been endowed, like each poet in his own right, with the gift of poetic creation (the imperishable favor of the Muses), Goethe-the-poet is able to transcend the laws of time and space to which the rest of the world is heir: 'Let Beginning now and Ending join and fuse into One! Let yourself soar past more swiftly than the objects! Be grateful that the Muses' favor promises Imperishability: the meaning in your heart and the form in your mind'.

> Laß den Anfang mit dem Ende
> Sich in Eins zusammenziehn!
> Schneller als die Gegenstände
> Selber dich vorüberfliehn!
> Danke, daß die Gunst der Musen
> Unvergängliches verheißt,
> Den Gehalt in deinem Busen
> Und die Form in deinem Geist (ll. 34–41).

Just as Goethe takes things experienced in the mutable world and converts them into the immutable structure of this poem, so too can all poets find and make permanence in change.

Through the mediation of the change-permanence synthesis achieved through cycle rhetoric, the seeming local chaos of one poet's death is transformed into a grand order of royal successions of holders of the office of Poet. When Goethe dies, another poet is ready to take up the place left vacant—in what might be termed a bardic apostolic succession; the name may change but the calling of the Poet remains the same. This strange refusal to be the single, solitary poet, and instead, the need to find a metaphor that will offer a context of process and community in or through time—all this we will see again and again throughout a significant number of the poets in the full range of (Neo)Classical-Realistic-Romantic culture. What is it if not the textual traces of a dis/*ease* with the alienation created by capitalistic individualism? And what is this recurrent need to see beneath change to some underlying permanent simplicity, but anxiety in the face of the industrial and the political revolutions? Perhaps, all that can be said at this point with any certainty, is that the theme of continuity in change embodies some half-hidden cultural transition and, above all, that water-symbols form the needed syntax to embody the metamorphic complexity of this theme.

Such is the theme of Goethe's 'Song and Form' / 'Lied und Gebilde', from the *West-Easterly Divan / West-östlicher Divan* (3: 296), which is arguably a more terse, quintessential expression of this theme. In the first stanza he sets forth the Greco-Roman classical aesthetic: 'May the Greek mold his clay into forms, taking ever-increasing delight in his own hands' son [= child or creature/creation]';

Mag der Grieche seinen Ton
Zu Gestalten drücken,
An der eignen Hände Sohn
Steigern sein Entzücken;

and in the second, he uses the river as a symbol for a system of aesthetics which emphasizes flux over stasis: 'but to us it is sheer bliss to grasp at the Euphrates and roam back and forth in the fluid element':

Aber uns ist wonnereich
In den Euphrat greifen,
Und im flüßgen Element
Hin und wider schweifen.

Goethe has again established his beloved polarity, and here it is his special preference: that between form and formlessness. Again, I appeal to that old metaphor (which by the way I have borrowed from Egyptology to show the pull between Seth, the god of chaos, and Horus, the god of cosmos) of the harmony gained by the tension between two opposed forces—like all metaphorical paradoxes, its essence can best be accessed and described in yet another metaphor.

Goethe, by means of his persona, describes the mission of the poet as the merging of these two aesthetics; this merger is especially interesting if one considers that Weimar Classicism was a melding of Early-Romantic Storm and Stress with Greco-Roman Classicism. This merging finds its expression in the symbol of the most mutable of substances being crystallized in the hands of the poet: 'When I thus have cooled the soul's ardor, the song will be sung; when the poet's pure hand creates something, water will take shape'.

Löscht ich so der Seele Brand,
Lied es wird erschallen;
Schöpft des Dichters reine Hand,
Wasser wird sich ballen.

Goethe's image for perfection is balanced on the apex between the chaos of formlessness and, on the other hand, the order of a crystalline form taking shape out of the water. These lines celebrate the precariousness of the act by which the image of pure change is welded by the magic of poetry into a symbol of permanence.

Water, as the formless, is the ideal challenge for the form-obsessed. And the prophet of Weimar Classicism was indeed form-obsessed. In a scientific treatise on meteorology, Goethe tells how in his testament of nature he seeks to set in opposition 'the sculpted life against the formless' / 'ein gestaltetes Leben dem Gestaltloser' (17: 642). This, then, is the goal; and for Goethe (as I will seek to demonstrate) the ultimate test lies in the nature of water phenomena. To return again to Paul de Man's words on Shelley: 'Water, which has no shape of itself, is molded into shape by its contact with the earth, just as in the sense of the water washing away the tracks, it generates the very possibility of structure, pattern, form, or shape by way of the disappearance of shape into shapelessness'.[44] Water then beckons to the poet or artist to return it to shape—or in Goethe's case to sustain it in that shape. He does not take the easy mode of representing form; it must be perched at the edge of a chaos of implied contradictions; once again he exalts in the use of a signifier that is fraught with the implications of flux and reflux to symbolize permanence. If the key element of Neoclassicism is stasis, then in Goethe's poem, it is something more: it is the sounding of a string held in tension by a pull in opposite directions. One of these directions is the Romantic element of a love of change; the other, of permanence. The harmony created is one of the glorious embodiments of Romantic culture.

With these lines in mind, I would like to requote from chapter one lines by Mélendez, where the poet has similarly employed the poetic strategy of multivalence in word usage: 'Oh, smiling little fountain, how my spirit takes delight!—my eyes become enchanted in your crystalline mirror!'

> ¡Oh! ¡cómo en tus cristales,
> Fuentecilla risueña,
> Mi espíritu se goza,
> Mis ojos se embelesan! (ll. 1–4).

Meléndez's phrase 'en tus cristales' carries yet another ambivalence beyond those examined in chapter one: 'crystal' or 'crystalline' means transparency—one of the water's manifestations; mirror creates a surface that destroys transparency but reflects—the same with water.

Romantic painters of particular gift could create the conflicting nature of water in motion and stasis—the fluid of its flow and the reflective powers of its still surface. The Stour runs through the center of *The Hay-Wain* and its flow is carefully rendered by the artist. The river's water is painted with an initial, unifying application of green, brown, or tan—blue-green for the area of water just above the hay wain—to represent the reflections of the trees, house, wain, and bushes. Above all, Constable uses impasto streaks of pure white to suggest a surface disturbed by the flow of the water. As one would expect, the facture representing water is more prevalent downstream from the wain. Graham Reynolds describes these effects as follows: 'In *The Hay-Wain* [...] the water is composed of dark-grey, yellowish-brown and thin white streaks. [...] [These] broken tints [were] often applied in small, crisp, dry touches, and sometimes with a vertical striation or fracture which gives further life to the surface texture of the paint'.[45] This stippling facture is the key to

44 Paul de Man, *The Rhetoric of Romanticism* (New York: Columbia UP, 1984) 107.

45 Graham Reynolds, *Constable: The Natural Painter* (St. Albans: Panther, 1976) 97.

Constable's Realistic-Romantic rendering of water, as well as—taking a broader point of view—of visually conveying the processes within nature.

There has appeared recently an impressive attack on Constable's fidelity to visible truth that first shook, then, refined my own ideas on the subject; this attack was made by William H. Galperin, in his *The Return of the Visible in British Romanticism.* Galperin writes that 'despite its uncanny abetting of a visible order, *The Hay-Wain* is opposed to the visible as I conceive it'.[46] The visible as Galperin conceives of it seems to be represented by the Diorama which renders (judging from the examples he offers) the visible with a painfully precise exactitude; they are the most absolute and elaborate forms of topographical rendering. The Dioramas are rendered with a frozen exactitude that allows it to bear only superficial resemblances to cinema; Dioramas' unrolling a changing scene makes the comparison superficial at best. The art of Diorama finds its comparison in the glossy photographic meticulousness of Friedrich.[47] This is not Constable's way. He does not paint with the sort of Realistic and oily-facture precision of a Friedrich—or, for that matter, of a Gustave Courbet; nor does he follow the intentions of the true Impressionists as Galperin suggests. The Impressionists sought to render reality as the cells of the cornea of the eye catches the light from a scene, viewed at a momentary glance. Constable would not have understood this; he also rarely paints objects and people in the foreground or middleground with the broken brushstroke technique (superficially termed Impressionistic because they used it). Nor does Constable apply primary colors with tiny adjacent comma-like brushstrokes; his broken brushstrokes are inevitably white stipplings, which suggested something beyond the visible phenomena—the noumena of recurring process.

Rendering *process* and *motion* is the goal of Constable's facture. He uses the white stipplings to suggest the elemental movement of phenomena: the passing, the transforming of clouds and the light effects created by their changing shadows on the ground below; the flutter of the leaves in the wind and uneven sunlight; and—as we have just seen—the stippling to create the flow of water. This is an accurate rendering of the visible—but not the frozen detail of objects in a Diorama or a Friedrich oil painting—but rather a facture that seeks to render the motions, the dynamisms of the visible world. This is why *The Hay Wain* seems, in the eyes of many, to shimmer. Constable was interested in rendering—in the words of Karl Kroeber—'the processes of nature as predominantly manifested in fluid, impalpable, unfixed phenomena'.[48] The broken hasty brushstrokes are attempts at creating a

---

46 William H. Galperin, *The Return of the Visible in British Romanticism* (Baltimore: Johns Hopkins UP, 1993) 84.

47 Galperin 207–43.

48 Karl Kroeber, *Romantic Landscape Vision: Constable and Wordsworth* (Madison: U of Wisconsin P, 1975) 19. This is an often insightful book; however, I have always been puzzled by this urge to compare Constable and Wordsworth, who seem to me to have had antithetical artistic sensibilities and interests. Constable preferred Hampstead Heath and the Stour River valley to the Lake Country. Constable seems to me closer to Jane Austen; Turner, to Wordsworth. The full-fledged Romanticism of *Hadleigh Castle* is an exception in Constable's oeuvre, and I find Kroeber's comparison of *Peele Castle* and *Hadleigh Castle* the most satisfying part of his study (44–60).

sensation of movement. It, and not the moving Diorama, is the Romantic equivalent of motion pictures, film.[49]

During one of the least frenetic moments in *Frankenstein,* Mary Shelley has her protagonist travel down the Rhine in which the humanized landscape still had patches of the wild interspersed:

> This part of the Rhine, indeed, presents a singularly variegated landscape. In one spot you view rugged hills, ruined castles overlooking tremendous precipices, with the dark Rhine rushing beneath; and, on the sudden turn of a promontory, flourishing vineyards with green sloping banks and a meandering river and populous towns occupying the scene (150).

It must be said that, in some outward details, the Rhine and its 'rugged hills' or 'tremendous precipices' is more in keeping with the content preference of Turner rather than Constable; however, as Shelley describes the scene, it fits the latter in spirit. She has Frankenstein state a preference for this humanized landscape over the wild, untrammeled ones of the mountains:

> Oh, surely the spirit that inhabits and guards this place has a soul more in harmony with man than those who pile the glacier, or retire to the inaccessible peaks of the mountains of our own country (151).

Harmony is the key term, it seems: man can alter nature without violation, but rather through harmonic compromise. According to my reading of the text, here, it seems key to compare this humanized landscape with the wild-mountain landscape from this novel already examined in chapter one: the alpine setting of streams, avalanche, and glaciers, where Frankenstein confronts his creature, who is to be taken as a symbol of violated nature.

Frankenstein expresses this sense of human imperialism over nature when he begins his search for the knowledge and technology needed to create his monster:

> soon my mind was filled with one thought, one conception, one purpose. So much has been done, exclaimed the soul of Frankenstein,—more, far more, will I achieve; treading in the steps already marked, I will pioneer a new way, explore unknown powers, and unfold to the world the deepest mysteries of creation (47).

To enter a wilderness of 'unknown powers', as a violating 'pioneer' is not only something not to be celebrated, it is rather the cardinal sin of the novel's ethical system—which I cannot help viewing as a key (proto)enviromental one. Entering the wild mountain landscape—the wilderness area, as it were—signifies Frankenstein's use of science and technology to create life. Shelley has carefully and aptly chosen wild-mountain peaks 'inaccessible' to most humans for locating the confrontation of the Monster with the Creator, and I think that it is no accident that, only a few pages after the close of the mountain-top encounter, Shelley has her protagonist-narrator wax lyrical on a humanized landscape of the Rhine valley where there is a harmony between man and nature.

Whereas Mary Shelley, Goethe, Austen, and Constable preferred the quiet streams that allowed for human activity to prosper at their sides, a considerable number of their contemporaries showed a predilection for the surge and rush of rapids and falls

49 Cf. Galperin 1–13.

(which we will examine later under their dominant image, the waterfall). John Martin, the very popular Romantic painter, rendered an extraordinary mountain torrent that fills the bottom half of his painting *The Bard* and is lit by a bright unseen moon. The streaks and swirls signifying the water are echoed by those representing the stormy clouds above and those of the bard's windswept cloak. On a crag opposite the bard, a castle precariously and impermanently perches—as if soon to be undermined by the water eating at the cliffs below.

Like Percy Shelley, Martin and Coleridge juxtapose the flux of the torrents set in high mountains or otherwise extreme landscapes against the impermanent works of man. The resulting torrents from the melting snow form a mighty stream offering a foil for human power and inventions:

> The race
> Of man flies far in dread; his work and dwelling
> Vanish, like smoke before the tempest's stream,
> And their place is not known. Below, vast caves
> Shine in the rushing torrents' restless gleam,
> Which from those secret chasms in tumult welling
> Meet in the vale, and one majestic River,
> The breath and blood of distant lands, for ever
> Rolls its loud waters to the ocean waves,
> Breathes its swift vapours to the circling air ('Mont Blanc' ll. 117–26).

One significant aspect that lends a relative power to the water phenomena is that the power and force of several quarters of the cycle are appealed to in this passage. The stream and sky quarter are mixed in this phrase—'like smoke before the tempest's stream'. The link between the stream and the ocean quarter is appealed to in these—'one majestic River [...] / Rolls its loud waters to the ocean waves'. Permanence is lent to the latter lines by the intensifying 'for ever'. Shelley's juxtaposing the relative insignificance of the human handiworks compared to the mighty flow of waters bears a striking resemblance to the contrast in the first two stanzas of Coleridge's 'Kubla Khan'.

These two stanzas of Coleridge's poem (1798; 1: 295–98) are constructed of many recurring motifs which form two major symbol complexes. The first of these is the vision of 'the stately pleasure-dome' which Kubla Khan decreed his men to build (ll. 1–2) and its walled gardens which have been carefully laid out in a controlled landscape. The natural certainly intrudes as 'forests ancient as the hills'—i.e., they were there before man (l. 10)—and in the form of 'rills' whose sinuosity (l. 8) suggests the 'meandering' of the larger stream (l. 5); however, the whole vision seems to have been brought safely under the control of human gardeners and of architects—and ultimately, of the monarch who commands them all. 'Kubla imposes his forms', as Marjorie Levinson expresses it, 'upon naturally occurring materials whose own properties are thereby modified, appropriated, or eliminated'.[50] There are, however, many elements appearing throughout the poem that fall well beyond

50 Marjorie Levinson, *The Romantic Fragment Poem: A Critique of a Form* (Chapel Hill: U of North Carolina P, 1986) 105.

Kubla's power; having said this, it must be admitted, with Levinson, that his impress upon the landscape is striking.

This whole landscaping process is suggested by the verse form which slowly changes from lines distinguished by varied length and enjambments to more controlled blank-verse lines with punctuated end-stops:

> So twice five miles of fertile ground
> With walls and towers were girdled round:
> And there were gardens bright with sinuous rills,
> Where blossomed many an incense-bearing tree;
> And here were forests ancient as the hills,
> Enfolding sunny spots of greenery (ll. 6–11).

Secondly, in opposition to this manmade order there is the natural landscape of 'Alph, the sacred river' and 'the caverns measureless to man' into which this river falls 'in tumult' (ll. 3–5, ll. 25–28). This cavern is variously described as a 'deep romantic *chasm*' (l. 12, l. 17) and as '*caves* of ice!' (l. 34, 36).

I disagree with Levinson when she says that Coleridge presents his dream vision is such a way that we would believe that 'Kubla subdues the profound and generative natural principles figured by the chasm, river, and cavern—yonic and psychic imagery—in the interests of civil and domestic order and by way of asserting his dominance'.[51] The poem opens temporally with Kubla's having commanded the building of the dome, rather than by saying 'there are pyramids in Egypt'. Unlike the Pharaohs' great building projects, however, Kubla's dome and the pleasure ground no longer exist: the coda states that they have to be rebuilt by poetic imagination. Furthermore, the water phenomena are portrayed dynamically, indeed pyrotechnically—the poet's descriptions of Kubla's humanized landscape pale by comparison. Most of all, the river symbol, running through the first 36 lines, in sense, sound, and structure, often permeates and undermines Kubla geometry.

Yet, so as not to oversimplify a complex poem, let me say that the river has two contradictory states: one that flows placidly and parallels, at least temporarily, the building of human structures and the ordering of landscape; the second that defies it with a threat of chaos. Before falling into the caverns, the river Alph's flow is Neoclassically expressed by 'stately cadences' of regular spondaic and iambic meters reminiscent of Denham's Thames and by the alliteration of the liquid consonant *m*:

> Five miles meandering with mazy motion
> Through wood and dale the sacred river ran,
> Then reached the caverns measureless to man [...] (ll. 25–27).

This marks only one of two essential moods of the river within the poem. After the fall, the meter is shattered like the flow of the stream:

> And sank in tumult to a lifeless ocean [...] (l. 28).

The hint of the Popesque couplet of lines six through eleven with its suggestion of order has already been undercut in the first stanza where the river spills from one Romantically enjambed line to the next:

---

51 Levinson 105.

[...] Alph, the sacred river, ran
Through caverns measureless to man
  Down to a sunless sea (ll. 3–5).

For the waters of this river, even in a state of repose, are beyond the control of man. The river is a sacred law unto itself; the walls of man can never gird it around. Where it later seems that seismic forces effect the river, the river water is tossed and turned:

A mighty fountain momently was forced [...] (l. 19).

These are the forces of disorder, not order; it is no surprise that Kubla's edifice stands no more. Yet, it must be noted that the river is controlled by its own laws and, even at the height of Kubla's building projects, its flow breaks out into seeming chaos. The river is a foil for Kubla the architect and landscape gardener; even its Neoclassical moods, when seen in context of the more Romantic, are clear ironic mocking of Kubla's khanate—which may have hegemony over humans, but not over nature.

Although these works of the two English Romantics resemble Goethe's 'Permanence within Change' and 'Song and Form' in that they employ streams as symbols for abstract ideas, the streams of the latter poems are brought under control (even if it is celebrated as a precarious and temporary control sustained in defiance of the changeful qualities of the signifier), whereas in Shelley and Coleridge, they represent things beyond the control of human power. In this respect the English poems more resemble the overflowing winter-streams found in *Werther* and 'To the Moon'.

The Stream in Turmoil: The Waterfall, Chaos, and Unrelieved Change

As we saw in the first chapter, the poets treated in this study often show their springs falling precipitously from their source in high mountain landscapes. With similar frequency, they show their brooks and rivers at their fall-zones dropping steeply down cliffs and boulders.

From among the water poems composed within the time-frame of this study which bear a strong stamp of previous topographical and eighteenth-century descriptive poetry, I have chosen one example from the last quarter of the eighteenth century, from 1787. In the tradition of topographical and descriptive poetry is a twelve-line fragment (285)[52] by Robert Burns—'Written with a Pencil, Standing by the Fall of Fyers, near Loch-Ness', which, as the title indicates, was written at the precise locus of Fyers Falls (also the central water image of the poem) and which is intended as nothing more than a descriptive verbal sketch:

Among the heathy hills and ragged woods
The roaring Fyers pours his mossy floods;
Till full he dashes on the rocky mounds,
Where, thro' a shapeless breach, his stream resounds.
As high in air the bursting torrents flow,

52 *Burns: Poems and Songs*, ed. James Kinsley (London: Oxford UP, 1969).

As deep recoiling surges foam below,
Prone down the rock the whitening sheet descends,
And viewless Echo's ear, astonish'd, rends.
Dim-seen, through rising mists and ceaseless showers,
The hoary cavern, wide-surrounding, lowers.
Still thro' the gap the struggling river toils,
And still, below, the horrid cauldron boils [...].

Although Aubin, in his *Topographical Poetry in XVIII-Century England*, calls this work 'pure description',[53] the water images that make up this poem are not merely described; for the persona intervenes into his conveying of the scene with adjectives, nouns, and verbs such as 'astonish'd', 'the struggling river toils', and 'the horrid caldron boils'. It must be noticed, too, that the title of Burn's poem puts us in the realm of the Romantic sketch of the visual arts—in music, the impromptu—the dashed off, incomplete, fragmented thing. The poet suggests the visual arts where sketches are the norm.

In rendering the cataclysmic rush, froth, and spray of the waterfalls, Turner was to perfect the technical innovations introduced by the late eighteenth-century painters. *The South Wales Sketch Book* offers a representative example of Turner's early topographical interests in loci with water, particularly falls, as well as an indication of a development of techniques in rendering water and foam. In this collection, thirty-nine of the ninety-six sketches treat water phenomena; more significantly, of the seventeen which were touched up with color, eleven are of water, and generally, it is the water alone which is colored. From among these early topographical sketches, 'Melincourt Fall, Vale of Neath'[54] may be singled out as representative as well as being an interesting work in its own right. At first glance, the sketch seems incomplete: for the trees, rocks, and the human figure (a man in traveling clothes with a cane at the bottom right of the falls) are done merely in pencil.

The sketch—in accordance with a Romantic traveler's interest—focuses on the waterfall as well as the cliffs and the boulders next to the falling water; these alone are painted. The rocks, especially those of the lower cliff, have been done with careful detail. Turner has rendered the cliffs in the same subdued, light ultramarine blue with which he painted the waterfall—suggesting that they are covered with spray from the waterfall and that only those parts covered by this spray were of interest to the artist. The rocks have a thicker application of paint. The paint of the lower cliff on the left has been very heavily applied to the point of opacity—here, the rocks have been carefully modeled with bluish-black paint. The top of the waterfall, where it spills over, has been rendered with stipple strokes—leaving the white of the paper showing through in order to suggest rapids. The fast-falling water of the waterfall itself is suggested by a quick dry brush application with very long, light strokes; here, the paper beneath is more visible and the paint application is

53 Aubin 237.

54 Andrew Wilton gives this work and its rending of water phenomena some attention; he titles it *Melincourt Fall near Abergarnedd*, 'Turner: Topographer of Sublimities', *J.M.W. Turner: 'That Greatest of Landscape Painters', Watercolors from London Museums* (Tulsa: Philbrook Museum of Art; Seattle: U of Washington P, 1998) 1–14.

more transparent than in the rendering of the surrounding cliffs, a contrast that suggests the lightness and transparency of the falling water—reduced as it is almost to a spray. At the point where the falling water hits some jutting rocks and at the foot of the falls, the strokes are even lighter and more transparently applied—suggesting a greater amount of mist and spray at these points. The rapids below are rendered like those above the falls. A pool, seen obliquely, is suggested by a dark blue stroke. In short, the sketch is a complete work in the sense that its color adds emphasis to the most important aspect—the waterfall—and subordinates the topographical details of the surroundings.

As with Burns' 'Fall of Fyers', which purports to be a sketch 'Written with a Pencil' *in situ,* Turner's 'sketch' goes beyond mere topography: both works show a concentration on and delight in viewing, then describing, the flux and flow of the water as it 'dashes on the rocky mounds', foams in 'deep recoiling surges', and '[p]rone down the rock the whitening sheet descends'. In both works, we see the artists breaking away from the conventions of topographical exactitude and polish; this happens, I think, because of the countervailing tradition of the sketchbook—perfect finish and precise detail can be completely ignored—allowing the process of phenomena to predominate over the rendering of place.

Loutherbourg's *The Falls of the Rhine at Schaffhausen* is an example of a topographical portrayal of a waterfall—but, the laborious attention given to the technical rendering of the water shows the painting to be a transitional work. There is an increased intensity in simulation. For while the water of the falls is set picturesquely in its landscape of trees, human figures, and buildings, the water itself has an importance at least equal to these. The foam of the falls is rendered and emphasized by a heavy impasto. The pool below is rendered with blue and green streaks. The water's tremendous flux conveys a strong sense of the grandeur of nature. Such striking use of water and foam is even more impressive in Koch's *Waterfall / Wasserfall.* The waterfall fills the whole left-center of the canvas. It spills over jagged outcroppings of rocks. Hans Engelhard writes that the entire composition rests on the formal contrast between the dull browns of the cliff and the silver-grey of the foam of the waterfalls.[55] The waterfall itself is rendered by a light application of shades of grey, from dark to very light, almost to white. Although there is no way of telling for certain, it would seem that the paint used to represent the foam was dappled on with a sponge. Be that as it may, the brush has been used in such a way as to make the water seem transparent. The striking effects of placement and technique are marshalled by Loutherbourg in order to achieve, as Engelhard expresses it, 'the sense of the elemental power of falling water'.[56]

To add to this impressive effect of the waterfall, the two human figures (one in the foreground and a much smaller one in the background) and the tiny goats

---

55 In the author's words: 'Die ganze Komposition ist auf den Gegensatz zwischen den zackigen braunen Felsen und der silbrig grauweißen Gischt des Wasserfalles angelegt' (Hans Engelhard, 'Kunstgeschichte und Geologie: Der Wasserfall in Gemälden des 17, bis 19. Jahrhunderts', diss. Cologne, 1975, 137).

56 My translated phrase comes from this passage: 'Koch kommt es darauf an, das Enge, Bedrohliche und Ausweglose der Schlucht und der Elementargewalt des stürzenden Wassers so eindrucksvoll wie möglich zu machen' (Engelhard 137).

grazing on the overhanging ledge are dwarfed proportionally so that the waterfall is made to seem very large—indeed, somewhat overlarge relative to the amount of water flowing in the foreground. This discontinuity of scale seems intentional: the artist may have consciously sought to make the waterfall appear inordinately large relative to the human figures so as to underscore their insignificance in the face of nature's magnitude. (This is a convention used but a few decades before by Piranesi for ancient ruins, but this is the most extreme application of it to water that I know of.) There is a true revelling by the artist in the titanic power of this scene of falling water. Only the classically costumed, posed, and rendered shepherd reminds one that the painting was done while Davidian Neoclassicism was at its height. Indeed, the art historian Fritz Novotny categorizes the painting as 'Classical'. Novotny goes on to make a point about what I term the (Neo)Classical *componentization* of the subject that may bear upon the presentation of Koch's careful use of quarters of the water-cycle. The painter carefully renders clouds and mist upon the mountain; the run-off from glaciers, the terrestrial source; the high-mountain falls; lower falls carrying the water below the tree line, and into the foreground as a whitewater torrent. Novotny says that Koch emphasizes these 'zones' and that 'the principle of the addition of part to part being carried to such a pitch that the picture offers no single uniform focal point. This method of construction, naive and artificial as it is, conveys almost better than any other an impression of the immensity of the world of high mountains'.[57] The sum effect is like that of the Chinese mountain landscape, as developed by the Northern Sung painters: they have a perspective consisting of multiple focal points; it is as if the viewer were to imagine himself *inside* the picture able thereby to move from position to position to see each landscape zone to its best advantage, which he could not possibly do if background views were subordinated to the single viewpoint of the foreground (the Italian Renaissance mode).[58] In Koch's picture, then, this allows the viewer to see the source quarter of the background (as well as the sky quarter) with the same clarity as the stream quarter of the foreground.

Storm and Stress poets, like the painter Koch, present the waterfall as a symbol of overweening nature, especially in its most destructive guise. The falling stream of water is always seen as a negative agent acting upon things alongside. It is a sign for unleashed forces—born of the social frustration of an underemployed middle class in aristocratically dominated realms, but not yet articulated into a political enunciation. Stolberg's 'The Rock Stream: Immortal Youth' / 'Der Felsenstrom: Unsterblicher Jüngling' (1: 104–06) presents just such a vision of destruction. The powerful rush of water destroys trees and cliffs around it as it comes roaring down from its mountain birthplace—its source. The persona of the poem makes it clear that he is envious of the stream's destructively powerful flow: 'How lovely you are with your silver locks! How dreadful you are amid the thunder of the echoing crags around you! Before you, the firs tremble. You uproot and dash down the firs! The rocks flee

57 Fritz Novotny, *Painting and Sculpture in Europe, 1780–1880*, 2nd. ed. (Harmondsworth: Penguin, 1971) 70.

58 William Watson, *Style in the Arts of China* (Harmondsworth: Penguin, 1974) 85.

from you; you catch the rocks and mockingly roll them down as if they were pebbles'.

Wie bist du so schön
In silbernen Locken!
Wie bist du so furchtbar
Im Donner der hallenden Felsen umher!

Dir zittert die Tanne.
Du stürzest die Tanne
Mit Wurzel und Haupt!
Dich fliehen die Felsen.
Du haschest die Felsen,
Und wälzest sie spottend, wie Kiesel, dahin! (ll. 8–17).

The 'silver locks' represent the white water as the stream spills down in rapids and falls. A further proof of the peculiar appeal of this phenomenon as a symbol for the German poets of the Storm and Stress might lie in the fact that Coleridge omitted this symbol from his own version of 'On a Cataract', which was inspired by that of Stolberg. There were other Stormers and Stressers who use this symbol of the irresistible flow of a downward rushing stream with waterfalls.

In *Faust, Part One* (5: 141–288), Goethe makes use of the falls as a symbol for a pessimistic world view based entirely on change and insecurity; however, in the later *Part Two,* he uses the rainbow with the falls to symbolize a more optimistic world view of permanence amid change. The first of these is to be found in the 'Wald und Höhle' / 'Forest and Cave' scene, written in the Storm and Stress days of the seventies (passage composed between 1773 and 1775). In this scene, Faust begins to feel remorse for his mistreatment of Margarete and a reluctance to inflict any further injury upon her. The devil introduces the image of a flooding mountain brook coming from melting glaciers and ice fields—here, form reflects content when the freshet spills out beyond the surrounding pentameter lines and draws back in again: 'Her love for you is overpowering. First came the spate of your mad passion like a brook that floods when snows are melting; you poured it into her heart, and now your brook is a shallow brook again'.

Sie hat dich übermächtig lieb.
Erst kam deine Liebeswut übergeflossen,
Wie vom geschmolznen Schnee ein Bächlein übersteigt;
Du hast sie ihr ins Herz gegossen,
Nun ist dein Bächlein wieder seicht (5: 3306–10).

Here, Goethe uses the annual cycle of the flow of the stream—where at certain seasons the stream is in full spate and at others the stream dries up. The tragedy here is that what is signified—the affection in two human hearts—does cycle unseasonably, out of sync with one another: Faust's lust ebbs, while Margarete's love is in full flow.

Later in the scene, Faust expresses his guilt: 'Don't I always sense her misery? Aren't I the fugitive? the homeless?'

Fühl ich nicht immer ihre Not?
Bin ich der Flüchtling nicht? der Unbehauste? (ll. 3306–48).

This rush of questions picks up the theme of the flooding stream, introduced by the devil, and converts it into a simile for a destruction-wreaking waterfall in the mountains—'the aimless and restless monster has roared, in a greedy rage, like a waterfall from cliff to cliff down toward the precipice? And at the side, she—with her childishly befogged brain—in a cabin on a tiny alpine meadow, and all her domestic potential—circumscribed within a tiny universe!'

> Der Unmensch ohne Zweck und Ruh,
> Der wie ein Wassersturz von Fels zu Felsen brauste,
> Begierig wütend, nach dem Abgrund zu?
> Und seitwärts sie, mit kindlich-dumpfen Sinnen,
> Im Hüttchen auf dem kleinen Alpenfeld,
> Und all ihr häusliches Beginnen
> Umfangen in der kleinen Welt! (ll. 3349–55).

The speaker is so overwrought with guilt and caught up in the stream and falls metaphor that his emotions and desire to express the destructive chaos of the phenomenon cause the final sentence, here, to break into non-grammatical fragments.

Faust describes the cataclysmic change which his love has wreaked upon her, by using the symbol of a waterfall undermining the side of the mountain on which is located a cottage—symbolic of Margarete's heretofore tranquil life: 'And I, the despised of God, was not satisfied to seize and crush the rocks into gravel: did I have to undermine her and her tranquillity!'

> Und ich, der Gottverhaßte,
> Hatte nicht genug,
> Daß ich die Felsen faßte
> Und sie zu Trümmern schlug:
> Sie, ihren Frieden mußt ich untergraben! (ll. 3356–60).

Goethe is using the waterfalls to convey the effect of one human destiny (that of Faust) upon another (that of Margarete); more often the poet employs the falls to symbolize the effect of misfortune (signified by the falls) upon the course of a human life (signified by the river).

As we learned from Segel, the Baroque poets anticipated Romantic poets in the use of the waterfalls—generally in connection with a river—as a symbol for misfortune in the course of a human's life. Such an allegorical usage is found in Hölderlin's 'Hyperions Schicksalied' (2: 143, 3: 143), from his novel *Hyperion* (1797–99).[59] This lyric opens with two stanzas describing the life of the Immortals who dwell in the sky (ll. 1–15) and closes with a stanza offering contrast to the lives of flux and uncertainties of those who dwell below: 'Yet it is our fate never to rest in any place; we, miserable humanity, dwindle and fall blindly from one hour to the next',

> Doch uns ist gegeben,
> Auf keiner Stätte zu ruhn,
> Es schwinden, es fallen
> Die leidenden Menschen

59 Johann Christian Friedrich Hölderlin, *Sämtliche Werke: Stuttgarter Hölderlin Ausgabe*, ed. Friedrich Beissner (Stuttgart: Kohlhammer-Cotta, 1943–53).

Blindlings von einer
Stunde zur andern (ll. 16–21).

Then, the persona finds an apt simile for his view of the human condition in the tumultuous flux of a waterfall: 'for years hurled like water from cliff to cliff down into uncertainty'.

Wie Wasser von Klippe
Zu Klippe geworfen,
Jahr lang ins Ungewisse hinab (ll. 22–24).

The 'Jahr lang' of this concluding line implies a sense of duration—but not of Eternal Continuity, for this belongs to the Immortals. The deathless fate of the gods is clocked in eternal terms ('ewig' in ll. 11–5) and the fateful destiny of mankind, in terms of years. In Hölderlin's poem, the fascination with the rush of the falls is simply one of fear and despair because water/man, in the mortal world of continuous mutability, is subject to constant shifting and can find no permanence ('Auf keiner Stätte zu ruhn'). There is no duality in the persona's attitude toward change and no sense of a human participation in its continuity; such participation is reserved for the gods. There is only a profound dread of change. Humans' fate is not part of a divine process, which is restricted to a higher level beyond their earthly station. In terms of human life, there is for Hölderlin no permanence, and this belief makes this poem more pessimistic—both in its vision of change and usage of water imagery to symbolize it—than we shall find in any of the poems under examination.

Although certainly not pessimistic to the same degree as in the Hölderlin lyric, a similar gloomy vision of change conveyed by means of a waterfall appears in Lenau's 'Niagara' (1: 258–59).[60] This is one of several poems on this subject that the poet completed under the inspiration of his travels in America. Above the falls, the Niagara River flows with all the stolid placidity of Denham's Thames: 'Clear and like a cheerful youth, and as if murmuring a sweet dream, Niagara moves farther along the green fringes of the primeval forest; and, with joyful leisure, it reflects the forest's splendor and the stars of a still night'.

Klar und wie die Jugend heiter,
Und wie murmelnd süßen Traum,
Zieht der Niagara weiter
An des Urwalds grünem Saum;

Zieht dahin im sanften Flusse,
Daß er noch des Waldes Pracht
Widerstrahlt mit froher Muße
Und die Sterne stiller Nacht (ll. 1–8).

The last stanza introduces the key signifier of the lyric: the stream's creative capability to reflect, recombine, and juxtapose the forest and the stars on its placid surface. According to my primary reading of Lenau's open symbolism, the river and its waterfalls are two extreme aspects of Romantic creativity: tranquility and frenetic dynamics. The fracturing of the reflections will prove central to the poem.

---

60 Nikolaus Lenau, *Sämtliche Werke und Briefe*, ed. by Walter Dietze after Eduard Castle, 2 vols. (Leipzig: Insel, 1970).

In the middle stanzas of the lyric, the placidity is broken by its steady approach to the falls. Here, too, the wanderer is struck by the contrast between the placid flow he sees and the torrential roar he hears; with a close juxtapositioning of adjectives ('ungestört / Und erstaunt'), the poet conveys the paradoxical feeling created by the contrast: 'Thus the waves glide softly so that the traveler—undisturbed and surprised—hears the roar of the mile-wide cataract'.

Also sanft die Wellen gleiten,
Daß der Wandrer ungestört
Und erstaunt die meilenweiten
Katarakte rauschen hört (ll. 9–12).

In the fourth stanza, the waterfall is endowed with human characteristics which intimate the metaphorical meaning of the last stanzas of the lyric: 'Where the course of the Niagara River moves nearer to the cataract, the stream has been seized with a wild presentiment of its fall'.

Wo des Niagara Bahnen
Näher ziehn dem Katarakt,
Hat den Strom ein wildes Ahnen
Plötzlich seines Falls gepackt (ll. 13–16).

In the fifth stanza, the reflective property of slow moving water is shown, but with the added irony that it is in the process of being shattered by the rapids just above the falls: it is as if the poet were describing the image in such a way as to suggest an apocalyptic vision of the destruction of heaven and earth: 'Unconcerned about heaven and earth, it now moves in a mad rush—it has destroyed its lovely reflection that it had so cheerfully carried'.

Erd und Himmels unbekümmert
Eilt er jetzt im tollen Zug,
Hat ihr schönes Bild zertrümmert,
Das er erst so freundlich trug (ll. 17–20).

Although this stanza is describing the rapids above the falls in order to prepare for the great downfall, Lenau shows a distinct predilection for a sudden change of mood from friendly to destructive.

The precise moment at which the stream plummets is conveyed with particular relish, which is suggested by an exuberant usage of verbs. For Lenau, a single verb is not enough to describe the water in the process of falling; he uses four— 'seethe' / 'stürzen', 'tumble' / 'schießen', 'thunder' / 'donnern', and 'jerked along' / 'hingerissen'. The phrase 'as if by a longing' / 'Wie von Sehnsucht' offers yet another suggestion—that the poet's consciousness has become identified with the fall of the stream—which, in turn, is described in such a portentous phrasing ('for the great downfall' / 'Nach dem großen Untergang') as to suggest a longing for death: 'The rapids seethe, tumble, and thunder loudly with a wild impulse, as if carried away by a longing for the great downfall'.

Die Stromschnellen stürzen, schießen,
Donnern fort im wilden Drang,
Wie von Sehnsucht hingerissen
Nach dem großen Untergang (ll. 21–24).

The theme of the apocalypse is clarified in the final two stanzas by reference to a prophet. This is conveyed by means of the waterfall sounding like thunder from afar—an effect which is lost when the hearer is close to the waterfall: 'What the wanderer can hear in the distance—Niagara's deep plummet—he cannot hear up close because the clamor of the surge is too deafening. And in vain will he listen who approaches the falls; however, the prophet could in the distance hear the sound of the future'.

Den der Wanderer fern vernommen,
Niagaras tiefen Fall
Hört er nicht, herangekommen,
Weil zu laut der Wogenschall.

Und so mag vergebens lauschen,
Wer dem Sturze näher geht;
Doch die Zukunft hörte rauschen
In der Ferne der Prophet (ll. 25–32).

'Transposing the concept of space into one of time' and a play 'on the symbolic connotations of the great 'fall',' according to Hugo Schmidt, both endow the prophet and the 'downfall' he predicts with symbolic significance and 'implications that reach far beyond the falling of water'.[61] The same might be said of the play of 'distance' terms: *fern* and *in der Ferne*.

Lenau was not the only one to take Niagara Falls as the archetypical waterfall and symbol of pessimistic change—indeed, it resulted in a delightful series of multicultural linkages among Europe, North America, and South America. This was initiated by Chateaubriand's description of the great falls in *Atala*[62]:

We soon arrived at the edge of the cataract, which announced itself by a frightful roar. The falls is formed by the Niagara River, which flows out of Lake Erie and flows into Lake Ontario; its perpendicular height is one hundred and forty-four feet. Between Lake Erie and the plunge, the river runs down a course with a quickly falling elevation, and at the moment of the fall, the torrents of the stream—less stream than sea—rush precipitously toward the gaping mouth of the chasm.

Nous arrivâmes bientôt au bord de la cataracte, qui s'annonçoit par d'affreux mugissements. Elle est formée par la rivière Niagara, qui sort du lac Érie, et se jette dans le lac Ontario; sa hauteur perpendiculaire est de cent quarante-quatre pieds. Depuis le lac Érie jusqu'au Saut, le fleuve accourt par une pente rapide, et au moment de la chaute c'est moins un fleuve qu'une mer, dont les torrents se pressent à la bouche béante d'un gouffre.

To lend intensity to the vision of this titanic water phenomenon, Chateaubriand feels compelled to draw in and link to the cycle. He emphasizes, with an unpoetic matter of factness, geography and hydrology by noting that this short river links two of the world's largest freshwater lakes—Lake Erie with Lake Ontario. He further draws in the sea quarter by way of comparing it to the vast amount of water pouring over the falls.

---

61 Hugo Schmidt, *Nikolaus Lenau* (New York: Twayne, 1971) 11; the German scholars and critics tend to ignore or at least slight Lenau's Niagara series.

62 Françoise-René de Chateaubriand, *Atala: Edition Critique*, ed. J.M. Gautier, Textes Littéraires Française (Geneva: Droz, 1973).

The two branches of the falls, created by an island just above the edge, receive meticulous attention: 'The cataract is divided into two branches and bends like a horseshoe. Between the two falls, an island extends out, hollowed underneath, and balanced with all its trees over this chaos of waters. The bulk of the river rushes southward, rounds itself into a vast cylinder, then unrolls in a blanket of snow and, in the sun, shimmers forth all the colors possible. The stream that falls to the eastward descends into a terrifying realm of shadows; it was like a column of water at the time of the Flood'.

> La cataracte se divise en deux branches, et se courbe en fer à cheval. Entre les deux chutes s'avance une île creusée en dessous, qui pend avec tous ses arbres sur le chaos des ondes. La masse du fleuve qui se précipite au midi s'arrondit en un vaste cylindre, puis se déroule en nappe de neige et brille au soleil de toutes les couleurs. Celle qui tombe au levant descend dans une ombre effrayante; on dirait une colonne d'eau du déluge (143).

Chateaubriand is at great pains here and elsewhere throughout the description to convey the sense of the great arch or horseshoe of the falls. The white water is likened to a great field of snow. To further convey the 'chaos of water', the writer compares it to 'a column of water during the Flood'.

Both the spiraling clouds of water droplets and the rainbow, which they help create, are central to his description: 'A thousand rainbows curve and crisscross over the abyss. Striking a fallen rock, the water rebounds in whirlpools of foam, which rise above the forests like the smoke of a vast conflagration. Pines, wild walnut trees, and rocks sculpted into fantastic shapes all add to the scene. Eagles, drawn by the current of air, spiral downwards into the depths of the chasm, and the badgers suspend themselves by their flexible tails from beneath a low-hanging tree in order to seize from the abyss the broken cadaver of an elk or bear'.

> Mille arcs-en-ciel se courbent et se croisent sur l'abîme. Frappant le roc ébranlé, l'eau rejaillit en tourbillons d'écume, qui s'élèvent au-dessus des forêts, comme les fumées d'un vaste embrasement. Des pins, des noyers sauvages, des rochers taillés en forme de fantôme, décorent la scène. Des aigles entraînés par le courant d'air descendent en tournoyant au fond du gouffre; et des carajous se suspendent par leurs queues flexibles au bout d'une branche abaissée pour saisir dans l'abîme, les cadavres brisés des élans et des ours.

The description is often more of a travelogue describing a conventionally Sublime scene summed up by the narrator, who says he contemplated the scene 'with a pleasure mixed with terror' / 'avec un plaisir mélé de terreur' (143–44). The final description, however, of animal corpses killed and broken by the fall floating at the foot of the falls, suggests something more. This sense of animal pathos prepares the way for the culminating recapitulation of the human tragedy of the prose tale *Atala,* which it concludes.

A site overlooking the waterfalls is the place where the final strands of the tragic histories of Atala, Chactas, and René are related to the narrator, who upon hearing all of it, cries out: 'Thus passes away upon the earth all that was good, virtuous, full of feeling! Man, you are nothing but a fleeting dream, a painful dream; you only exist for misfortune; nothing else compares with the sadness of your soul and the eternal melancholy of your mind!'

Ainsi passe sur la terre tout ce qui fut bon, vertueux, sensible! Homme, tu n'es qu'un songe rapide, un rêve douloureux; tu n'existes que par le malheur; tu n'es quelque chose que par la tristesse de ton âme et l'éternelle mélancolie de ta pensée! (147)

It is this ability to extract from an exotic scene the last element of melancholy that made Chateaubriand so influential in France; it must also be added that this influence was also, in part, due to Chateaubriand's particular knack for choosing the scene—a gigantic waterfall, for example—that would capture the imaginations of his contemporaries by embodying precisely those qualities and values that fascinated them—that is, by embodying their world view. Although his influence elsewhere in Europe seems limited when compared to that of Byron or Staël, Chateaubriand made a deeper impression in Latin America.

Chateaubriand influenced one of the most frequently anthologized of Romantic Latin American lyrics—'Niágara' by the Cuban poet José Maria de Heredia.[63] According to a scholar of Heredia's poetry: 'It is certain that Heredia read *Atala*. This is confirmed by the letter that he wrote to his mother, addressed Brooklyn May 1824 and concluded with: "Greeting to Ignacia [his sister] who has read in *Atala* the description of the cataract of Niagara which I am going to see"'.[64] As with Chateaubriand, Heredia seeks and finds Judeo-Christian associations in Niagara Falls. Heredia finds in the roar, the origin, the destination, and large place of the waterfall within the water-cycle, an emblem for the continuity of eternal change and the mystery of divine origins and ends—the mystery of God; in this regard, the poem has much more to do with the Baroque outlook of Henry Vaughan than Lenau and the Romantic poets we have just considered. When, however, Heredia treats the destiny of man, the same view of gloomy change appears: 'I see your tireless water run, like the long flood of centuries sweeping through eternity! thus the flowering days of a human's life pass by, and he awakes to sorrow'.

¡Miro tus aguas que incansables corren,
como el largo torrente de los siglos,
rueda en la eternidad! así del hombre
pasan volando los floridos días,
y despierta el [al] dolor [...] (ll. 105–09).

William Cullen Bryant wrote an admirable, if somewhat free translation of Heredia's poem: 'Ode to Niagara'. He renders the above lines as:

I see thy never-resting waters run
And I bethink me how the tide of Time
Sweeps by eternity. So pass, of man,—

63 Heredia wrote two very different versions of 'Niágara', *The Literature of Spanish America*, ed. Angel Flores (New York: Las Americas, 1966); in his revised version, these lines were completely transformed, but I have adhered to the version translated by William Cullen Bryant, cf. Heredia, *Poesías líricas*, ed. Elías Zerolo (Paris: Garnier, 1893) 225–29.

64 'Es cierto que Heredia leyó *Atala*. Está comprobado por la carta que escribió a su madre, fechada en Brooklyn en mayo de 1842, en la que al final dice: 'A Ignacia [su hermana] que lea en *Atala* la descripción de la catarata de Niágara que voy a ver" (Francisco Gonzalez Del Valle, 'El Niágara de Heredia y el de Brainard', *Lyceum* 4:4 [1939]: 6–18).

Pass, like a noonday dream—the blossoming days,
And he awakes to sorrow [...] (ll. 71–87).

Byrant actually wrote a considerable number of poems centering on water images and symbols; however, only in this translation does the versifier—at least to my mind—become, by proxy, a poet.

It also seems that Heredia's poem inspired the Columbian Poet Rafael Pombo to write his own lyric to Niagara—which is less Baroque, as well as quite a bit more Romantic, innovative, and interesting—'On Niagara: Contemplation' / 'En el Niágara: Contemplaci ón'.[65] Pombo, unlike Lenau and Heredia, sees little gloom in Niagara Falls; for him it is a celebration of change and of the larger continuity of change inherent in Nature: 'Always new, brilliant, in constant motion; always abundant, mighty, and forceful like the living stream of the boiling sap which pours from the shining breast of the great mother of all nature, you manifest your magnificence through your fall and you raise from that abyss up to the firmament a hymn of power and of life'.

Siempre nuevo, brillante, en movimiento;
Siempre fecundo, poderoso y fuerte
Como el vivo raudal de hirviente savia
Que de los pechos deslumbrantes brota
De la madre común naturaleza,
Despliegas tu grandeza en tu caída,
Y alzas de aquel abismo al firmamento
El himno de la fuerza y de la vida (ll. 215–29).

This and the other three celebrations of a gigantic mass of water spilling over the arcing cliff closely parallel in words the rendering of the falls by two North American painters—Thomas Cole and, above all, Frederic Edwin Church.

Cole's *Niagara* (1830) contains a facture carefully rendered to suggest the flow of water over the falls, viewed from the front. Cole takes care with the surrounding tree-covered hill. This painting—with its Native Americans placed on the central axis of the painting on a jutting rock just beneath the portrayed arc of the falls—looks for all as if it is an illustration for Chateaubriand's passage. Cole could also give a pessimistic vision of the destructive implications of the change involved in a river spilling over a falls, as he does in the adulthood or third panel of the *Voyage of Life,* which we will consider along with other holistic water-cycle visions later in this chapter. Compared to this panel of the voyage, *Niagara* strikes one as being understated—the rush of the falls itself rivaled in importance by the hills, trees, and human figures.

A contrast to Cole's treatment of the falls and a comparison with that of Pombo's poem is offered by the *Niagara* of Church. Church has cut out most of the environs, and he has devoted almost three-fifths of the entire canvas to the rush and flow of the rapids above and waterfall itself. Church has consciously framed his subject so that its bowed shape (on the Canadian side) creates a serpentine form across the full

---

65 *Poesías de Rafael Pombo,* ed. Antonio Gomez Restrepo, 2 vols. (Bogota: Imprenta Nacional, 1916). Heredia's poem had such a strong influence on Pombo's that the Columbian poet mentions the Cuban's poem in lines 132 and 150.

center of the canvas. The facture used by the artist to create the sense of flow is a white stippling surpassing that of Cole and rivaling that of Constable, with its careful realism, capturing the visible sense of phenomena in motion. Also, unlike Cole, Church—rather than giving importance to the mist below—concentrates on the rapids above, where the sense of a real flow is not lost in the chaos of foam and spray. As Franklin Kelley expressed it: 'Church managed to create one of the most convincing depictions of water ever to appear on canvas'.[66] The remaining two-fifths of the canvas' expanse is filled almost completely with clouds, including a rain shower on the upper left. Between sky and falls, there is the thinnest line of trees along the horizon, and even this is sliced in the left center by an upstream view extending almost to the actual horizon. Most dramatic of all, he makes a complete break with the Neoclassical landscape of Poussin and Claude by his omission of any framing foreground: the spectator is thrown right into the immediate experience of the rush of water. Church's painting is remarkable for the overwhelming emphasis he gives the clouds, rain, river, and falls.

Waterfall with Mist and Rainbow: Permanence in Change

The waterfall taken by itself, then, gives a vision of unrelieved change, but when the poets add mist and the sun shining through that mist to create a rainbow, the sense of permanence may be added. It is for this reason that I feel that the symbolic significance of the falls carries even greater weight in another poem, 'Multivalence' / 'Verschiedene Deutung' (1: 258), where Lenau employs the same central symbol—that of Niagara—for the same theme, that of human fate. The lyric consists of the two stanzas which, without redundancy, mirror one another in a four-part structure: (1) the river's deep, dark, reflecting waters (2) being dashed to pieces by the fall (3) and reunited in the glorious light/color of the rainbow, (4) which the poet sees as a symbol for the destiny of our consciousness. The first opens with a generalized description of the Niagara River as it comes to the Falls:

> See, as the Niagara's waves in a thunderous fall are disintegrated into spray, and as they now, fragmented into spray, receive golden rays of sunshine, and across the abyss, beautifully paint the rainbow so rich in color. Oh friend, we too are darkling waves, and our consciousness—it must disintegrate. Only if it is dissolved into spray into the air, it will receive the radiance of the rainbow.

1

Sieh, wie des Niagara Wellen
Im Donnerfall zu Staub zerschellen,
Und wie sie, sprühend nun zerflogen,
Empfangen goldne Sonnenstrahlen
Und auf den Abgrund lieblich malen

66 Franklin Kelly, 'A Passion for Landscape: The Paintings of Frederic Edwin Church', *Frederic Edwin Church*, ed. Franklin Kelly (Washington: National Gallery of Art and the Smithsonian Institution, 1989) 32–75, qtd. from 52; see also Jeremy Elwell Adamson et al. *Niagara: Two Centuries of Changing Attitudes, 1697–1901* (Washington: Corcoran Gallery of Art, 1985; David Huntington, 'Frederic Church's "Niagara": Art as the Nation's Type', *Texas Studies in Language and Literature* 25 (1983): 100–38.

Den farbenreichen Regenbogen.
O Freund, auch wir sind trübe Wellen,
Und unser Ich, es muß zerschellen,
Nur stäubend in die Luft zergangen,
Wird es das Irislicht empfangen.

The '-en' rhyme is not merely a bit of poetic virtuosity; it lends a sense a forward impetus to the stream. The cesura—'and our consciousness—it must disintegrate' / 'Und unser Ich, es muß zerschellen'—is a very telling rendering of the water's fall or the death of the self. Again, let me stress that such touches are not mere formalism: the poet is expressing a world view encompassing change, death, and art. The Niagara lyrics of Lenau can be read as having an underlying politically unconscious subtext in Fredric Jameson's sense.[67] In the first stanza, the poet may well be demonstrating the political anxiety a conservative from Metternichian Central Europe would feel in democratic America.

In stanza two, however, the flow is described in more detail: here, too, the key idea is that the water has not achieved its higher destiny as long as it flows, 'dark, colorless' / 'trüb, farblos', in the stream bed; the vision is revolutionary: out of horror is produced beauty—out of the darkness, light in the form of a rainbow: 'Dark, colorless were these waters, as long as they still flowed in the stream; they had to be cleaved into multiplicity in order that they could bloom in a glow of colors. Now each solitary drop flies alone, each a pathetic consciousness, yet together they shine in the bright, heavenly light of the rainbow's harmony of colors'.

2

Trüb, farblos waren diese Fluten,
Solang sie noch im Strome wallten;
Sie mußten vielfach sich zerspalten,
Daß sie aufblühn in Farbengluten.
Nun fliegt ein jeder Tropfen einsam,
Ein armes Ich, doch strahlen sie
Im hellen Himmelslicht gemeinsam
Des Bogens Farbenharmonie.

The first stanza seems to be the statement of one persona; the second, a reply in agreement with the first by another persona (perhaps the 'Freund' referred to in stanza one). The poet establishes no clear symbolic relation at this point, but it seems safe to take the stream as a metaphor for the dreariness of life as seen by Lenau—one of the most overt sufferers of *Weltschmerz.*[68] Transfiguration from this stasis is achieved by the symbolic death of the plunging water, which is described in relatively straightforward topographical terms in the first stanza: 'in a thunderous fall are disintegrated into spray' / 'Im Donnerfall zu Staub zerschellen', or put in more abstract and—as we shall see—more symbolic terms in the second: 'they had to be cleaved into multiplicity' / 'Sie mußten vielfach sich zerspalten', in order that it can be transfigured into a rainbow: 'that they could bloom in a glow of colors' /

67 Fredric Jameson, *The Political Unconscious: Narrative as a Socially Symbolic Act* (Ithaca: Cornell UP, 1981).

68 Schmidt 75.

'Daß sieh aufblühn in Farbengluten'. It is significant also that both stanzas end with four lines which have a threefold purpose: (1) they recapitulate the three stages of the lyric 'event': (2) they share images which closely parallel each other (e.g., 'Wellen'//'Tropfen', 'und unser Ich'//'Ein armes Ich', 'in die Luft'//'Im hellen Himmelslicht', 'das Irislicht'//'Des Bogens Farbenharmonie'); (3) finally, they set forth the relationship—by means of simile in the first and metaphor in the second—between the destiny of the water and the poet. The water the poet metamorphoses into radically different states: from unity in flow to fragmentation to unity in light/color; from darkness to light; from the nameless mass to the solitary consciousness of the individual ('Ich') to some sort of transcendental unity in death. It should be noted too that just as the light of the moon in Goethe's 'To the Moon' and the mountains and snow of Shelley's 'Mont Blanc' all offer images of permanence presiding over scenes of flow and change, so does the continuity—at least, during the moment of perception—of the rainbow over the falls in 'Multivalence' / 'Verschiedene Deutung'. Moreover, what is an apparently still phenomenon of light/color is caused by the reflection from a swirling mass of tiny water drops. Thus the image of the rainbow lends permanence to what is in actuality only change.

Almost half a century after Storm and Stress, writing *Faust, Part Two* (5: 289–526), Goethe returns to the image of the waterfalls but is no longer content with the simple image of cataclysmic change (passage composed in 1800 and 1801). For his more mature symbolic purposes, more complexity in the nature of the symbol itself is needed, in addition to the straightforward relationship between the symbol and the symbolized: all this is necessary in order to make the symbol sufficiently complex to give expression to the paradox of permanence in change. Just as with Lenau's 'Multivalence', the manifold meaning cannot be expressed by the symbol of the falls per se and must be enriched by the image of the rainbow, so, too, with Goethe's waterfall passage in the 'Charming Region' / 'Anmutige Gegend' scene which opens part two. Faust is in a mountain meadow recovering from the guilt and shock of the Gretchen episode and is storing up energy to act out the last and greater part of his divine tragicomedy. Although he wants to look directly at the sun, he cannot: it is too godlike to be experienced by the naked eye of man. Thus, he turns away from it and says: 'Let the sun, then, stay behind me!'

> So bleibe denn die Sonne mir im Rücken! (l. 4715).

He now looks at a waterfall—which, like the earlier one, is located in the high mountains: 'I behold with growing rapture the waterfall that cleaves the rocky ridge. From plunge to plunge, it rolls down, now pouring forth into one thousand—then, into many thousands of streams—blowing spume upon spume, high in the air'.

> Der Wassersturz, das Felsenriff durchbrausend,
> Ihn schau ich an mit wachsendem Entzücken.
> Von Sturz zu Sturzen wälzt er jetzt in tausend,
> Dann abertausend Strömen sich ergießend,
> Hoch in die Lüfte Schaum an Schäume sausend (ll. 4716–20).

But out of this image of chaotic change rises a contrasting image—that of the still, unmoving rainbow which is created out of the ever-moving mass of droplets thrown off by the falls: hence the rainbow has the paradoxical quality of 'changeful-permanence' / 'Wechseldauer': 'Yet how splendidly—arising from out of this storm—vaults the many-colored rainbow's changeful-permanence—this moment, clearly drawn—the next, dispersing in thin air, casting about a fragrant-refreshing shower!'

> Allein wie herrlich, diesem Sturm ersprießend,
> Wölbt sich des bunten Bogens Wechseldauer,
> Bald rein gezeichnet, bald in Luft zerfließend,
> Umher verbreitend duftig-kühle Schauer! (ll. 4721–24).

The readers of this poem are faced with a tension between the symbol of a changeful phenomenon to represent permanence. We must hold in our mind a series of contradictory epistemes: the swirling mass of water droplets; the fact that the rays of the sun pass through them collectively to create a seemingly static rainbow; the apparent stasis of the latter is to be imagined against an ever-changing rush and spray of water.

This symbol of the seemingly permanent rainbow, more permanent than the whirring water droplets that cast it, refers ahead to the whole essence of act 1 of *Part Two* of *Faust*. This act exposes that things like wealth are only a figment of meaning cast by the human mind upon gold or upon paper currency: value is a myth. Beyond this there is the transcendent realm of the Mothers, who keep all the archetypes. By going to them to seek their aid, Faust brings back Helen and Paris, in a sort of proto-holograph—a fancy lightshow.

In the sunlight reflecting in light/color from these swirling water drops, Goethe finds a symbol for human endeavor: 'This mirrors aspiring human endeavor. Ponder over it and you will comprehend more accurately: In colorful refraction, we find life'.

> Der spiegelt ab das menschliche Bestreben.
> Ihm sinne nach, und du begreifst genauer:
> Am farbigen Abglanz haben wir das Leben (ll. 4725–27).

In addition to the miniature rainbow created by the waterfall, Goethe makes use of one further aspect of the waterfall, which lends itself as a symbol for permanence in change: this is the water-cycle in miniature which, like the rainbow, is created by the mist cast out from the waterfalls.

Goethe employs a somewhat similar symbol in his 'Song of the Spirits over Water' / 'Gesang der Geister über den Wassern' (1: 306–07). In this lyric, he makes use of a stream spilling over a waterfall, with its undulating cloud of mist, as his central image in order to represent the destiny of man: 'The soul of man resembles water':

> Des Menschen Seele
> Gleicht dem Wasser:

thereby involving not only the destiny of one man's life but also that of all men. The symbol, which has dual significance at once microcosmic and macrocosmic, implies the greater water-cycle while graphically describing a water-cycle in miniature, as suggested by the motion made by the mist from the falls. Goethe, by means of prepositions, offers a vivid description of the water's cyclic motion: 'It comes from heaven, it rises to heaven, and down again to earth it must go, eternally changing'.

Vom Himmel kommt es,
Zum Himmel steigt es,
Und *wieder nieder*
Zur Erde muß es,
Ewig wechselnd (emphasis mine, ll. 1–7).

The exact philosophical—or, perhaps, more precisely, theological—meaning here is not clear. It would seem that transmigration of souls is being referred to.[69] What is certain is that in line seven of the above passage we again find the motif, continuity of flux, expressed by the phrase, 'eternally changing' / 'Ewig wechselnd'. The paradox inherent in this phrase, in the context of the poem, is much more than a mere figure of speech; for in the context of a poem on the motion of the soul/water, it is clearly an expression of the continuity of change inherent in the nature of both. The discovery of a miniature cycle with the outward-seeming chaos of the falls present a vision of transcendent order.

In the next stanza, Goethe presents another precise description of the waterfall and the miniature water-cycle: 'If the pure jet streams from the lofty vertical rockface, it then sprays beautifully, in billowing clouds, to the smooth rock, and lightly received, it rolls down to the depths, becoming hazy and softly murmuring'.

Strömt von der hohen,
Steilen Felswand
Der reine Strahl,
Dann stäubt er lieblich
In Wolkenwellen
Zum glatten Fels,
Und leicht empfangen
Wallt er verschleiernd,
Leisrauschend
Zur Tiefe nieder (ll. 8–17).

---

69 James Boyd, *Notes to Goethe's Poems* (Oxford: Blackwell, 1944) 1: 156–57. Boyd first offers these interpretations: 'Stanza I raises a problem: are we to accept the parallel of the soul and the water literally and in its entirety? or only that part of it where the experiences of the soul and water seem to correspond or overlap. The former would give us a conception of the soul's single descent from and return to heaven as illustrated by the eternal movement of water in nature'. He then rejects the first interpretation on the grounds that it is uncharacteristic of Goethe's ideas as expressed throughout his *oeuvre*: 'Nowhere in his works is there the slightest indication of such a belief'. Boyd accepts the second interpretation; but also goes on to put it into a larger context: 'We are thus forced to accept the more limited simile, namely, that the sojourn of man's soul on earth finds a parallel in water's descent from and ascent to the skies. But humanity does not end there: It is [...] an unending chain, and, as the water falls again, so is the soul of man in a mysterious way replaced and the race renewed'. If the latter interpretation is correct, this poem bears a strong resemblance to Shelley's 'The Cloud' and Hugo's 'Dicté en présence du glacier du Rhône' where the individual finds immortality or, at least, a sense of permanence, being a part of a larger and collective process. Nevertheless, a widely different approach is set forth in the following quotation from Richard M. Müller, *Die deutsche Klassik: Wesen und Geschichte im Spiegel des Strommotivs* (Bonn: Bouvier, 1959) 63: 'Wie das Wasser, kann auch der Mensch keine Ruhe finden, weder im Himmel des Ideellen, Geistigen, noch auf der Erde mit ihren griefbaren Realitäten, und so wechselt er wie jenes im ewigen Auf und Ab [...]'.

The mini-water-cycle includes water droplets collecting into clouds, which spin and fall in a vortex of mist. The description remains graphic and objective until the third stanza, where—especially with the word *unmutig*—one gets a volitional element added to the water's nature: 'If crags project against the fall, it foams indignantly step by step into the abyss'.

Ragen Klippen
Dem Sturz entgegen,
Schäumt er unmutig
Stufenweise
Zum Abgrund (ll. 18–22).

Chapter three will treat the lake imagery of the lines that follow—according to the natural progression of the highland stream, oscillating from turbulence to calm.

As with these German poets, the waterfall and its corollary phenomena found favor with at least two important English Romantic poets—Wordsworth and, above all, Byron. Not only did they use the same images, but they used them to symbolize comparable and often identical concepts. The alpine-falls passage in the sixth book (ll. 58–72) of *The Prelude,* which we examined in the last chapter, bears an interesting parallel to Lenau's use of the waterfall in 'Niagara' for suggesting Apocalyptic finality or to Wordsworth's 'Characters of the great Apocalypse, / The types and symbols of Eternity, / Of first and last, and midst, and without end'. In a similar way, Byron evokes the Apocalyptic vision with a falls in an Alpine landscape in his poetic drama *Manfred* (4: 70):

A lower Valley in the Alps.—A Cataract.
Enter MANFRED
It is not noon—the sunbow's rays still arch
The torrent with the many hues of heaven,
And roll the sheeted silver's waving column
O'er the crag's headlong perpendicular,
And fling its lines of foaming light alone,
And to and fro, like the pale courser's tail,
The Giant steed, to be bestrode by Death,
As told in the Apocalypse [...] (*Manfred* 2. 2. ll. 1–8).

As for the scene which this symbol of the ultimate end or limit of human history precedes, it is a discussion between Manfred and the Witch of the Alps concerning the limitations of human endeavor.

For a pessimistic use of the waterfall employed in isolation from the water-cycle, one should turn to the arch-pessimist among Romantic painters: Turner. His *The Bay of Baiae, with Apollo and the Sybil* contains what I consider to be a significant example of his symbolic use of waterfalls, although the falls seem, at first glance, an unimportant detail; also, it is small and without a rainbow. The midpoint of this painting lies halfway between the sea above and the waterfalls below. The latter is rendered with streaks of light blue and white. These falls spill over numerous rocks and from parts of ruined buildings. It seems as if the falls spring from various sources all over the whole lower center area of the painting. A close examination of this area of the painting reveals that the land is made by the artist to seem insubstantial, waterlogged, and eroded by the many falls. Water seems to flow from and over almost all the land in this spot. This insubstantial *terra firma,* which appears to be eaten away by water, offers a symbol which is comparable in its pessimistic overtones to that of the snake about to consume a rabbit in the foreground of the canvas.[70] Presumably, to underscore the interconnection of symbols, Turner places the falls just above these two animals. Similarly, as if to emphasize the symbol of falling water's gnawing away at the land, there is a small waterfall of a rivulet located in the lower left of the canvas. This miniature waterfall receives a certain emphasis by its being located just over the slab of marble bearing a Latin inscription of the painting's title. Finally, in this ideal landscape—over which the artist had complete control in his positioning of the natural phenomena constituting the composition—the placement of these things is not accidental: so that chaos can contrast with and even undercut order, Turner seems to have quite intentionally juxtaposed the torrential waterfalls with the placid sea, just as Byron places a passage treating the wild Velino waterfall next to a passage conveying the placid Clitumnus.

The Velino-Clitumnus passages are found in canto four (1812) of Byron's *Childe Harold* (2: 120–86). The waterfall is that of the Cascata del Marmore on the Velino River near Terni. While most Byron scholars (even Yax, in his treatment of water imagery) pass over this passage without mention, Marjorie Hope Nicolson sees in it the quintessential Byron: 'Sound and motion and elemental energy—even elemental fury—led Byron to exultation, even more than to exaltation'.[71] The Velino passage is complex because the description and symbolic implications contain almost every aspect which we have found in the waterfalls of the German poets in a new and enriching combination. In addition, Byron—like Turner, loving contrast and contradiction—juxtaposes the description of the turbulent waterfall with a Classical stream, Clitumnus, which placidly flows like Denham's Thames:

—the purest god of gentle waters!
And most serene of aspect, and most clear;

thus, then, the topography contributes to the Classical aura and the water's Neoclassical flow:

And on thy happy shore a temple still,
Of small and delicate proportion, keeps,
Upon a mild declivity of hill,
Its memory of thee; beneath it sweeps
Thy current's calmness [...] (stanza 66–67).

This scene has a soothing effect upon the persona, Harold, who by looking

on the heart the freshness of the scene
Sprinkle its coolness, and from the dry dust

70 Concerning the snake and rabbit symbolism, see Clark, *Romantic Rebellion* 238.

71 Marjorie Hope Nicolson, *Mountain Gloom and Mountain Glory: The Development of the Aesthetics of the Infinite* (1959; New York: Norton, 1963) 385.

Of weary life a moment lave it clean
With Nature's baptism,—

feels a 'suspension of disgust', of his *Weltschmerz* (stanza 68).

This Neoclassical section (stanza 66–68) offers an antinomy with the Romantic falls passage (stanza 69–72) that opens with exclamation-marked excitement and tumultuous imagery:

The roar of waters!—from the headlong height
Velino cleaves the wave-worn precipice;
The fall of waters! rapid as the light
The flashing mass foams shaking the abyss;

here, the verbs, adjectives, and nouns—such as 'roar', 'headlong height', and 'flashing mass', 'foams', 'shaking the abyss'—are close in tone and content to such words and phrases in Goethe's 'Song of the Spirits' / 'Gesang der Geister', which bears requoting here: 'If the pure jet stream from the lofty vertical rockface' ('Strömt von der hohen, / Steilen Felswand / Der reine Strahl') and 'If crags project against the fall, it *foams* indignantly step by step into the *abyss*' ('Ragen Klippen / Dem Sturz entgegen / *Schäumt* er unmutig / Stufenweise / Zum *Abgrund*'). Of equal importance are similar phrases found in Lenau's 'Niagara': '*The rapids seethe, tumble, and thunder* loudly with a wild impulse, as if carried away by a longing for *the great downfall*' ('*Die Stromschnellen stürzen, schießen, / Donnern* fort im wilden Drang, / Wie von Sehnsucht hingerissen / Nach *dem großen Untergang*' [emphasis mine]).

Byron's passage, however, does not use symbols to allegorize abstract concepts, as does Goethe's; rather, like Lenau's it uses symbols to externalize consciousness; for, Byron has his persona-narrator in *Childe Harold* describe the waterfall as

The hell of waters! where they howl and hiss,
And boil in endless torture; while the sweat
Of their great agony, wrung out from this
Their Phlegethon, curls round the rocks of jet
That gird the gulf around, in pitiless horror set [...] (69).

In this demonic image of a river which represents so many tortured souls howling in Hell, we have what might be considered a projection of certain Satanic elements of the so-called Byronic personality—or certainly one of many aspects of his manifold personality.[72] Goethe must also have seen something Satanic in a cascade because, along with water jets, the Mephistopheles desire to have cascades in his ideal garden landscape: 'cascades combined through rocks with rocks':

Kaskadensturz, durch Fels zu Fels gepaart [...] (l. 10,165).

I think it is no accident that this German and this British poet, who so much admired each other's works, should have been fascinated in the falls and their manifold phenomena, which in their turn flicker around the main sign (the cascade) like

72 See Mario Praz, 'Metamorphoses of Satan', in *The Romantic Agony*, trans. Angus Davidson, 2nd. ed. (London: Oxford UP, 1951) 67–76.

modifying signifiers that alter its meaning: at one moment the falls is pure change; the next it is permanence in change.

Like Goethe's falls in 'Song of the Spirits' / 'Gesang der Geister', Byron's creates a miniature water-cycle giving forth an 'unceasing shower' symbolizing a continuity of change, as the thrown-off water

> mounts in spray the skies, and thence again
> Returns in an unceasing shower [...]

Here, too, an implicit sense of permanence embodied in the mini-cycle is made sharper in Byron's poem by the contrasting flux of lines that follow (stanza 70). There is, also, a sudden change of tone—from that of the Hell of the falls to the April of the cycle—conveyed by images such as 'gentle rain' and, more dramatically, by the shift from the predominance of sibilants in the lines above (averaging roughly two to three per line) to few or none below; those above best describe the noise of the fall; these, the noiselessness of the cycle,

> which round,
> With its unemptied cloud of gentle rain,
> In an eternal April to the ground,
> Making it all one emerald [...] (stanza 70).

Here, the recurrent process is expressed in the fact that the cloud is 'unemptied'; this is true because its moisture is replenished at the same time as it discharges itself in a shower (stanza 70). The reference to April showers adds the seasonal cycle to the mini-water cycle, and yet the showers of the falls are 'eternal'. The permanence is within the change.

In the lines that follow, the gentle aspect is thrown off and the water is described as 'the Giant Element', which moves with 'fierce footsteps' (cf. 'Führertritt' of Goethe's 'Song in Praise of Mohammed'):

> —how profound
> The gulf! and how the giant element
> From rock to rock leaps with delirious bound,
> Crushing the cliffs, which, downward worn and rent
> With his fierce footsteps, yield in chasms a fearful vent
>
> To the broad column which rolls on [...] (70–71).

It is verses like this of Byron, falling so close to those of Goethe's Storm and Stress style and attitude, that attracted the older poet and made him look upon the younger as an immature version of his own genius.

In passing, Byron brings together two opposite quarters of the cycle by describing how the falling stream of water

> shows
> More like the fountain of an infant sea
> Torn from the womb of mountains by the throes
> Of a new world, than only thus to be
> Parents of rivers, which flow gushingly,
> With many windings, through the vale [...] (71).

Out of this vision of change comes a symbol for Eternity:

—Look back!
Lo! where it comes like an eternity
As if to sweep down all things in its track,
Charming the eye with dread,—a matchless cataract,

Horribly beautiful! (71–72).

These oxymoronic concepts of something at once horrible and beautiful, at once 'an Eternity' and the agent of cataclysmic change, lead into the Titanic section below. Here, as with Lenau's 'Multivalence' / 'Verschiedene Deutung' and Goethe's falls passage in *Faust, Part Two,* the mini-rainbow lends a sense of permanence, or at least tranquillity to the rush of the water:

but on the verge,
From side to side, beneath the glittering morn,
An Iris sits, amidst the infernal surge,
Like Hope upon a death-bed, and, unworn
Its steady dyes, while all around is torn
By the distracted waters, bears serene
Its brilliant hues with all their beams unshorn:
Resembling, 'mid the torture of the scene,
Love watching Madness with unalterable mien (72).

The symbolic relation of rainbow//Love to cataract//Madness bears comparison with a strikingly similar relationship in Goethe's 'To the Moon'—moonlight representing the beloved contrasts with the rushing water symbolizing the sorrows of life: in both cases, something above and beyond the stream of water must give a sense of permanence to the change.

Victor Hugo's *Legend of the Centuries / La Légende des siècles* (6: 31–33), a vast compendium of poetry composed and published in parts between 1859 and 1883, seeks to survey the history of Western humanity to reveal it as one great collective entity; moreover, *The Legend of the Centuries* contains some fine shorter pieces, including one of the poet's most anthologized and written-on lyrics, 'Booz endormi' / 'The Sleeping Boaz'. In 'The Waterfalls: Rivers and Poets' / 'Les Chutes: Fleuves et Poetës' from this vast compendium, the falls are shown to be fateful vengeances wreaked on the rivers, symbolizing—as the subtitle implies—the misfortunes which threaten to overwhelm poets during the course of their lives: 'The mighty Niagara falls, the Rhine plummets; the monstrous abyss seeks to be a grave, it hates the giant river, and says, "I will devour"'.

Le grand Niagara s'écroule, le Rhin tombe;
L'abîme monstrueux tâche d'être une tombe,
Il hait le géant fleuve, et dit: J'engloutirai (ll. 1–3).

The river does not submit meekly to its fate—described, as in Byron, in terms of human damnation: 'And the river—like a lion enticed into the unexpected cave of a thousand-headed hydra—struggles with all its cries and all its storms'.

Et le fleuve, pareil au lion attiré
Dans l'antre inattendu d'une hydre aux mille têtes,
Lutte avec tous ses cris et toutes ses tempêtes (ll. 4–6).

Nature is a place of impermanence and abrupt change: 'What! immense nature is thus an uncertain place'.

> Quoi! la nature immense est donc un lieu peu sûr! (l. 7).

At such injustice, the falls 'rears in revolt, it resists the dark precipice, it foams and boils, and—white and black like marble—it grabs at the rocks, it clutches at tree trunks, it inclines, and—as if struck with a curse—rolls over, spinning like the undying Ixion'.

> Il se cabre, il résiste au précipice obscur,
> Bave et bouillonne, et, blanc et noir comme le marbre,
> Se cramponne aux rochers, se retient aux troncs d'arbre,
> Penche, et, comme frappé de malédiction,
> Roule, ainsi que tournait l'éternel Ixion (ll. 8–12).

The reference to eternal torment suffered in Hades by Ixion does offer a vision of the perpetual recurrence of change; but the vision is very pessimistic.

The tone deepens even more into what appears for the moment to be a nihilism worthy of Byron, Turner, or Lenau: 'Twisted, broken, conquered—God himself is responsible—the wild river submits to its cruel torture. The abyss desires its death; but the effort of its scourges, intended to cause extinction, only create chaos; the frightful pit of hell opens its funereal jaws and roars'.

> Tordu, brisé, vaincu, Dieu même étant complice,
> Le fleuve échevelé subit son dur supplice.
> Le gouffre veut sa mort; mais l'effort des fléaux
> Pour faire le néant, ne fait que le chaos;
> L'affreux puits de l'enfer ouvre ses flancs funèbres,
> Et rugit (ll. 13–18).

The mist thrown off from the falls, referred to as a sinister cloud, is compared to the smoke from a volcano; the paradoxical relation of creation and destruction which will be the celebratory culmination of the poem is introduced, but then the tone of negativity temporarily returns:

> What effort in order to create darkness! It is jealousy, it is wrath, it is night; and destruction—that is what it creates. Like the smoke that pours from Vesuvius, a sinister cloud hangs over the gigantic vat and hides the tormented betrayal of the mighty river. Why all this hatred against the life-giver? What has it done to the forests, to the sublime mountain, to the green field, that they should push it into the abyss? Its force, its splendour, its beauty, its goodness are all destroyed. [...] What an ambush and what baseness!

> Quel travail pour créer les ténèbres!
> Il est l'envie, il est la rage, il est la nuit;
> Et la destruction, voilà ce qu'il construit.
> Pareil à la fumée au faîte du Vésuve,
> Un nuage sinistre est sur l'énorme cuve,
> Et cache le tourment du grand fleuve trahi.
> Lui, le fécondateur, d'où vient qu'il est haï?
> Qu'est-ce donc qu'il a fait aux bois, au mont sublime,
> Aux prés verts, pour que tous le livrent à l'abîme?
> Sa force, sa splendeur, sa beauté, sa bonté,
> Croulent. ... Quel guet-apens et quelle lâcheté! (ellipsis in original, ll. 18–28).

The sense that somehow the landscape is the cause of the river's great cataclysmic drop is a rather novel concept. Scientifically speaking, it is true that rivers shape the land, yet by the same token, the land directs and shapes the rivers. Symbolically speaking, it would seem that the shape and contour of the land are used by Victor Hugo to signify a shaped and developed destiny of the poet/waterfalls. The disastrous changes wrought by a volcano is appealed to by Hugo, who seems set on offering an encyclopedic vision of all such changes.

As with the poets whom it symbolizes, the river gives vent to its agony in the form of sounds:

> The water swells like the leather bags where the Boreades murmur, and the horror disperses in despairing voices; all is downfall, shipwreck, engulfing, night, and one might say that a base laugh is to be heard in the noise; nothing is spared, nothing survives, nothing floats; the river in its struggle with the horrible mesh, falls, agonizes, and casts into the distant sky a long, low murmur.

> L'eau s'enfle comme l'outre où grondent les Borées,
> Et l'horreur se disperse en voix désespérées;
> Tout est chute, naufrage, engloutissement, nuit,
> Et l'on dirait qu'un rire infâme est dans ce bruit;
> Rien n'est épargné, rien ne vit, rien ne surnage;
> Le fleuve se débat dans l'atroce engrenage,
> Tombe, agonise, et jette au lointain firmament
> Une longue rumeur d'évanouissement (11. 29–36).

Here, again we have examples of Hugo's elemental all-inclusiveness: wind and seastorms presented as allusions to Homer and Virgil. The constant reiteration of 'rien' / 'nothing' rhetorically drives home the nihilisms.

This nihilism, however, has only been a prologue to the final vision. Out of the horror is produced beauty; out of the darkness, light—in the form of a rainbow; and, if I have read the symbols right, out of the agony of their life comes the poets' art: 'All at once, above the chaos of despair, there appears—created from all that is hideous, violent, torrential in the abyss—a stately shimmer, the rainbow; the plot is vile, the rock is traitorous, the water is black, and out of this frightful gloom you spring, oh glory!'

> Tout à coup, au-dessus de ce chaos qui souffre,
> Apparaît, composé de tout ce que le gouffre
> A de hideux, d'hostile et de torrentiel,
> Un éblouissement auguste, l'arc-en-ciel;
> Le piège est vil, la roche est traître, l'onde est noire,
> Et tu sors de cette ombre épouvantable, ô gloire! (ll. 37–42).

While this glorious rising of the triumphant rainbow out of the mist of cataclysm reminds us of the Classical vision of *Faust, Part Two,* and indeed both Goethe and Hugo create a symbol for permanence and change, the French lyric is almost otherwise devoted to the vision of disruption and change, and it thereby lies closer to the works of Byron and Lenau examined in this chapter. In their poems, the falls have generally implied an interruption, a fateful intrusion in the course of life—indeed, often, a transfiguration from life to another state; behind this is the supposition that the river's flow before and after the falls is a symbol for the course of human existence; in the poems treated in the section that follows, this notion is not a corollary but a central symbol.

The Course of Life: The River in the Water-Cycle

The poetic employment of the course of a river as a symbol or allegory for human life was certainly not first used by the (Neo)Classical-Realistic-Romantics of Romantic culture; however, no one ever appealed to it with such frequency or used it so expansively as did they. It represented life in all the ways that would give expression to their world view. It represented a single human life as a becoming thing. It allowed—usually by unspoken implication—for the inclusion of this river, which often symbolizes the life course of a human, in the hydrological-cycle, which can represent the birth and death cycle. Sometimes, this life cycle can include the Christian conception of the resurrection.

The persona in Meléndez Valdés' ode 'The Brook' / 'El arroyuelo' (45–47) describes a brook—of which the opening stanzas offer sufficient examples: 'With what placid ripples, you tranquilly glide—oh, graceful brook—through the flower-filled valley! How your clear fluid—free now from fetters of January's cold—now please my ear! How with twisting and laughter, and in your cheerful roar, the fresh mint gets splashed with dew!'

> ¡Con cuán plácidas ondas
> Te deslizas tranquilo,
> Oh, gracioso arroyuelo,
> Por el valle florido!
> ¡Cómo tus claras linfas,
> Libres ya de los grillos
> Que les puso el enero
> Me adulan el oído!
> ¡Cuál serpean y ríen,
> Y en su alegre bullicio
> La fresca yerbezuela
> Salpican de rocío! (ll. 1–12).

The persona intrudes only by describing the stream in adverbs and adjectives which remind the reader that the human senses are being appealed to by means of the images: 'tranquilo', 'gracioso', and, above all, in the line in which the rushing of the newly thawed water is commanded by the persona: 'Me adulan el oído!' / 'now please my ear!'

It is important to point out that the aspect of water emphasized by Meléndez is the changing, liquid motion of the water—placid (e.g., 'plácidas ondas / Te deslizas tranquilo'). This emphasis alone makes it an artifact of Romantic culture. The persona, however, in Meléndez Valdés' ode describes the water image with no symbolic end in mind. On the other hand, such a symbolic purpose is inherent in the passages from the following lyrics by Lamartine, Goethe, Byron, and Zorilla, where the stream becomes a symbol around which the poet builds a lament for the speedy passing of one's allotted lifetime. These mournful refrains are taken from Zorilla's 'Stream, it is very sad' / 'Arroyo, es muy triste' in his series *Stream Side* \ *La margen arroyo* (1: 160–63)[73]: 'Stream, it is very sad to ponder by your side that this is the way that life rolls to its end!'

¡Arroyo, es muy triste
pensar, junto a ti
que así van las vidas
rodando a su fin! (ll. 1–4).

This stanza is repeated for emphasis, as for music in lines 18–19. The persona's sad thoughts are on mutability and courage/ignorance in the face of it: 'Stream, in just this way those live—who enjoying drunkenly the happy moment—even though they are about to die!'

¡Arroyo, así viven
los que han de morir
gozando embriagados
el tiempo feliz! (ll. 25–28).

There is here neither delight in the motion of the water or consolation in change. The change is seen as death. Zorilla offers no mediating way of viewing the symbol in a way that might offer a larger context of permanence.

This larger context is to be found in a comparable work by Larmartine, but only after he has treated us to a melancholy opening. 'The Valley' / 'Le Vallon', from the *Poetic Meditations / Méditations poétiques* (19–20), like the Zorilla, uses the stream to signify human existence. Here, the persona laments the brevity of life's pleasures, but the added motif of the anonymity of death sounds a new note of pessimism. The waves and murmuring of the lake are replaced by those of two hidden streams, which meet to form one. The streams flow from their respective sources only to lose their individual identities by merging with each other: 'There, two streams hidden under bridges of greenery trace the valley's serpentine contours; they mix for a moment, they mingle their waves and their murmur, and not far from their source they are namelessly lost'.

Là, deux ruisseaux cachés sous des ponts de verdure
Tracent en serpentant les contours du vallon;
Ils mêlent un moment leur onde et leur murmure,
Et non loin de leur source ils se perdent sans nom (ll. 9–12).

The connection is then established with the source quarter and with the process of waters of one brook mingling with another—usually an image reserved for flow into the ocean. In keeping with the sorrowful tone, the stream is cast as a simile for the person's lost youth: 'The source of my days like the streams has flowed away; it has passed without a sound, without a name, and without any hope of return'.

La source de mes jours comme eux s'est écoulée,
Elle a passé sans bruit, sans nom, et sans retour (ll. 13–14).

To sound, at this point in the poem, an unhappy note, Larmartine has for the moment bracketed the water cycle out of the stream symbol. He also does this because he wishes to concentrate on the ongoing flow of the stream itself.

73 José Zorìlla y Moral, *Obras completas*, ed. Narciso Alonso Cortes (Valladolid: Santaren, 1943).

Lamartine, however, sets himself off from most poets in that he also refines his analogy by pointing out a difference between the symbol (the two streams) and the thing symbolized (human life): 'But their water is limpid, and my troubled soul is not able to reflect the clarity of a beautiful day'.

Mais leur onde est limpide, et mon âme troublée
N'aura pas réfléchi les clartés d'un beau jour (ll. 15–16).

Here, Lamartine employs one of his favorite metaphors: the capacity of sounds in nature to lull the senses and to heal hurts. These effects he sought, often with great success, in his lyrics. In 'The Valley', the persona says of the two streams: 'The freshness of their grassy spots, the shadows that crown them keep me all day long on the border of the brooks; like an infant rocked by a simple chant, my soul is assuaged by the murmur of the waters'.

La fraîcheur de leurs lits, l'ombre qui les couronne,
M'enchaînent tout le jour sur les bords des ruisseaux;
Comme un enfant bercé par un chant monotone,
Mon âme s'assoupit au murmure des eaux (ll. 17–20).

The music of these lines, with their *n* and the *m* sounds, creates precisely the soothing effect that they describe.

The water of the brooks and—in the lines which follow (ll. 37–48)—other images taken from nature are used to symbolize the transience of human life; paradoxically, for the poet, nature has the ability to console humans because, although subject to change in its parts, it is permanent in its totality: 'While everything changes for you, nature is the same':

Quand tout change pour toi, la nature est la même (ll. 51).

Indeed, behind the nature of change is what most thinkers have taken for the quintessential permanence—God: 'God created intelligence so that we could conceive of Him; in nature, discover at last its author!'

Dieu, pour le concevoir, a fait l'intelligence;
Sous la nature enfin découvre son auteur! (ll. 61–62).

For Lamartine, there is no consolation in change as manifested in natural phenomena, like the running water of a stream—except if it be accompanied by the thought that it partakes of some greater mystic whole. The full implications of all this can only be comprehended by comparison of this with another poem by Lamartine.

This deep awareness of permanence and change that ran through his poetic vision also infused his later political vision, as expressed in 'The Revolutions' / 'Les Révolutions' (510–19). All creation, for the poet, is subject not to divine stasis but rather to divine evolution. Under the direction of this God, the history of mankind is a course of changes, of rises and falls of empires: 'All the course [of this history] is marked out only by the debris of nations':

Toute la route n'est tracée
Que des débris des nations [...] (513).

The poet catalogs a variety of both religious and political forms that have been invented and discarded along the roadway of human history: 'Thrones, altars, temples, porticos, cultures, kingdoms, republics are the powder covering the roadway';

> Trônes, autels, temples, portiques,
> Peuples, royaumes, républiques,
> Sont la poussière du chemin [...] (513–14).

Lamartine presents one of his comprehensive, universal visions of the history of Western civilization since ancient Egypt. His conception is that of an evolutionary progression of human ethics and social structures—each is valid only for a single epoch: '"Advance!" Humanity does not live by a single idea! Each night it extinguishes the candle that has guided it, it lights another from the eternal torch:'

> Marchez! l'humanité ne vit pas d'une idée!
> Elle éteint chaque soir celle qui l'a guidée,
> Elle en allume une autre à l'immortel flambeau [...] (516).

It is in light of this creative celebration of the development of all human culture that I feel we must interpret Lamartine's stream symbolism in 'The Valley'.

Lamartine is the one poet treated in this section who feared sudden change—political, aesthetic, or otherwise. He liked, or rather repected change; still, he remained, as we have seen, a gradualist. In the following poems by Hölderlin and Byron, the poets welcome change without reserve.

I would like to observe how the fatalism and the symbol embodying Lamartine's poem, of the flowing together of the two streams (particularly in these lines—'There, two streams hidden under bridges of greenery trace the valley's serpentine contours; they mix for a moment their waves and their murmur, and not far from their source they are namelessly lost') both appear in a passage from Hölderlin's *Hyperion*. The eponymous hero and first-person narrator has long been attracted to another young man, Alabanda. Hyperion describes the meeting in a way transcending the sentimental male bonding so common in German literature of the period and seeming to be an instance of the homoerotic, or at least homosocial: '"Good evening!" exclaimed the charmingly hale and hardy one and looked at me with his wild, yet tender eyes, while his sinewy hand grasped mine so that the touch of it penetrated to my inmost being. O now the meaninglessness of my existence was at an end!'

> Guten Abend! rief der liebe Rüstige, sah mit zärtlich wildem Blicke mich an und drückte mit seiner nervigen Faust die meine, daß mein Innerstes den Sinn davon empfand.
> O nun war mein unbedeutend Leben am Ende! (3: 25).

First, Hyperion describes how they had checked into a nearby caravansary, had sat arm and arm by an open window overlooking the sea, 'while thunderclouds lulled themselves to sleep in the bed of the ether',

> indeß die Donnerwolke sich wiegt' im Bette des Aethers [...].

A person fearless or even hopeful of change can be lulled by the possibility of storms.

Hyperion, then, embodies the coming together of the two friends and its fatalism by means of the two streams metaphor:

Our souls were impelled toward each other all the more strongly because they had come together against our will. We met like two brooks that, pouring from a mountain, fling off their burden of earth and stone and rotted wood and the whole sluggish chaos that holds them back, determined to clear the way to each other, to burst through until, clasping and clasped with equal force, they set out, united in one majestic stream, on their long journey to the sea.

Unsere Seelen mußten um so stärker sich nähren, weil sie wider Willen waren verschlossen gewesen. Wir begegneten einander, wie zwei Bäche, die vom Berge rollen, und die Last von Erde und Stein und faulem Holz und das ganze träge Chaos, das sie aufhält, von sich schleudern, um den Weg sich zu einander zu bahnen, und durchzubrechen bis dahin, wo sie nun ergreiffend und ergriffen mit gleicher Kraft, vereint in einen majestätischen Strom, die Wanderung in's weite Meer beginnen (3: 26).

Homosexuals or bisexuals of the Romantic culture—like any other clear socio-political outcast—would clearly celebrate a change in a status quo where they were so oppressed.

Although I have yet to find a comparable symbol in any of the homoerotic or what Louis Compton and others say are homosexual poems of Byron, one finds that the same sense of the fatal acceptance of the condition of eternal change in the following lines of an 1816 fragment (4: 29), which expresses consolation before the process of change and uses the natural phenomenon of water as the expressive symbol:

Could I remount the river of my years
To the first fountain of our smiles and tears,
I would not trace again its stream of hours
Between its outworn banks of withered flowers,
But bid it flow as now [...] (ll. 1–5).

In a world where death was the possible sentence for same-sex love, Byron has nothing to lose in change—even cataclysms.

Byron was not the only bisexual among the great poets of the Romantic culture. In 1790, Goethe wrote but did not publish this epigram: 'I might well have loved boys too, but girls are what I prefer. If I tire of one as a girl, I can still enjoy her as a boy'.

Knaben liebt ich wohl auch, doch lieber sind mir die Mädchen;
Hab ich als Mädchen sie satt, dient sie als Knabe mir noch (2: 183).

And it is this same Goethe who has his persona or speaker in 'To the Moon' (1: 71–72) bid the river-as-simile-for-life to flow: 'Flow on, flow on, beloved stream! Never will I be full of joy, so did jest and kiss pass away, and so too did fidelity'.

Fließe, fließe, lieber Fluß!
Nimmer werd ich froh,
So verrauschte Scherz und Kuß,
Und die Treue so (ll. 13–16).

As in the Lamartine lyric, there is a quality in nature that offers consolation for Goethe's persona: the ability to externalize one's sorrows by finding symbols for them in nature—thereby objectifying them. In this way, water offers a perfect objectification and consolation for the mutability of things. Water offers an accompaniment to the speaker's lament on the nature of things: 'Rush, stream, along the valley, without rest and peace, rush! and—to accompany my song—whisper melodies when on a winter's night in your fury you overflow, or when you flow around the spring's splendor of fresh buds'.

> Rausche, Fluß, das Tal entlang,
> Ohne Rast und Ruh,
> Rausche, flüstre meinem Sang
> Melodien zu,
>
> Wenn du in der Winternacht
> Wütend überschwillst,
> Oder um die Frühlingspracht
> Junger Knospen quillst (ll. 21–28).

Thus, the persona *qua* poet finds consolation in his song, which at once finds its key symbol as well as its accompaniment in the stream's flow. Also, it should be noted that, as in Lamartine's lyric, there is in Goethe's lyric a sense of permanence behind nature's superficial mutability: the seasons (ll. 25–28 above) suggest a cyclic continuity.

In his *Ring* Cycle, Wagner includes the sky and river quarters of the hydrological-cycle. It opens literally in the depths of the Rhine and closes with a cosmic flood. Wagner in a letter to Liszt dated 11 February 1853 writes of the poetic text for the *Ring* that it contains 'the world's beginning, and its destruction'.[74] The conclusion, where the great river inundates and destroys the world of human beings, can be read as the end of one cosmic cycle. The opening of the *Ring*, then, can be read as the origin of life from the Rhine. Of this opening, Robert Donington writes:

> The *Ring* opens quietly, but with an effect which in the context of harmonized music is apparently unique. For a very long passage there is not only no modulation but no change of chord. A chord of E flat major builds up: first the tonic sounds in the abysmal depths; next a fifth is added; then an arpeggio movement on the complete triad, calm but swelling, an embryonic motive which becomes partly stepwise to give us the first genuinely melodic motive. But still the chord does not change. We become increasingly aware of the modulation which is not happening. A sense of timelessness sets in.[75]

The effect, then, is the metamorphosis of the music out of a single prolonged chord—as Donington observes the effect of staying in one harmony for so long had the effect of arresting time. Time seems to become geological, evolutionary. The slow metamorphosis of the music signifies the metamophosis of life out of the water. So for Wagner, the Rhine River becomes symbolic of the course of human life in accordance with the creation-destruction cycle of Norse myth. The sustained chord has one other key effect. There must be a fixed point to allow for the sensation of something flowing past: The sustained $E^{\flat}$ offers the fixed point over which the rest flows. This does not, in the least, detract from the sense of the duration of the flow. The rich counterpoint of the eight- and sixteenth-notes moves up and down, as well as over the top of of one another.

---

74 Qtd. by Robert Donington, *Wagner's 'Ring' and its Symbols*, 3rd ed. (London: Faber, 1974) 35.

75 Donington 35.

Although Wagner, Zorilla, Lamartine, Hölderlin, Byron, and Goethe each set up a unique and only partly comparable relation between the simple image and that feeling or concept for which it is made to serve as a symbol, nevertheless the following generalization can be made: in the above works, the stream represents the passage of human life and the course of human existence and experience. Although they are backed up by the suppositions of beginning and end, of source and sea, and of birth and death, these are not elaborated on in the poems treated above as they are in those below.

The Course of Life into Death: The Flowing of the River into the Receptacle Quarter of the Water-Cycle

In the lyrics I have just examined, the poets only alluded to the step in the water cycle from stream to sea. In chapter one during my examination of the opening lines of 'Song in Praise of Mohammed' / 'Mahomets Gesang' (1: 304–05), I emphasized the way Goethe uses the river as the allegorical representation of the passage of man's life from infancy to old age. We will now, as it were, pick up where we left off: the point at which the prophet/river was still a young-leader/a rivulet. The speaker of the poem tells us that the religious leader spreads his ideas as a river waters and fertilizes the soil. The indomitable movement of the leader is expressed by the unrestrainable drive of the water downward and seaward. As we shall see, this demonic drive dominates Goethe's poems that employ man/stream metaphors. The stream's (the prophet's) creative acts, which are now appropriate to the small stream, will amplify as the stream grows: 'Below in the valley flowers grow beneath his footprint, and the pasture thrives from his exhalation'.

Drunten werden in dem Tal
Unter seinem Fußtritt Blumen,
Und die Wiese
Lebt von seinem Hauch (ll. 18–21).

Temptations of leisure and pleasure—which might divert and endanger the prophet's great movement—are expressed with slightly more anthropomorphic terms; in keeping with this, I continue to translate the 'it' of the stream rather as the 'he' of the prophet Mohammed—the Germanic gendering of the pronoun *it* for *the stream* adds an ambivalence that cannot be Englished simply: 'Yet no valley full of shade can hold him—no flowers entwining his knees and coaxing him with lovelorn eyes: winding snakelike, he presses on his course toward the flatland'.

Doch ihn hält kein Schattental,
Keine Blumen,
Die ihm seine Knie umschlingen,
Ihm mit Liebes-Augen schmeicheln:
Nach der Ebne dringt sein Lauf
Schlangenwandelnd (ll. 22–27).

Goethe is drawing on the Rococo vocabulary of imagery of his previous poetic style, only to have his Storm and Stress persona, as well as subject, reject them and what they signify.

In the next stanza, the motif of the stream's unrestrainable motion—introduced in lines 13 through 17—is employed as the voices of disciples/streams are heard requesting to join the movement of the prophet/river toward God/the sea. The metaphorical relationship between the army of Mohammed and the streams joining the river becomes so intricate that it is hard to determine which is the metaphor for which: 'Freshets snuggle up to him for company'.

Bäche schmiegen
Sich gesellig an (ll. 28–29).

Let me requote from De Man's reading of the source passage in Shelley's *Triumph of Life*: 'The property of the river that the poem singles out is its sound; the oblivious spell emanates from the repetitive rhythm of the water, which articulates a random noise into a definite pattern'.[76] Notice in the following lines the recurrent 'In die Ebne', 'Und die Ebne', 'von der Ebne', and such reiteration of words, phrases, and syntactic patterns:

Now, silver-sparkling, he steps into the flatland, and with him, the flatland sparkles, and the rivers of the flatland and the brooks from the mountains cheer him and call: Brother! Brother, take your brethren with you, with you to the ancient father, to the everlasting ocean, who awaits us with outstretched arms—opening them, alas, in vain to embrace us who yearn for him [...]'.

Nun tritt er
In die Ebne silberprangend,
Und die Ebne prangt mit ihm,
Und die Flüsse von der Ebne
Und die Bäche von den Bergen
Jauchzen ihm und rufen: Bruder!
Bruder, nimm die Brüder mit,
Mit zu deinem alten Vater,
Zu dem ewgen Ozean,
Der mit ausgespannten Armen
Unser wartet,
Die sich, ach! vergebens öffnen,
Seine Sehnenden zu fassen [...] (ll. 29–41).

The lines that follow explain why these streams 'despair' that—on the level of natural phenomena—they might not partake of the water-cycle process: 'for the greedy sand of the barren desert devours us; the sun above sucks our blood';

Denn uns frißt in öder Wüste
Gierger Sand; die Sonne droben
Saugt an unserm Blut [...] (ll. 42–44).

This touch is particularly apt because he is referring to Middle-Eastern streams that never reach the sea but evaporate or end in marshes and ponds.

We find again the theme of a thwarted river that is found in the later poem 'Overwhelming Astonishment': 'a hill dams us up into a pond! Brother, take with you the

---

76 Paul de Man, 'Shelley Disfigured', *The Rhetoric of Romanticism* (New York: Columbia UP, 1984) 107.

brethren of the flatland, take with you the brethren of the mountains—to your father'.

ein Hügel
Hemmet uns zum Teiche! Bruder,
Nimm die Brüder von der Ebne,
Nimm die Brüder von den Bergen
Mit, zu deinem Vater mit! (ll. 44–48).

The progress of Mahomet/the river continues unchecked and is described as the impulse behind the creation of an empire: 'Come, all of you!—and now he swells more gloriously; a whole race raises on high the ruler! And with triumphant impetus, he names countries; in his track, cities appear. Unstoppable, he rushes forth—leaves behind the flaming tops of towers, houses of marble, his creative plenitude'.

Kommt ihr alle!—
Und nun schwillt er
Herrlicher; ein ganz Geschlechte
Trägt den Fürsten hoch empor!
Und im rollenden Triumphe
Gibt er Ländern Namen, Städte
Werden unter seinem Fuß.

Unaufhaltsam rauscht er weiter,
Läßt der Türme Flammengipfel,
Marmorhäuser, eine Schöpfung
Seiner Fülle, hinter sich (ll. 49–59).

In the next chapter, we will follow the stream flowing into the next quarter—the ocean as receptacle.

English Romantic poetry offers a direct imitation of Goethe's allegorical usage of streams representing both a prophet and his movement, when around 1799, Coleridge and Robert Southey jointly planned and individually wrote what were to remain fragments of an unfinished epic called *Mahomet*. Coleridge's fragments (1: 329–30) are not only significant as an English tribute to German poetry and thought, but also—for all their roughness—they are remarkable bits of verse that capture, with phrasing and punctuation, as well as imagery, the sensation of the forward rush of a great movement of waters/peoples, suddenly obstructed, shattered, and united in their course—

the people with mad shouts
Thundering now, and now with saddest ululation
Flew, as over the channel of rock-stone the ruinous river
Shatters its water abreast, and in mazy uproar bewilder'd,
Rushes dividious all—all rushing impetuous onward (ll. 10–14).

Is it stretching too much to see, in these lines, English ambivalence, but mainly fear, of the potentiality of the army of the French Revolution and—more and more—the army now lead by Napoleon Bonaparte? I suspect it is not.

Similarly, in a sonnet composed around 1802 or 1803 (*Poetical Works* 3: 117), Wordsworth uses the image of a river as the metaphor for the destiny of a nation, rather than that of man:

It is not to be thought of that the Flood
Of British freedom, which, to the open sea
Of the world's praise, from dark antiquity
Hath flowed, 'with pomp of waters, unwithstood',
Roused though it be full often to a mood
Which spurns the check of salutary bands,
That this most famous Stream in bogs and sands
Should perish; and to evil and to good
Be lost for ever [...] (ll. 1–9).

The celebration of governmental restraining power over the energies of the people hints at the rightward turning of Coleridge and his fellow Lake Poets.

Several Romantic works, especially *River Duddon* of Wordsworth, allow for the possible inclusion of the rich variety of natural and human life along the course of the river. The flow of the river, especially in its water-cycle context can offer a full poetic scheme of relatedness. And here, of course, I will appeal repeatedly to the complexity theory and praxis of Lansing. After all, the particular subject of his study treats the social and religious structures found all along the Bali streams as they move seaward.

Goethe suggests the complexity of life on the banks of a stream in his *Werther*; here, the stream induces in the narrator this primal state of mind(lessness) of the protagonist which is achieved through the contemplation of the flux and reflux of water. This vision of flow is manifested by a small stream—which we have already seen in chapter one and in the present chapter is used by Romantics to signify childhood in the river of life—accompanies the narrator, Werther's, return to the scene of his childhood:

I followed the river downstream to a certain farm—a walk I always took, and up to the spots where we practiced causing flat stones to make as many skips as possible as possible when we were boys. I remembered so vividly how I would sometimes stand there gazing at the flowing water, following it with my mind full of wonderful notions, imagining the exciting regions where it would fly to, and all too soon finding the limitations of my imagination; and yet it had to go on and on, until I was quite lost in the contemplation of an invisible distance.

Ich ging den Fluß hinab, bis an einen gewissen Hof; das war sonst auch mein Weg, und die Plätzchen, wo wir Knaben uns übten, die meisten Sprünge der flachen Steine im Wasser hervorzubringen. Ich erinnerte mich so lebhaft, wenn ich manchmal stand und dem Wasser nachsah, mit wie wunderbaren Ahnungen ich es verfolgte, wie abenteuerlich ich mir die Gegenden vorstellte, wo es nun hinflösse, und wie ich da so bald Grenzen meiner Vorstellungskraft fand; und doch mußte das weitergehen, immer weiter, bis ich mich ganz in dem Anschauen einer unsichtbaren Ferne verlor.

Notice the careful layering of the two different spots of time where the narrator as a child and as a young man has fallen into meditations over the water. These meditations bring to Werther's mind 'romantic notions' following the expected flow of the stream through places downstream from where he is standing.

Childhood becomes closely linked with the pristine age of human history—of Homer: '—Look, my dear fellow, that is how the magnificent patriarchs were: so circumscribed and yet so happy; their feelings and poetry so childlike! When Odysseus speaks of the measureless sea and the boundless earth, it is all so true, so human, so inwardly, so closely felt, and so mysterious'.

—Sieh, mein Lieber, so beschränkt und so glücklich waren die herrlichen Altväter, so kindlich ihr Gefühl, ihre Dichtung! Wenn Ulyß von dem ungemess'nen Meer und von der unendlichen Erde spricht, das ist so wahr, menschlich, innig, eng und geheimnisvoll (7: 454).

By and large, in this passage from Goethe's novella, the scenes along the stream are intuited by Werther through a sort of meditation. Like Lansing's complexity study of the Balinese farmers, this passage from *Werther* teaches us a crucial ecological lesson: a stream—visually and aurally—shapes human consciousness along its banks. Wordsworth's *River Duddon*—which goes further in anticipating Lansing's complexity approach to rivers and anthropology—shows how a stream's effect extends into shaping lifestyles.

In the *River Duddon* sonnets, Wordsworth is able to develop this idea of life along the shores of a stream with great actuality and at greater lengths than does Goethe. The Duddon passes a

Cottage rude and grey;
Whose ruddy children, by the mother's eyes
Carelessly watched, sport through the summer day,
Thy pleased associates;—light as endless May
On infant bosoms lonely Nature lies (sonnet 5).

So the river is a sort of educative agent of Wordsworth's great teacher, Nature; also, implied in these lines is the running theme of the river's relation to the cycle of human life.

The cycle of human prehistory is also touched on, at least by way of imagination:
WHAT aspect bore the Man who roved or fled,
First of his tribe, to this dark dell—who first
In this pellucid Current slaked his thirst? (sonnet 8).

This sonnet series is different from the 1802 sonnet in that in *River Duddon* Wordsworth departs from the central allegorical relation of river/one man. The above is Wordsworth's evocation of the primordial man who first sampled the water of the Duddon. In this passage, too, we have one example of Wordsworth's occasional departure from the beginning-to-end, temporal pattern of the river's flow. This pattern is often interrupted for a moment when the poet suggests the continuity of the river's history by recalling 'a spot of time' in the river's past. (This continuity, as we shall see in the next chapter, is further elaborated by the 'After-thought' sonnet, which is the last in the series.)

After the fashion of Coleridge, Wordsworth finds the entire history of humankind along the shore to be paltry compared to the flow of the stream which has carved 'deep-worn' channels which seem

Bright liquid mansions fashioned to endure
When the broad oak drops, a leafless skeleton,
And the solidities of mortal pride,
Palace and tower, are crumbled into dust!—
The Bard who walks with Duddon for his guide,
Shall find such toys of fancy thickly set [...] (sonnet 12).

As in the case of the rivers in Bali described by Lansing, the stream not only passes through the lives of the people along the way, it also provides energy for their technology—here a water mill the 'spouting' chute above is touched on:

> HAIL to the fields—with Dwellings sprinkled o'er,
> And one small hamlet, under a green hill
> Clustering, with barn and byre, and spouting mill! (sonnet 13).

Even a pool in the stream is used for sheep washing (23) and for watering and providing pasturage for cattle (28).

In sonnet nine, already touched on in chapter one, the reader is returned abruptly to the river's present, to its movement in time and to its allegorical representation of one man's life—as he 'grows':

> THE struggling Rill insensibly is grown
> Into a Brook of loud and stately march [...] (sonnet 9).

Wordsworth himself appeals to the musical analogy, and one of the great comprehensive cultural scholars of Romanticism, H.G. Schenk does compare Wordsworth and Beethoven—*The Prelude* and the 'Pastoral' Symphony, the brook music from the second movement. While drawing directly upon Schenk's analysis and comparison, I would like to replace *The Prelude* with *The Duddon River*. It is crucial that the scholar rejects the other famous nature reference Beethoven makes to bird calls in order to seize on what Schenk rightly deems is a more profound consideration: the musical signs in the score that represent water and that he takes to be an accurate rendering of the flow of water, heard before his growing deafness had begun. Schenk points out 'the composer's musical rendering, in quavers or semiquavers on the lower strings, of the murmur of the brook which, it seems, he had acoustically observed some five years earlier while his outward sense of hearing was still acute'.[77] The time for the movement is 12/8, an unusual time that blurs the metrical emphasis—adding to the flowing sensation. The movement opens with eight-notes (quavers) moving up and down single scale intervals, with the exception of a single third-interval step. Against this flowing movement are sustained notes of the horns: there must be, as I have observed before, some stable point before the flow can be recognized as such. Then later at the fifth measure, we find the sixteenth-notes (semi-quavers) figures often oscillating between notes in intervals of thirds and fourths, the fixed point of the horns, and now the first theme, fragments of which had already begun to sport in measure one like whirlpools spiraling off from the main current—represented throughout by two solo 'cellos.

Actually, all of these effects—organized into a full sonata form—make perfect sense as pure music. By entitling the movement 'Szene am Bach' / 'Scene by the Brook', however, Beethoven has asked us to see the movement also on the level of representation of the natural phenomenon itself and, more perhaps, of the feeling the composer has for it. And if we are really honest about it, the words of poetry hardly do more in the way of representing; they are mere signs for the thing itself. After all, what does Wordsworth mean in the passages cited above by 'a Brook of loud and

---

77 H.G. Schenk, *The Mind of the European Romantics* (Garden City: Anchor-Doubleday, 1969) 167.

stately march'? How can a flowing brook either look like humans marching or sound loudly like the music to accompany them? Such language shatters signification beyond all precise recognition, and it certainly does not recreate the natural phenomenon. Musicologists who are so eloquent at what music fails to do should question the very words they use to say so.

Wordsworth sustains the unstable, metaphoric relation between a youth/brook through sonnet twenty, where the poet traces the stream during the placid part of its course and describes

> the entrancement that detains
> Thy waters, Duddon! 'mid these flowery plains;
> The still repose, the liquid lapse serene,

but the poet endows the stream with a human-like desire to break chains and rebel against the restraints which offer a safe, placid existence:

> But these chains
> Will soon be broken;—a rough course remains,
> Rough as the past [...] (sonnet 20).

This same motif of the unrestrainable desires of youth appears in Goethe's 'Song in Praise of Mohammed' in a passage which it seems valuable to requote here: 'Yet no valley's shade can hold him, no flowers entwine his knees and coaxing him with lovelorn eyes' (ll. 22–27), and in 'Overwhelming Astonishment', the stream 'sweeps irresistibly valleyward and onward' (l. 4). Thus, we have in these poems the motif of a stream being endowed with a rebellious impatience to hurl forth and spill out of old bonds—symbolized by valleys and by meadows. This theme, as well as that of the flower-covered landscape, recurs later in *River Duddon* as it continues its destined course:

> Not hurled precipitously from steep to steep;
> Lingering no more mid flower-enamelled lands
> And blooming thickets; nor by rocky bands
> Held [...] (sonnet 32).

Here, Wordsworth has added his own particular complexity to the image of the hindrances to the course of the brook: sometimes the brook/youth is impatient to hurl forth; sometimes given to moments of lazy, directionless ease. Only adulthood's maturity offers the axis point upon which to center a real sense of purpose. In river terms, it offers a clearer course.

By now the Duddon has passed into its 'middle age', as it becomes a magnificent river, soon destined to pour into the sea:

> in radiant progress toward the Deep
> Where mightiest rivers into powerless sleep
> Sink, and forget their nature—*now* expands
> Majestic Duddon, over smooth flat sands
> Gliding in silence with unfettered sweep!
> Beneath an ampler sky a region wide
> Is opened round him [...] (the emphasis is Wordsworth's; sonnet 32).

If we remember that this is a hike made during the cycle of a single day, the persona of the poem is getting tired and thinking of the coming night's rest. Out of this mood grow thoughts of the sleep at the end of the human cycle of life. Offering itself as a ready symbol is the river water losing its identity as such by merging with the sea, in the course of its great movement within the larger cycle, which in the next chapter we will see Wordsworth view from various metaphysical angles.

Both Turner, in his *Coniston Fells,* and Koch, in his two versions of the *Schmadribach,* imply the water-cycle context of the stream by relating the latter, shown in the foreground, to the spring-falls, glaciers, and clouds, in the background. The place of the stream in the cycle is even clearer in Friedrich's *Cycle of Life's Ages / Lebensalter-Zyklus,* where both the 'Summer' / 'Sommer' panel and the 'Autumn' / 'Herbst' contain river symbols. The former represents youth. Above the two young lovers, the artist has placed two rivers, which merge. The apparent flow of these rivers is from left to right. On the left and in the background, smaller streams can be distinguished flowing down from the mountains and into the rivers in the middle ground. According to Börsch-Supan, 'The river, which has grown wider, crosses a cultivated landscape and serves as an image of the interest directed towards earthly existence at this time of life'.[78] 'Autumn' treats adulthood. The river has now become large enough to bear ships. At least, one sailing ship can be discerned. The allegorical meaning is to be gained indirectly from a member of Friedrich's circle, G.H. von Schubert, who observes that 'the impulse of passions within us has turned into a river carrying ships down its way'.[79] Both Goethe, in 'Song in Praise of Mohammed' and Wordsworth, in *River Duddon,* employ the navigable river as an allegorical symbol for the shouldering of adult responsibilities.

Thomas Cole—in his own application of the water-cycle to the life cycle—uses a river, as well as a waterfall, for symbols in two panels of his *Voyage of Life*. For the second panel, 'Youth', Cole wrote this program to the allegorical significance:

> The stream now pursues its course through a landscape of wider scope, and more diversified beauty. Trees of rich growth overshadow its banks, and verdant hills form the base of lofty mountains. The Infant of the former scene is become a Youth, on the verge of Manhood.
>
> He is now alone in the Boat, and takes the helm himself, and, in an attitude of confidence and eager expectation, gazes on a cloudy pile of Architecture, an air-built Castle, that rises dome above dome in the far-off blue sky (29).

This 'cloudy pile', of course, looks forward to the sky quarter; in the fourth chapter, we will consider the application of Romantic fancy to clouds in more detail. Its appearance in this painting appears to be outrageously allegorical; it is interesting, however, that there is one passage in *Faust, Part Two* (5: 289–526) paralleling Cole's cloud city: 'Haven't you heard of those wisps of fog that drift along off the coasts of Sicily? There—with swaying clarity, in broad daylight, lifted up into mid-air, mirrored in a particular haze—appears a strange apparition: There cities are

---

78 Helmut Börsch-Supan, 'Caspar David Friedrich', *Caspar David Friedrich, 1774–1840: Romantic Landscape Painting in Dresden,* eds. William Vaughan, Helmut Börsch-Supan, and Hans Joachim Neidhardt (London: Tate Gallery, 1972) 51–101, qtd. from 82.

79 Qtd. in *Friedrich* 106.

swaying back and forth, and gardens are floating up and down, as image upon image pierces through the ether':

> FAUST. Vernahmst du nights von Nebelstreifen,
> Die auf Siziliens Küsten schweifen?
> Dort, schwankend-klar, im Tageslicht,
> Erhoben zu den Mittellüften,
> Gespiegelt in besondern Düften,
> Erscheint ein selsames Gesicht:
> Da schwanken Städte hin und wieder,
> Da steigen Gärten auf und nieder,
> Wie Bild um Bild den Äther bricht (10, 584–92).

This is Thomas Cole's version:

> The Guardian Spirit stands upon the bank of the stream, and, with serious, yet benignant countenance, seems to be bidding the impetuous Voyager Godspeed. The beautiful stream flows for a distance, directly toward the aerial palace; but at length makes a sudden turn, and is seen in glimpses beneath the trees, until it at last descends with rapid current into a rocky ravine, where the Voyager will be found in the next picture. Over the remote hills, which seem to intercept the stream, and turn it from its hitherto direct course, a path is dimly seen, tending directly toward that cloudy Fabric, which is the object and desire of the Voyager.
>
> The scenery of the picture—its clear stream, its lofty trees, its towering mountains, its unbounded distance, and transparent atmosphere—figure forth the romantic beauty of youthful imaginings when the mind elevates the Mean and Common into the Magnificent, before experience teaches what is the Real. The gorgeous cloud-built palace, whose glorious domes seem yet but half revealed to the eye, growing more and more lofty as we gaze, is emblematic of the day-dreams of youth, its aspirations after glory and fame; and the dimly-seen path would intimate that Youth, in its impetuous career, is forgetful that it is embarked on the Stream of Life, and that its current sweeps along with resistless force, and increases in swiftness, as it descends toward the great ocean of Eternity (29).

Here are references to the ocean quarter and the link/sky quarter of the water-cycle; so Thomas Cole is not just content with having the cycle move progressively through the series of panels: he sets out to include as much of all the cycle-quarters as is meaningfully possible in each panel.

In analyzing 'Youth', one sees to the right of and in opposition to the 'cloud-built palaces' a barren, rocky mountain that is solid and real. It represents earthy reality rather than heavenly idealism. It is in the direction of this barren rocky landscape that the youth's boat will carry him. There are two rivers in the painting. On the left of the canvas at the upper edge of the real river, there is an imaginative river (or 'path' as the painter says) that seems to move in the direction of the cloud city; however, it moves uphill—obviously to us, not a possibility. River or path—that is the direction where the naïve youth thinks he will travel. At the same point, the real river turns sharply to the right. Here, too, the river's placid, reflective surface becomes a cataract: the placidity of youth is about to be shattered by the trials of adulthood. To the upper right, the barren lands are seen in opposition to the lush vegetation in the foreground from whence he is drifting. There is, unlike in the Friedrich, no sense of season—the vegetation seems generically subtropical, with a date palm on the lower right. The light is softened by the remaining haze of morning. The placid water is streaked in zigzag patterns with white to suggest sparkles; the rapid water of the cataract is rendered with heavy impasto application.

The third 'Manhood' panel takes up the story after the bend in the river, when it is channeled through the narrows between the mountains, and just before the falls:

> Storm and cloud enshroud a rugged and dreary landscape. Bare, impending precipices rise in the lurid light. The swollen stream rushes furiously down a dark ravine, whirling and foaming in its wild career, and speeding toward the Ocean, which is dimly seen through the mist and falling rain. The boat is there plunging amid the turbulent waters. The Voyager is now a man of middle age: the helm of the boat is gone, and he looks imploringly toward heaven, as if heaven's aid alone could save him from the perils that surround him.

The conventional symbolism is drawn from the linking, sky quarter:

> The Guardian Spirit calmly sits in the clouds, watching with an air of solicitude, the affrighted Voyager: Daemon forms are hovering in the air.

The waterfall is being used—just as in Hölderlin, Lenau, and Hugo—for the period moments of adversity in the flow of one human's life; here, too, the next stage of the water-cycle, the ocean as the end of all, is anticipated:

> Trouble is characteristic of the period of Manhood. In childhood, there is no carping care: In youth, no despairing thought. It is only when experience has taught us the realities of the world, that we lift from our eyes the golden veil of early life; that we feel deep and abiding sorrow: and in the Picture, the gloomy, eclipse-like tone, the conflicting elements, the trees riven by tempest, are the allegory; and the Ocean, dimly seen, figures the end of life, which the Voyager is now approaching. The demon forms are Suicide, Intemperance and Murder; which are the temptations that beset men in the direst trouble (29).

The appearance of 'Suicide' in this vision of potential shipwreck of the boat in the picture conveys the power that this allegorical symbol had to the Romantics.

In her very late and very personal *Reflections on Suicide,* Germaine de Staël uses the shipwreck and the river of life symbols. While I admit they might be read as rhetoric rather than poetics and thereby might embody a conventional figure rather than a distinctively Romantic image, they present a vision that has a power rivaling that of Cole: 'life often seems like one long shipwreck—with friendship, fame, and love as the debris. The riverbanks along the course of our own generation are littered with them; but if we have salvaged our inner harmony from the wreck, we may yet be able to enter in communication with the works of the Divine Being'.

> la vie ne semble souvent qu'un long naufrage, dont les débris sont l'amitié, la gloire et l'amour. Les rives du temps qui s'est écoulé pendant que nous avons vécu en sont couvertes; mais si nous en avons sauvé l'harmonie intérieure de l'âme, nous pouvons encore entrer en communication avec les oeuvres de la Divinité (2: 182).

This abstract reference to God is found in Cole:

> The upward and imploring look of the Voyager shows his dependence on a Superior Power; and *that faith saves him* from the destruction that seems inevitable (emphasis is Cole's, 29–30).

'Manhood' is indeed rendered to give a sense that destruction is beyond doubt. The clouds swirl in a Turner-like vortex, and they join with the rocks to form a cavern-like shape. The clouds open to show the ocean; indeed, in the sketch for the final version, one can see where the river flows into the sea; however, in the final version the artist seems to have literally shrouded it with clouds, perhaps for the sake of mystery. The clouds also seem to form into vague demon shapes in the upper middle of the canvas. Here, as predicted by the composition of the previous panel, we are in a barren, rocky landscape, and with what trees there are blasted by the elements. Here, we get a hint of the seasons because the leaves seem to be turning an autumnal red. A sunset over the sea is at exactly dead center of the canvas; this is also the heart of the cloud, mist, and waterfall vortex. The facture to create the water of the cataract is the greatest application of impasto in the entire series. Some white stippling creates the main flow and eddies swirling out to the side. The mist is suggested by what seems to be a facture using finger or sponge and spreading a very thin coat of light paint, perhaps, over an already dried surface.

In these works of Staël and Cole, then, the landscape with water-cycle is—as in Caspar David Friedrich—being used allegorically to convey Christianity, or at least a Theism. There is a directness of approach in Staël and Cole that, as with Friedrich, sets our Postmodernist teeth on edge; Friedrich even faced contemporaneous criticism from his German countrymen for his expressive directness.[80] But in this chapter we have seen a wide range from the subtle intricacies of Goethe, Constable, and Lamartine to the blunt directness of Cole, Staël, or Friedrich. Many in the Romantic culture valued clarity and directness; these frank values were beginning to come under attack, but they would not be totally discounted until the Symbolists. And the flow quarter of the water-cycle had by now become familiar as a source of a common fund for the rhetoric of birth, life, and death.

For my summary of this chapter, I am going to use a comprehensive, inclusive lyric that contains most of the themes contained in this flow chapter.

A compatriot of Thomas Cole, Philip Freneau, wrote a poem included in the second volume of *A Collection of Poems on American Affairs and a Variety of Other Subjects Chiefly Moral and Political* (1815). 'The Brook of the Valley' (2: 31–33)[81] not only presents a fuller vision of the stream in the cycle than the other poets we have considered, but the poet also enlarges its symbolic import to offer it as a political celebration of the permanence of change. Indeed, the historical-political situation is presented at the very outset of the poem:

> The world has wrangled half an age,
> And we again in war engage,
> While this sweet, sequester'd rill
> Murmurs through the valley still (ll. 1–4).

The poet, for the fleeting moment of a stanza, contemplates the peace and security offered by stasis:

> All pacific as you seem:
> Such a gay elysian stream;—
> Were you always thus at rest
> How the valley would be blest (ll. 5–8).

---

80 William Vaughan, 'Caspar David Friedrich', *Friedrich* 9–44, qtd. from 26–27.

81 Philip Freneau *A Collection of Poems on American Affairs and a Variety of Other Subjects Chiefly Moral and Political*, ed. Lewis Leary (Delmar: Scholars Facsimiles, 1976).

Setting the river's cycles of high and low flow levels within the cycle of the seasons, the poet realizes that this stasis is undesirable and change is necessary for sustaining life:

But, if always thus at rest;
This would not be for the best:
In one summer you would die,
And leave the valley parch'd and dry (ll. 9–12).

Then follow a series of questions very like those posed, in a Baroque Christian context, by Henry Vaughan in the poem with which this chapter opened and with an air of anachronism rather like that seen earlier in Heredia (and Bryant):

Tell me, where your waters go,
Purling as they downward flow?
Stagnant, now, and now a fall?—
To the gulph that swallows all (ll. 13–16).

The density of ideas in these lines is impressive; gravity works with a universal destiny toward an end that is not an end. The waterfall is some sort of break in the continuity as we have seen in Hölderlin, Lenau, and Hugo—something to be feared as much as wished for; indeed, it is in this that they instinctively assert their European attitude toward change. The citizen of the new United States celebrates the falling that prevents stagnation: discontinuity and change are nothing but good.

Since, as I will discuss later, some see this poem as pessimistic, I will need to look outside the work itself to show how it is—if not entirely optimistic—resigned; indeed, I would go so far as to say it is a qualified celebration of change, which is in turn central to the political world view of Philip Freneau and of Thomas Jefferson. Freneau was indeed the key figure in Jefferson's media war against the Federalists, who sought to retain as much of the English system as possible and prevent as many revolutionary transformations as possible in the new American government; indeed, Jefferson, in a letter to George Washington discussing his connections with Freneau, terms him 'a man of genius' (462). Jefferson is probably the most thorough-going believer in change of all political thinkers. All the great eighteenth- and nineteenth-century thinkers, Hegel and Marx included, always saw some goal or end—therefore, ending—of that change. Jefferson saw no need for such an end. In his great 6 September 1789 letter to James Madison, he rejects the binding force of political traditions or even constitutions by saying that 'no society can make a perpetual constitution, or even a perpetual law. The earth belongs always to the living generation' (449). In the face of the terror of the French Revolution, he never turned into a faint heart; for him the terror and the fate of its victims never called the revolutionary changes into question: 'In the struggle which was necessary, many guilty persons fell without the forms of trial, and with them some innocent. These I deplore as much as any body, & shall deplore some of them to the day of my death. But I deplore them as I should have done had they fallen in battle' (465). Only his friends Thomas Paine and Freneau looked with a similar unflinching stare at the sadder implications of political change.

Freneau certainly is fearless in the face of the Heraclitean paradox. It is an answer to a deeper riddle of origins; it is expressed with exhilaration; and it celebrates Heraclitos' vision of unrelenting change:

> Flowing, peaceful, from your urn
> Are your waters to return?—
> Though the same you may appear,
> You're not the same we saw last year (ll. 17–20).

It is true certainly that there is no admixture of pessimism in Goethe's use of this 'never the same river twice' concept in his 'Permanence and Change'; but, as we have seen earlier in this chapter, Goethe was seeking within the realm of art an escape from the fearful implications of change. And he certainly did not throw open the image to a cosmic tracing of the other years' water throughout the vast reaches of earthly space, as does the poet of the new United States:

> Not a drop of that remains—
> Gone to visit other plains,
> Gone, to stray through other woods,
> Gone, to join the ocean floods! (ll. 21–24).

While it may not be pessimistic; such a view cannot be called optimistic. 'In this poem', writes Mary Weatherspoon Bowden,

> the poet takes care to impress us with the cycles of the brook—and, by extension, with the cyclical nature of man's history. Acceptance of the adage 'history repeats itself' carries the implication of lack of progress; for, if man goes in circles, he is not advancing directly toward a goal. Freneau's fondness for cyclical theories here betrays his desire to see the progress of man.[82]

While Bowden does bring to the surface the ahistorical implications of the cycle, she draws out of it ideas held by Heraclitos or his philosophical heir Nietzsche, but not by a Freneau or his mentor Jefferson.

Notice in the stanzas we have just considered that the cycle is never the same for each human viewpoint. The cycle as described by Freneau is not some recurrence of a vast waste of sameness. The same water does not recycle through the source (urn); the river is never precisely the same. Here, we have in Freneau's lyric the paradoxical combination of the arrow and the cycle described by Gould. Freneau sees the paradox and delights in it. The river's flow is shown by him to move arrow-like from point A to B; yet, almost as if it were conceived of as being superimposed on the river vision, there exists the larger cycle vision. At least according to my reading in opposition to that of Bowden, Freneau is avoiding the pessimism of an endless eternal return to the very same point. To take an obvious example, while Freneau and Jefferson accepted the recurrence of revolution, they would not feel that each duplicated the other. Their progressive faith in the change of the 1789 revolution was to be realized in the progressive move through 1830 to 1848 to 1871 and so on. Freneau's tone throughout is resignation to, indeed celebration of permanence in

---

82 Mary Weatherspoon Bowden, *Philip Freneau* (Boston: Hall, 1976) 166. Other important studies on Freneau neglect this poem; see, for example, Philip M. Marsh, *Philip Freneau: Poet and Journalist* (Minneapolis: Dillon, 1967).

change. Freneau's complex cycle partakes of the larger cycle rhetoric discussed at the end of the introduction: it is the upward spiral of Vico and Hegel. Cyclic recurrences happen in the spiral, but at a higher level of development. Freneau is just looking at the unfortunate recurrences with the clear eye of recognition, not with the cynical eye of a Turner or Byron. In the spiral, recurrence includes rather than precludes progress. Bowden is correct in saying that the cyclic image is a dangerous one because of its embodiment of endless recycling; however, that is only one symbolic implication of the water-cycle. Complexity theory warns against one-way readings.

Freneau's vision is worldly; the lacuna left for the Christian Mystery in Vaughan and, it seems, that in Heredia (Bryant), are filled with science—all questions are answered by the water-cycle. Notice, too, how form embodies content when the only enjambed stanzas introduce the water-cycle and its crucial processes of evaporation and precipitation, which will be the focus of the next chapter. It, however, would be wrong to split up the poem according to the quarters of the cycle; indeed, it is one of the glories of this poem that the aspects of the water-cycle are bound into such a great unity.

> Yes—they may return once more
> To visit scenes they knew before;—
> Yonder sun, to cheer the vale,
> From the ocean can exhale
>
> Vapors, that your waste supply,
> Turn'd to rain from yonder sky;
> Moisture, vapors, to revive
> And keep your margin all alive (ll. 25–32).

In none of the river poems so far has the hydrological-cycle been so explicitly presented. The varied and intricate phenomena of a stream—which is shown moving in obedience to the seasonal cycling from summer quiescence (the drought is hypothetical) to a winter spate—is viewed in its full environmental complexity.

The cycle is present to frame the vision of unrest and flux of storm and flood, which in their turn stand as symbols of war and stress in human history. 'The events surrounding the War of 1812', writes Bowden, 'inspired Freneau to write many reflections in which he attempts to adjust to his disappointment at the seeming reversal of man's progress—the best of which is "The Brook of the Valley"'.[83] I, however, fail to see it as the completely pessimistic poem of a disappointed man. Flood/war dominates the climatic stanzas before it cycles back to peace/quiet flow:

> But, with all your quiet flow,
> Do you not some quarrels know!
> Lately, angry, how you ran!
> All at war—and much like man (ll. 33–36).

The most pessimistic aspect of all is that the warlike state of the river is caused by the workings of the hydrological-cycle. Flood/war is for the river part of the workings of nature; hence, by analogy so is war for humans:

---

83 Bowden 165.

When the shower of waters fell,
How you raged, and what a swell!
All your banks you overflow'd,
Scarcely knew your own abode!

How you battled with the rock!
Gave my willow such a shock
As to menace, by its fall,
Underwood and bushes, all [...] (ll. 37–44).

War and peace are part of an ongoing cycle in which one temporally falls hard upon the other—an attitude made clear by the close placement of the two in a single stanza:

Now you are again at peace:
Time will come when that will cease;
Such the human passions are;
—You again will war declare (ll. 45–48).

Freneau goes on to leave us in no doubt that the stream and its seasonal cycling of moods are emblematic of human history:

Emblem, thou, of restless man;
What a sketch of nature's plan!
Now at peace, and now at war,
Now you murmur, now you roar;

Muddy now, and limpid next,
Now with icy shackles vext—
What a likeness here we find!
What a picture of mankind! (ll. 49–56).

Freneau is here emphasizing change. The important point is that he is like Stäel, Turner, and Byron: he looks change fearlessly in the face. He is totally accepting of the fact that change can lead to worse and much as better, but he is willing to take the chance. The caution of a Goethe or a Lamartine is not for him.

As we have just seen, Freneau is in the company of many poets and artists, working between 1760 and 1870, who use the river as a symbol for human destiny. Unlike him, many of them employ the waterfall as a point of discontinuity. Byron celebrates this supreme emblem of change, for change's sake. Lenau and Hugo use it as a symbol of the vicissitudes of life, out of which great art is made. The massive falls of Niagara were a catalyst of great creative efforts among Western painters and poets of three continents. Generally the (Neo)Classical-Romantic-Realist did not need the titanic flow of a cataract; they rather found in the simple flow a signifier concentrated enough to covey their particular approach to permanence and change. Those creators drawn to the (Neo)Classical vein in Romantic culture—like the young Constable, Meléndez, Beethoven, and the Goethe of Weimar Classicism—found a tranquility forged from the material of change and a sense of order drawn from the actuality of the natural world. More romantically inclined poets, Percy Shelley and Coleridge, use flow symbols as foils for human attempts at permanence and order; their apperceptions found a visual-arts parallel in a rare landscape by Goya. A younger Goethe, a Byron, or a Hölderlin could celebrate change as a departing of an embittered past. Wordsworth's neglected *Duddon River* offers a rich diversity of human experiences lived along the river—so rich, indeed, that this chapter could only offer a taste. Mary Shelley, Austen, Constable, Goethe, and Crévecoeur offer a Realistic-Romantic view of a humanized landscape that seems to correspond with that posited by Marx and found as existing today in Bali by Lansing.

As will be demonstrated in the next chapter, from among these poets, Goethe, Wordsworth, and Byron connect the river with the next quarter of the cycle—the sea. In addition, it will be discussed how Stolberg links the river with the lake. In the continuing analyses of works selected from the *oeuvre* of these poets, it will be seen how some employ the sea as an allegorical symbol for stagnation and death—in short, the end of a life's course.

CHAPTER THREE: RECEPTACLE

Like the flow phenomena treated in the last chapter, the lakes, seas, and oceans shaped the complexity of life in much of America and Europe, especially at their margins. Unlike flow, the receptacle is not so apparently bound up in the hydrological-cycle. The *flux et reflux* of lakes, seas, and oceans were for Romantic culture, however, such a compelling metaphor for change and, even paradoxically, for permanence that it will be as extensively treated as were streams—where the water phenomena flows from one place to another. Except at flood time, the linear pattern of the stream's flow from headwaters A to ocean B offers a vision of order. Orderliness, however, is not what first comes to mind in mid-ocean. The vast receptacle of water is often used in Romantic culture to represent order balanced on the edge of chaos. From the glacier or the ground, water has gushed forth and streamed down; now, in the form of streams, we will see it flow out into lakes and seas—the two types of receptacles that comprise this water-cycle quarter. Here the water reaches a state of equilibrium—or so it seems. Nevertheless, the flow, or what we might better call *flux,* continues even when the water seems to be held in its receptacle. Actually, this water is not received, let alone trapped for very long, because evaporation continuously moves the water of this quarter on to the next—the sky. As always, this view of the receptacle quarter is a theoretical simplification of the actual complexity of science and nature: water can evaporate from the ground and especially from trees—this is how a rain forest can generate so much humidity and, even, rain in its own micro-water-cycle. Even the artists allow the receptacle quarter its own complexity; they find in the lake its own peculiar trait that distinguishes it from the sea: a lake, as well as ponds, can be viewed not as a finality, but as a stage in the process of the main stream that feeds it from above and that, in turn, feeds down below. Whereas the lake usually represents placidity, the ocean represents seething turmoil—the size adding, of course, to its impact.

Placid or storm-tossed, this concentration of liquidity, this flux and reflux within one often vast expanse—this ocean or lake—particularly attracted the attention of painters and musicians, perhaps even to as great an extent as that of poets. Music clearly has the potential to suggest the flux and reflux of the sea.

Toward a Romantic Model

The first composer to combine an interest in the elemental vision of the world, to include prominently water in that vision, and most significantly, to have a rich harmonic and timbric pallet to render the effects necessary was Rameau. French opera, with its emphasis on scenic spectacle, was rich in occasion for such elemental visioning; and, importantly, before Rameau even set his first note to an opera, he had published several revolutionary books in music theory full of harmonic and melodic innovations—a useful reserve from which to draw his colors or vocabulary for recreating the elemental in aural terms. In describing one of Rameau's earthquake and volcanic eruptions in *Les Indes galantes,* Cuthbert Girdlestone writes:

> One cannot but be struck by the modernity of much of the writing. It is, indeed, in descriptive passages of battles, storms, monsters, 'frémissements des flots', that Rameau is most modern and

that we find strains and progressions that carry us on fifty years or more. The piano version of the opening of the earthquake music might almost be mistaken for a page from Liszt. The practice of description, even more than the desire to depict new emotional states, impelled eighteenth-century musicians to develop the expressive resources of their art.[1]

It might be added to the latter point that in the hands of a Rameau, a Beethoven, a Rossini, or a Liszt (as I will hope to later indicate) these descriptions of natural events correspond to emotional events. Also, I would like to take, as representative of Rameau's sea storm, the tempest from *Les Indes galantes*. In addition, I would like to employ as a model for the interpretation of such musical signs for storm elements the interpretation by Girdlestone: 'The storm scene in *Le Turc génereux* begins with a startling change from D minor to B flat'—this is a change from a key with only one flatted note to another key of five flatted notes, which is much further around the cycle of keys arranged in fifths—conceptualized on the basis of Rameau's theoretical writings. This cycle—so important to our understanding of the works of not only Liszt, but also Rameau—has been introduced and illustrated in the introduction. After this breathtaking modulation of High Romantic proportions, Rameau presents 'a description *symphonie* of tremolos and up-and-down rushes on the common chord, with some dramatic changes of key towards the end'. Following the opening measures, the heroine, standing on the shore, begins an impassioned commentary on the fate of the ship floundering in the storm offshore and of its sailors, which she declaims 'against the background of tremolandos and scales—the key is now G minor—interpret the fury of the elements as a symbol of her own agitation'. A chorus, representing the sailor on the distressed ship and singing offstage, 'adds physical terror to elemental fury and moral perturbation'. 'In the recitative', observes Girdlestone, 'which marks the subsiding of Nature's anger, all the string parts except the basses are in double stopping, a device which Rameau is said to have been the first to use in orchestral writing'. He then goes on to discuss an opera of 1774 that Rameau's storm scene influenced: 'A movement like the overture and opening chorus of Gluck's *Iphigénie en Tauride* is a descendant of this tempest of Rameau's where mingle descriptive *symphonie,* recitative and chorus'.[2]

Since the larger operatic context of the elemental passages is pure Rococo, Rameau cannot be claimed as a Romantic; however, the term 'Protoromantic' for these passages would serve to point how their composer developed for the Romantics the vocabulary to embody the receptacle quarter. Music, as well as literature, was throughout the rest of the century to celebrate the flux and reflux of the sea. In the lake and sea poetry, as with that of falls and streams, one finds the Romantic poets anticipating by half a decade the Symbolists' demand that poetry should aspire to the condition and quality of music. It is the reflective quality of water and the effect upon it of rich colors—particularly those of sunrise and sunset—that inspired the Proto-Impressionistic aspect of the Romantic painters Turner and, to a lesser extent, Delacroix. With subtle application of colors, but in particular, of varying brush applications of white paint, Constable sought on the sea coast precisely what he sought inland at the Stour Valley and Hampstead Heath—the nuanced evanescence

---

1 Cuthbert Girdlestone, *Jean-Philippe Rameau*, rev. ed. (New York: Dover, 1969) 337.

2 Girdlestone 328.

of nature-naturing. Poetry, especially that of Stolberg and of Heine, as we shall see, can also seek to exploit this reflective, coloristic side of still waters.

The sea in literature has received the most extensive scholarly and critical treatment; the sea in music and painting, much less. Most of the literary studies, however, treat the receptacle quarter as it is found in the works of individual authors or national literatures. The few that are comparative are of great value. Of these comparative studies, the most significant is by Paul Van Tieghem. Van Tieghem places considerable emphasis on the eighteenth-century neoclassical poets; he attributes to these poets the initial impetus—but not the full culmination—of sea poetry. This occurred when these poets began to imitate the description of sea storms of 'the Classical poet' / 'le poète Classique'. Van Tieghem expresses it as follows: 'It is through the description of tempests by which the sea makes its entrance into modern poetry'.[3] Surprisingly enough, however, in the descriptive-nature poetry of James Thomson and his imitators, Van Tieghem 'is surprised to see how rarely the sea occurs'. Even in Macpherson's *Ossian,* it is 'an element of landscape, without anything more'.[4] However, in other sorts of water images, both Macpherson and Thomson are very rich. Although Thomson's dates (1700–48) fall before the focal period of this study, I have found it helpful to draw from his *The Seasons* examples of descriptive verse in order to contrast his use of water with that of post-1760 poets.[5]

Van Tieghem finds in Stolberg's poetry a significant turning point in the poetic usage of the sea—from the description of the sea to a sincere, direct perception of and the resulting true feeling for the sea: 'Physical sensation is at the source of all actual feelings for nature; we can verify this once more by finding, in the works of Count Friedrich von Stolberg, the first texts where one can truly breathe the poetry of the sea'.[6] He uses the first stanza of Stolberg's 'To the Sea' / 'An das Meer' (1: 173–75) to support his argument: 'You holy and wide sea, what a glorious sight you are to me! I greet you in the early shaft of light which, with a shudder, kisses your lips!'

---

3 My translation of: 'C'est par la description des tempêtes que la mer fait son entrée dans la poésie moderne'. Paul Van Tieghem, 'Les Débuts de la poésie de la mer au XVIIIe siècle', *Neophilologus* 23 (1937–38): 437–43, qtd. from 438.

4 My translation of: 'est étonné de voir comme la mer tient peu de place'; 'un élément du paysage, sans plus' (Van Tieghem, *Neophilologus* 440).

5 I use examples of poetry from James Thomson as paradigms of descriptive verse advisedly, for as J.R. Watson observes: 'Even Thomson, however, could not sustain pure description' (*Picturesque Landscape and English Romantic Poetry* [London: Hutchinson, 1970] 33). See also Jean Hagstrum, *The Sister Arts: The Tradition of Literary Pictorialism and English Poetry from Dryden to Gray* (Chicago: U of Chicago P, 1958) 246.

6 My translation of: 'La sensation physique est à la source de tout sentiment vrai de la nature; nous le constaterons une fois de plus en trouvant chez le comte Friedrich von Stolberg les premiers textes où l'on respire vraiment la poésie de la mer' (Van Tieghem, 'Les Débuts de la poésie de la mer aux XVIIIe siècle', *Neophilologus* 23 [1937–38]: 437–43, qtd. from 441). This passage is reprinted with minor changes in vol. 4 of *Le Préromantisme* (Paris: Nizet, 1960) 206–07. See also Alfred Biese, 'Die Poesie des Meeres und das Meer der Poesie', *Preußische Jahrbücher* 88 (1897): 279–301, esp. 298–99.

Du heiliges und weites Meer,
Wie ist dein Anblick mir so hehr!
Sei mir im frühen Strahl gegrüßt,
Der zitternd deine Lippen küßt! (ll. 1–4).

Although he does not elaborate, Van Tieghem certainly is alluding to the enthusiasm of the persona, his expression of feelings, and his description of the seascape before him. Stolberg has certainly endowed these opening lines with an uncanny verbal music—created by a rocking and pounding iambic rhythm, by the recurrence of long vowel sounds (*ei, ie,* and *ü*), and by the half-rhyme echoes of 'Meer', 'mir', and 'hehr'. All this is clearly an attempt to recreate in verse the effect of the dynamic flux and reflux of the sea waves upon the speaker of the poem.

This swelling rhythm and these echoes continue in the following stanzas, where the hypnotic sights and sounds of the sea are put into words: 'Happy am I who, close to you, looked at you many thousands of times! Every time my glance is drawn back with the deepest feelings. I listen to you with drunken ear; my spirit rises up with you and sinks down with you into Nature's mysterious grave'.

Wohl mir, daß ich, mit dir vertraut,
Viel tausendmal dich angeschaut!
Es kehrte jedesmal mein Blick
Mit innigem Gefühl zurück.

Ich lausche dir mit trunk'nem Ohr,
Es steigt mein Geist mit dir empor,
Und senket sich mit dir hinab
In der Natur geheimes Grab (ll. 5–12).

This rising-up and sinking-down motion, here with an English Graveyard touch, is evident throughout this study so far and will be seen to an even greater degree later.

A strong Neoclassicism pervades the lyrics of Stolberg and the other leading Storm and Stress poets. In the former's 'To the Sea' / 'An das Meer', the persona remarks on the effect of sun-, moon-, and starlight on the sea, and does so with a four-squareness of construction. The two enjambments are false ones, each with a caesura at the end of the first line: 'When the sun bows down toward you and turning red comes to your bedside, then the roar of your waves sound the cradle song to the weary earth. The evening star listens to you and greets you at a distance; the moon smiles at you, when her light is refracted a million times on your surface'.

Wann sich zu dir die Sonne neigt,
Erröthend in dein Lager steigt,
Dann tönet deiner Wogen Klang
Der müden Erde Wiegensang.

Es lauschet dir der Abendstern,
Und winket freundlich dir von fern;
Dir lächelt Luna, wann ihr Licht
Sich millionenfältig bricht (ll. 13–20).

For Stolberg's persona-as-poet, the sea is a source of inspiration and images: 'Often I hurry out of the peaceful groves with great delight toward your waves, and plunge into you, and enjoy, refresh, invigorate myself'.

Oft eil' ich, aus der Haine Ruh',
Mit Wonne deinen Wogen zu,
Und senke mich hinab in dich,
Und fühle, labe, stärke mich (ll. 21–24).

This vision of the sunlit sea reminds the poet of Homer:

> The lord's spirit creates the poet, the earth gives him motherly nourishment; on the blue lap of your waves, his imagination rocks and waxes. The blind singer was standing by the sea; the waves roaring around him, and gigantic deeds of the golden age roared around him in holiday apparel. The inspiration came to him in the form of a melody on the wings of a swan, and the *Iliad* and *Odyssey* emerged singing from the sea. If he'd had eyesight, then heaven, earth, and sea would have disappeared around him; they sang in order to have heaven, earth, and sea come back before his blind eyes.

Der Geist des Herrn den Dichter zeugt,
Die Erde mütterlich ihn säugt,
Auf deiner Wogen blauem Schooß [sic]
Wiegt seine Phantasei [sic] sich groß.

Der blinde Sänger stand am Meer;
Die Wogen rauschten um ihn her,
Und Riesenthaten gold'ner Zeit
Umrauschten ihn im Feierkleid.

Es kam zu ihm auf Schwanenschwung
Melodisch die Begeisterung,
Und Ilias und Odüssee [sic]
Entstiegen mit Gesang der See.

Hätt' er gesehn, wär' um ihn her
Verschwunden Himmel, Erd' und Meer;
Sie sangen vor des Blinden Blick
Den Himmel, Erd' und Meer zurück (ll. 25–40).

The work is otherwise controlled and such images are scarce and diffusely scattered. No matter how strongly Romantic in certain respects, Stolberg—like many of his fellows of the Storm and Stress Movement—falls naturally and easily into a Classical appeal to Greco-Roman conventions—thereby, helping us to refine, even qualify, the otherwise useful Romantic/Classic dichotomy.

It is not the classical reference per se, but the sheer concentration of feeling and density that makes the difference. Stolberg's Classical-Romantic pacing of symbols becomes intensified in Coleridge's Romantic distillation. From these lines by Stolberg, Coleridge took the images for the following lines of the lyric, 'Fancy in Nubibus: Or the Poet in the Clouds' (1: 435). And it is in the intensity created by this selection and distillation of images that sets Coleridge's High Romantic imitation apart from its Early Romantic model. Coleridge concentrates on the aural aspects of the sea to the exclusion of the visual—appropriate to the blind poet, Homer, whose 'light' is 'inward':

Or list'ning to the tide, with closed sight,
Be that blind bard, who on the Chian strand
By those deep sounds possessed with inward light,

Beheld the Iliad and the Odyssee
Rise to the swelling of the voiceful sea (ll. 10–14).

What is diffuse in Stolberg's Early Romantic lyric takes on a density in Coleridge's High Romantic work.

There is a late eighteen-century example of this same density, but it is found not in verse, but rather in poetic prose: Rousseau's *Lac de Bienne* passage that we will examine extensively in the following section. Early as it is, it still qualifies as High Romanticism at least in its treatment of the lake symbolism. While, as we shall see, Rousseau at one point makes clearly symbolic use of water phenomena, this is the exception. Similar to Stolberg's nearly coeval lines but with a new concentration of expressive power, Rousseau's poetic prose combines the descriptive reproduction of the external phenomenon with the expression of the responding emotions of the poet.

The Lake Itself

Writing after Waterloo, Byron spends the greater part of canto III of *Childe Harold* delineating two mad geniuses whom he deems to be the most responsible for the madness his generation had just experienced and, indeed, thanks to attempts by Castlereagh and Metternich to turn back the political clocks of Europe, were still experiencing: first, representing 'Conquerors' is Napoleon; representing both 'Founders of sects and systems', as well as 'Bards', is Jean-Jacques Rousseau.[7] He sees in them, more than any other, the creators of the great age of chaos that Europe had just experienced.

Rousseau wrote a collection containing what might be termed prose-poems: *Reveries of the Solitary Walker / Les Rêveries du promeneur solitaire*. In the 'Fifth Promenade' / 'Cinquième promenade', sometimes referred to as the 'Fifth Rêverie'/ 'Cinquième Rêverie' (61–86), Rousseau describes the profound effect that the flow and flux of Lac de Bienne had once had upon his consciousness. He describes how he would leave the company of his wife and friends in order to seek the solitude and reveries which the lake offered:

> and while the others were lingering at table, I would escape and launch myself in a boat which I maneuvered to the middle of the lake where the water was calm, and there, lying full-length in the boat and looking up into the sky, I let myself go and slowly drift wherever the water carried me, sometimes for several hours, immersed in a thousand confused, but delightful reveries, which—although they had no clear or constant subject—were still in my eyes infinitely to be preferred to all that I had found most sweet in the so-called pleasures of life.

> et pendant qu'on était encore à table, je m'esquivais et j'allais me jeter seul dans un bateau que je conduisais au milieu du lac quand l'eau était calme, et là, m'étendant tout de mon long dans le bateau les yeux tournés vers le ciel, je me laissais aller et dériver lentement au gré de l'eau, quelquefois pendant plusieurs heures, plongé dans mille rêveries confuses mais délicieuses, et qui sans avoir aucun objet bien déterminé ni constant ne laissaient pas d'être à mon gré cent fois préférables à tout ce que j'avais trouvé de plus doux dans ce qu'on appelle les plaisirs de la vie (67).

---

7 W.J. Calvert, *Byron: Romantic Paradox* (Chapel Hill: U of North Carolina P, 1935) 146; Andrew Rutherford, *Byron: A Critical Study* (Standford: Stanford UP, 1961) 51–58.

Rousseau's repetitive rhythm of incantation certainly carries the passage beyond the merely descriptive. The French-language editor of my software terms this passage from Rousseau a 'phrase incessante', which is too long with too many conjunctions, and it advises a revision! Does Rousseau also mean to suggest the reiterated motion of the water with the resounding nasals? It is certain that this passage comes very close to what Irving Babbitt called 'Rousseauist revery' in which the persona 'would [...] become an oak tree and so enjoy its unconscious and vegetative felicity. [...] Rousseau's great discovery was revery; and revery is just this imaginative melting of man into outer nature'.[8] This state of 'Rousseauist revery' (reverie) of losing oneself in the contemplation of an external, natural object—here a lake—is to be found in another passage from the 'Cinquième promenade':

> When evening approached, I came down from the highpoints of the island, and I enjoyed going to sit at the edge of the lake, on the beach, in a secluded spot, there the sound of the waves and the agitation of the water captured my senses and chased from my soul all other agitation plunging me into such a delightful revery that night would steal on me unawares.
>
> Quand le soir approchait je descendais des cimes de l'île et j'allais volontiers m'asseoir au bord de lac, sur la grève, dans quelque asile caché; là le bruit des vagues et l'agitation de l'eau fixant mes sens et chassant de mon âme toute autre agitation la plongeaient dans une rêverie délicieuse où la nuit me surprenait souvent sans que je m'en fusse aperçu.

The nebulous, drifting quality of Rousseau's long sentence, with its continuously flowing final clause, perfectly captures the psychological state he is seeking to convey.

As Rousseau proceeds, the sense of permanence in change—conveyed by the ongoing process of the flux with its corresponding reflux and the continuous sound incessantly acting on the persona—seems, at first glance, to completely overwhelm him and draw him into the unity of things and to support Babbitt's view of Rousseau as being somewhat unconsciously Taoist: 'The flux and reflux of this water, its sound—continuous but oscillating—unrelentingly striking my eye and ear, supplanting the internal stirrings which reverie has extinguished in me and sufficing for making me sense with pleasure my existence, without active thought'.

> Le flux et reflux de cette eau, son bruit continu mais renflé par intervalles frappant sans relâche mon oreille et mes yeux, suppléaient aux mouvements internes que la rêverie éteignait en moi et suffisaient pour me faire sentir avec plaisir mon existence, sans prendre la peine de penser (68).

A careful examination of this passage reveals, however, that the described effect of the water does not induce an unconscious state of reverie but rather resonates off something inside Rousseau, who because he keenly and deeply feels the external phenomenon, the water, senses at that moment that he indeed *is*—far different from the desired egolessness of Taoism.

---

8 Irving Babbitt, *Rousseau and Romanticism* (1919; Austin: U of Texas P, 1977) 210. While Babbitt's views of Rousseau and his place in the origins of Romanticism still plays too important a role in the literary fortune of the latter in Anglo-American criticism to avoid confronting them, French criticism has produced more sympathetic, more sophisticated, and perhaps, more accurate analyses of Rousseau, his reverie state, and his 'Lac de Bienne'; for example, see Robert Ricatte, *Réflexions sur les 'Rêveries'* (Paris: Corti, 1960) 81–86.

The Toaist would probably have the sound and flux of the water drop entirely from the very consciousness that it has helped to empty. The flux and reflux of the water is present, but it is transparent to the consciousness; like a thin veil before a scene that holds a reader's attention, it is there and not there at once. The sound of water is apprehended on some level of semi-consciousness. Paul de Man offers this description of Rousseau's state: 'The sound of the water that Rousseau perceives [...] is the source, the power that permits us to see objects in their true dependency in relation to being and in their authentic dialectical role as a protective but nevertheless transparent obstacle'.[9] The 'relation to being' for Rousseau and for many after him lies in the constant tension between change and permanence, which are not antinomies in the old sense of desirable set against undesirable; a state of balance created out of their dialectical tension and offering compromise somewhere between the two extremes is what Rousseau seeks. The ground of Rousseau's thought is founded on meditation about the way things were in the past, how they are different in Rousseau's present, and how they might be in the future. At this point in the 'Fifth Promenade', Rousseau is on the verge of meditation about change. Soon Rousseau is about to experience the often sublimated thoughts of all deep-ecological writers and artists of Romantic culture.

Rousseau, thus, has yet another reaction, a third and even less passive reaction than the other two concerning the movement of the lake water: in fact, he actively uses the water's motion as a symbol for change, for the insecurity of human existence: 'From time to time, there would arise some vague and passing reflection on the instability of things of this world for which the surface of the water provided me a metaphor':

> De temps à autre naissait quelque faible et courte réflexion sur l'instabilité des choses de ce monde dont la surface des eaux m'offrait l'image [...].

Then, however, the uniformity of the water's movement or change pulls him back into a more passive state: 'but soon these airy impressions dissipated in the uniformity of the continuous motion which rocked me, without any active effort on the part of my mind, that I could hardly bring myself to return home at the set time and signal':

> mais bientôt ces impressions légères s'effaçaient dans l'uniformité de mouvement continu qui me berçait, et qui sans aucun concours actif de mon âme ne laissait pas de m'attacher au point qu'appelé par l'heure et par le signal convenu je ne pouvais m'arracher de là sans effort (68–69).

This uniformity of movement, this flux and reflux, this glimpse of permanence, paradoxically, in an element which has only just offered itself so readily to the persona as an image of change and instability, draws him back into his reverie.

Rousseau goes on to elaborate on change, on permanence, and on the importance of the events (or lack of them) that he experienced at Lac de Bienne in the formulation of his world view:

> these short moments of abandon and of passion, however powerful they may be and perhaps due to that strength, occur all too seldom in life. They are too rare and too fleeting to constitute a real state of being, and a happiness deeply yearned for, could not be merely composed of

9 Paul de Man, *The Rhetoric of Romanticism* (New York: Columbia UP, 1984) 39.

fugitive instances but rather be a simple and permanent state, which seems the height of felicity because duration, rather than impact, added charm.

ces courts moments de délire et de passion, quelque vifs qu'ils puissent être ne sont cependant, et par leur vivacité même, que des points bien clairsemés dans la ligne de la vie. Ils sont trop rares et trop rapides pour constituer un état, et le bonheur que mon coeur regrette n'est point composé d'instants fugitifs mais un état simple et permanent, qui n'a rien de vif en lui-même, mais dont la durée accroît le charme au point d'y trouver enfin la suprême félicité (69–70).

What Rousseau has expressed through the symbolic use of water he goes on to expound in prose of a more expository nature:

All is in a state of continuous flux on the earth. Nothing there is able to sustain a solid and fixed form, and our own affections, which focus on external things, pass away and necessarily change like those very things. Always just ahead or just behind us, our affections recall the past which no longer exists or anticipate a future which may never come to pass: there is nothing substantial enough for the heart to latch onto.

Tout est dans un flux continuel sur la terre. Rien n'y garde une forme constante et arrêtée, et nos affections qui s'attachent aux choses extérieures passent et changent nécessairement comme elles. Toujours en avant ou en arrière de nous, elles rappellent le passé qui n'est plus ou préviennent l'avenir qui souvent ne doit point être: il n'y a rien là de solide à quoi le coeur se puisse attacher.

Here they are: the three modalities or referents of change—past, present, and future.

The dilemma of Goethe's *Faust*—even down to the very wording of his wager with Mephistopheles—is broached here: either permanence or change. Faust and Rousseau both see in permanence an undesirable trap leading to a kind of damnation; but with Rousseau, there is just the slightest, ambiguous sense that permanence is a desirable but unachievable aspiration:

Thus one only finds fleeting pleasures here below; as for the lasting happiness—I doubt it can ever be known. There is hardly in our most delightful experiences an instant where the heart could in all honesty say to us: *I would like for this instant to endure forever*; and how can one give the term joy to a fugitive state that leaves our hearts restless and empty, that leaves us regretting something that is past or desiring something that is yet to come?

Aussi n'a-t-on guère ici bas que du plaisir qui passe; pour le bonheur qui dure je doute qu'il y soit connu. A peine est-il dans nos plus vives jouissances un instant où le cœur puisse véritablement nous dire: *Je voudrais que cet instant durât toujours*; et comment peut-on appeler bonheur un état fugitif qui nous laisse encore le cœur inquiet et vide, qui nous fait regretter quelque chose avant, ou désirer encore quelque chose après?

Here is the advanced codification of the Romantics' paradoxical attitude toward permanence and change: singly, neither is pleasing; each is threatening by itself. It is remarkable to find this ambivalence in the works of the person who would be hailed as the prophet of that greatest of all changes: the French Revolution.

Rousseau goes on to suggest the possibility that there may be something that has duration through recurrences, which might reconcile the opposites of permanence and change and offer thereby happiness to the Romantic consciousness:

But if there is a state where the mind can find a place of rest secure enough for complete repose and concentration, without needing to remember the past or anticipate the future; a place where time has no significance, where the present runs on indefinitely but without any sense of the duration or succession of moments, without any other feeling of loss or of joy, neither of pleasure or pain, of desire or fear—only the feeling of existence, a sense that fills our soul entirely; only so long as this state lasts, we can call ourselves happy, not with an incomplete, impoverished, and relative joy such as we find in the pleasures of life, but instead with a sufficient, matchless, and complete joy, which leaves in the soul no void to be filled.

Mais s'il est un état où l'âme trouve une assiette assez solide pour s'y reposer tout entière et rassembler là tout son être, sans avoir besoin de rappeler le passé ni d'enjamber sur l'avenir; où le temps ne soit, rien pour elle, où le présent dure toujours sans nèanmoins marquer sa durée et sans aucune trace de succession, sans aucun autre sentiment de privation ni de jouissance, de plaisir ni de peine, de désir ni de crainte que celui seul de notre existence, et que ce sentiment seul puisse la remplir tout entière; tant que cet état dure celui qui s'y trouve peut s'appeler heureux, non d'un bonheur imparfait, pauvre et relatif, tel que celui qu'on trouve dans les plaisirs de la vie mais d'un bonheur suffisant, parfait et plein, qui ne laisse dans l'âme aucun vide qu'elle sente le besoin de remplir (70–71).

What is the solution? It is simply a moment of contemplation about natural phenomenon that takes us out of ourselves. Schopenhauer would find this release in music, and his disciple Wagner would create music aspiring to the state of water, by means of the patronage of that great lover of fountains Ludwig II. Earlier, however, Wordsworth, Beethoven, and Shelley would find it in the nature of water itself. Rousseau set the themes for future Romantic variations: consolation lying in the permanence-of-recurrence found in the changeful processes of a water phenomenon.

The author/persona goes on to reiterate how he received intimations of this state of godlike permanence from the water of the agitated lake; also, he compares the flux of the latter to the flow of brooks and rivers: 'Such is the state which I often found in my solitary reveries on the Island of Saint-Pierre, where seated in my boat, I surrendered myself to the current, or seated on the shore of the wind-tossed lake, or otherwise, at the side of a beautiful river or of a brook burbling over pebbles'.

Tel est l'état où je me suis trouvé souvent à l'île de Saint-Pierre dans me rêveries solitaires, soit couché dans mon bateau que je laissais dériver au gré de l'eau, soit assis sur les rives du lac agité, soit ailleurs, au bord d'une belle rivière ou d'un ruisseau murmurant sur le gravier (71).

Of course, there is a tremendous dichotomy and potential tension between the psychological state based on stasis, on the one hand, and the locus of lapping waves and a rocking boat where the state is induced, on the other. Paradoxically, it is precisely this tension between opposites that induces the sense of permanence; the sense of a pivot point is made clearest when the wheel which rests on it begins to turn.

Water phenomena offered Rousseau what, in a later passage, he describes as the continuous, uniform movement to induce the placid state which intimates godlike permanence:

There must neither be an absolute stasis nor too much motion, but rather a uniform and moderate movement without jolts or disruptions. Without movement life is only a lethargic existence. If that movement is uneven or rough, then it arouses; in thus making us again aware of our surroundings, it destroys the charm of revery, and tears us from our inner self only to put us suddenly under the yoke of fortune and of men and makes us feel the sting of our sorrows.

Il n'y faut ni un repos absolu ni trop d'agitation, mais un mouvement uniforme et modéré qui n'ait ni secousses ni intervalles. Sans mouvement la vie n'est qu'une léthargie. Si le mouvement est inégal ou trop fort il réveille; en nous rappelant aux objets environnants, il détruit le charme de la rêverie, et nous arrache d'au-dedans de nous pour nous remettre à l'instant sous le joug de la fortune et des hommes et nous rendre au sentiment de nos malheurs (72).

It seems no accident that this thinker whose ideas were celebrated as harbingers of political change by the French Revolutionaries should feel the demand for change.

According to my reading of the Lac de Bienne passage, it is as much a pronouncement of Rousseau's ideological mind—if certainly not precisely a political statement, as is the *Social Contract*. Or to put the matter another way—this celebration of flux and reflux reveals the ideological matrix that contains the specifics of Rousseau's politics. I am reading not merely the surface of a cultural artifact, but rather, the 'subtext' of Rousseau and, even, of his culture. 'The type of interpretation here proposed', writes Fredric Jameson,

> is more satisfactorily grasped as the rewriting of the literary text in such a way that the latter may itself be seen as the rewriting or restructuration of a prior historical or ideological *subtext*, it being always understood that that 'subtext' is not immediately present as such, not some common-sense external reality, nor even the conventional narratives of history manuals, but rather must itself always be (re)constructed after the fact.[10]

Where better for the creator of a revolutionary world view to explore what Jameson terms 'the political unconscious' than in a rocking boat?

The next step is to see how widespread the various signifiers of Rousseau's Lac de Bienne passage either spread out from it attracted by a common sympathy or grew out of the common cultural ground. Just as did Rousseau, Goethe suggests in several early lyrics that the flux and reflux of water can be used to access some secret part of the psyche, and it can stand in opposition to the rational adult intellect glorified by the Enlightenment. Goethe even includes the rocking boat since the poem comes from the ideological, if not precisely politically defined, Storm and Stress movement.

Even more to the point is another of Goethe's Storm and Stress works—'On the Lake: Early Version' / 'Auf dem See: Frühere Fassung' and 'Later Version' / 'Spätere Fassung' (1: 56–57); in this lyric, a state which offers remarkable parallels with that depicted in the Lac de Bienne passage is induced in the persona: here the persona is also adrift in a gently rocking boat; here, too, he is open to the manifestation of external nature—indeed, so passively that the sensation can only be evoked by the primal symbol of the navel and umbilical cord: 'I now imbibe at my umbilical cord nourishment from the world'.

> Ich saug' an meiner Nabelschnur
> Nun Nahrung aus der Welt.

I can think of no symbol that offers a better sense of the poet's Pantheistic connection to the Nature/God. This umbilical-cord symbol distills the essence of the *deus(a) sive natura* of Spinoza, whose *Ethics*—according to the sixteenth book of Goethe's biography (*Dichtung und Wahrheit*)—was such a revelation for the young poet. Conceived at Lake Zurich, these spontaneously direct lines, dashed down upon the pages of a journal covering a Swiss jaunt, were later polished by Goethe into: 'And fresh sustenance, new blood I imbibe out of a free world';

10 Fredric Jameson, 'The Political Unconscious: Narrative as a Socially Symbolic Act' (Ithaca and New York: Cornell, 1981) 81.

Und frische Nahrung, neues Blut
Saug ich aus freier Welt [...].

These revised lines, when compared to the original ones, seem neither as striking nor as expressive of the connectedness of the subject/object connection between poet and nature; however, for the rest of the poem, the later version either remains close to the 'On the Lake: Earlier Version' or even improves upon it: 'How sweet She is, and good, this Mother Nature who holds me close to her bosom! The wave gently rocks our boat upwards to the beat of the striking oar, and mountains, cloud-enshrouded, rising skyward, meet us in our path'.

Wie ist Natur so hold und gut,
Die mich am Busen hält!
Die Welle wieget unsern Kahn
Im Rudertakt hinauf,
Und Berge, wolkig himmelan,
Begegnen unserm Lauf (ll. 3–8).

Nature here also draws the speaker out of himself.

In Rousseau's prose-poem, we have found that the escape was from the uncertainties of human destiny; here in Goethe's early verse, it is perhaps from the unhappy memory of a distant loved one who is the subject of the 'Golden Imaginings': 'Eye, my eye, why are you drooping? Golden Imaginings—are you coming back? Leave me, you Dream! golden as you are; here also there is love and life'.

Aug, mein Aug, was sinkst du nieder?
Goldne Träume, kommt ihr wieder?
Weg, du Traum! so gold du bist;
Hier auch Lieb und Leben ist (ll. 9–12).

Thus it should be noted that these golden dreams are not the same as Rousseau's reveries: the *Träume* have content, substance which is imaged by the mind; the *rêverie* has nothing beyond simple consciousness of the ego. If one can tamper with the older Goethe's stricture on High Romanticism, one might read the above lines as: Making nature the object is health; making the subjective self the object is unhealthy.

Health is available to the persona by his rendering himself open to the influence of natural phenomena. This state of mind finds its corresponding symbol in the reflecting property of the lake. The mirroring of the surface of the lake reproduces the natural objects in the sky above and on the surrounding shore with the same clarity which the persona and, by extension, the poet create through the descriptive imagery: 'On the water, a thousand floating stars glitter; all around, soft mist drinks up the towering distance; the wind of morning flies around the shadow-dappled bay, and in the lake, ripening fruit casts reflections'.

Auf der Welle blinken
Tausend schwebende Sterne,
Weiche Nebel trinken
Rings die türmende Ferne;
Morgenwind umflügelt
Die beschattete Bucht,

Und im See bespiegelt
Sich die reifende Frucht (ll. 13–20).

The mental state which Goethe is describing in the above lines is not the inward concourses of Rousseau's reverie, where outside influences are reduced to the motion and sound of the waves; it combines a close attention to the persona's psyche with an out-looking appreciation and description of natural objects. The poet brings in the link or sky quarter with the image of the mist slowly forming and veiling distant objects—here, again a water phenomena intervenes and acts upon the consciousness of the poet. This time, however, the object is obfuscating rather than clarifying perceptions of the subject.

The piano accompaniment for Schubert's 1817 setting of 'Auf dem See' is based on what might be termed an inward motif, an outward one, and a synthesis of these two; all three are quasi-ostinato in their effect. The inward figure that opens the song is a generic rocking motif: as much the rocking of the persona in Mother Nature's breast as the rocking of the boat. Schubert set the piece in a highly chromatic C major. The composer conveys the sense of stasis (the mirroring placidity of Goethe's lake) by his slow block harmony, with chords of quarter-note value in the four-beat measure (even if the time is split). Schubert creates a sense of the flux and reflux of water currents each moving after its own way; over the top of this, he has quarter-note motifs running in one direction and over the top of that sixteenth-notes; these latter set up their own often contrary pattern; then, over the top of both, eight-note figures move yet another way.

So far in this section the works of literature have been from the eighteenth century, and Schubert has conveniently transported us into the opening decades of the nineteenth.

Another Romantic inspired by Rousseau is Lamartine, particularly in his early *Poetic Meditations*. The Lac de Bienne passage seems to have influenced 'Le Lac' / 'The Lake', the most famous and enduring work of this collection—indeed, Lamartine's most frequently anthologized poem. The essential theme of the poem—mutability in the light of the permanence of nature—is introduced in the first stanza. Here, the natural world is treated metaphorically—'the eternal night' / 'la nuit éternelle' (l. 2), 'time's sea' / 'l'océan de âges'(l. 3)—to suggest the uncertainty of human fate in the eternal flux. In short, these metaphors of the first stanza maintain the tradition of the sublime language that is at least as old as the first-century Greek critic Longinus. Beginning with the second stanza, however, a new, albeit tentative, tone is struck. Natural objects, ceasing to be metaphors, have an existence all their own and are conveyed to the reader with a directness that, until their composition date, had not been heard in French poetry since the early seventeenth century. Although nature for Lamartine certainly lacks the concrete immediacy already found in the poetry of Wordsworth or Goethe, the lake—addressed as it is by the persona—is a natural phenomena and not an imaginary shepherdess or actual patron of the poet. Now, nature—not a hackneyed poetic trope—is the focus. Stanzas two and three picture a time when the persona sat by an alpine lake with his beloved.

Stanza four deals with a third Wordsworthian spot of time: the persona recalls a night when the beloved and he were rowing on the lake. This complex layering of three events is characteristic of Lamartine's obsession with time. Stanza five introduces the motif of a Nature sympathetically resonant with human relations. The persona's beloved begins to speak, causing the shore to be spellbound and drawing the attention of the waves. The beloved's reply in stanzas six through nine is in imagery and symbols still in the eighteenth-century tradition; personified Time is now addressed far more conventionally than in the persona's earlier address to the lake. Stanza nine is a two-fold culmination of the beloved's address. First, it gives clear expression to the *carpe diem* theme to which it has all been leading: 'So let us love, so let us love! let us hasten, let us enjoy the fleeting hour':

> Aimons donc, aimons donc! de l'heure fugitive,
> Hâtons-nous, jouissons! (ll. 33–34).

Second, the stanza returns to the opening image of the lyric, time as an expanse of water without a harbor: 'Man has no harbor, time has no shore; It flows, and we pass on'.

> L'homme n'a point de port. le temps n'a point de rive:
> Il coule, et nous passons! (ll. 35–36).

Time is eternal in its flow, and our impermanence is juxtaposed against it.

In stanza ten, 'jealous time' / 'temps jealous' is addressed by the persona, who asks how time can take away the same 'moments of passion' / 'ces moments d'ivresse' that it gives (l. 37). In the last four stanzas, the persona addresses the 'lake! mute rocks, grottos! dark forest!' / 'lac! rochers muets! grottes! forêt obscure!' (l. 49) asking them to keep alive the memory of the young couple's night on the lake. With measured rhythms, the poet asks all the different sounds of nature 'everything that is heard, seen, or breathed, let all say: 'They loved!''

> Que tout ce qu'on entend, l'on voit ou l'on respire,
> Tout dise, 'Ils ont aimé!' (ll. 61–64).

This musical voice of nature is a poetic credo that Lamartine repeats again and again in his poetry; the limpid rhythm and assonance that embodies this natural music is the cause of the great popularity of this poem.

Written in 1816, within a year of Lamartine's lyric and Schubert's art-song, Byron's lines on the theme of Lake Leman from canto three (2: 76–119) of *Childe Harold's Pilgrimage* share many strong parallels with Rousseau's passage on Lac de Bienne and Goethe's lines on Lake Zurich. Like them, Byron experiences moments of the passive perception of external objects; like Rousseau, he finds momentary escape from thoughts of human existence in contemplating the lake's surface; like Goethe, he finds his perception of natural phenomena heightened by having it reflected in a lake's smooth surface, and, by extension, suggests the relation of this mirroring reality to a similar reproduction of reality achieved by poetry. L.D. Yax calls attention to the following lines from the third canto of *Childe Harold* and to their significance in this work:

> Nature's pages glass'd by sunbeams on the lake [...] (canto 2: stanza 13).

> Lake Leman woos me with its crystal face,
> The mirror where the stars and mountains view
> The stillness of their aspect in each trace
> Its clear depth yields of their far height and hue:

There is too much of man here, to look through
With a fit mind the might which I behold;
But soon in me shall Loneliness renew
Thoughts hid, but not less cherished than of old,
Ere mingling with the herd had penned me in their fold [...] (68).

I live not in myself, but I become
Portion of that around me [...] (72).

'Lake Leman is symbolic of a poet's total vision', according to Yax. 'The lake is the mirror of nature and the looking-glass of the imagination through which the inner spirit and the external world are harmonized by the vision of unity [...] which the poet perceives as the ultimate truth [...] of the universe'. In directing his discussion to stanza 68, Yax writes that the persona's 'quiet contemplation of the scene is disrupted by thoughts of mankind, and he is prevented from achieving spiritual fulfillment by what is apparently a heightened awareness of human contention in the world'.[11] Yax is fully justified in finding in the above lines by Byron a certain dissatisfaction with Lake Leman as a poetic image: Byron, found a much more congenial image/symbol in the Cascata del Marmore of Terni, and he will do so again in his extensive hymn to the ocean at the close of the fourth canto. Lake Leman as portrayed in this passage is too tame and its environs too settled to meet Byron's romantic demands—especially the expression of his Satanic personality. In treating the mood of this passage, Marjorie Hope Nicolson expresses it this way: 'this is not Byron's customary mood, nor are these quiet scenes the most recurrent in either *Manfred* or *Childe Harold's Pilgrimage*. [...] To Childe Harold and to Manfred, Nature was more often in motion than at rest'.[12] Jerome McGann sees similar contradictions between this passage and others in *Childe Harold*.[13]

Byron, in short, has no great love for calm waters—either his vision of nature or the world view it conveys. Constable, Goethe, and Austen would rarely say, 'There is too much of man here'. He sees the lake as being analogous to a pen for livestock. Containment is not to be preferred to the chaos of the great ocean.

Since Byron returns with greater relish to the ocean in the next canto—indeed, he concludes the whole of *Childe Harold* with it—the following rejection of the 'ocean's roar', or even a stormy alpine lake, for the Leman in a placid state is unconvincing:

Clear, placid Leman! thy contrasted lake,
With the wild world I dwelt in, is a thing
Which warns me, with its stillness, to forsake
Earth's troubled waters for a purer spring.
This quiet sail is as a noiseless wing
To waft me from distraction; once I loved
Torn ocean's roar, but thy soft murmuring

11 Lawrence Donald Yax, 'Ocean and Other Water Imagery in Byron's Poetry', Diss. Case Western 1971: 146–47, 156.

12 Marjorie Hope Nicolson, *Mountain Gloom and Mountain Glory: The Development of the Aesthetics of the Infinite* (1959; New York: Norton, 1963) 385.

13 Jerome McGann, *Fiery Dust: Byron's Poetic Development* (Chicago: U of Chicago P, 1968) 118–19.

Sounds sweet as if a sister's voice reproved,
That I with stern delight should e'er have been so moved (85).

Perhaps, instead, it is only partly convincing—as Byron says in a short prose passage (containing a sea reference): 'I doubt sometimes whether, after all, a quiet and unagitated life would have suited me; yet I sometimes long for it'.[14] The rejection of the roaring ocean for the quiet lake is perhaps that side of Byron that admired the Enlightenment-Neoclassical style of Alexander Pope.

One must not go too far in comparing the Romantic-Neoclassicism of that essentially, if not entirely, pessimistic poet Byron to the optimistic Pope (at least when he penned *Essay on Man,* if perhaps not the third book of the *Dunciad*): Byron uses Lake Leman in a much darker and richer context in another poem, *The Prisoner of Chillon* (4: 4–16):

Lake Leman lies by Chillon's walls:
A thousand feet in depth below
It massy waters meet and flow [...] (ll. 107–09).

Here, then, it is not the reflections on the surface that concern the poet, but rather flux of that water in its depths. For François Bonivard the political prisoner in the Chillon dungeon below lake level,

A double dungeon wall and wave
Have made—and like a living grave (ll. 113–14).

The free and open flux of the lake also offers a sardonic foil for the enclosed prisoners:

Below the surface of the lake
The dark vault lies wherein we lay:
We heard it ripple night and day [...] (ll. 115–17).

According to Andrew Rutherford, 'Byron makes good use of Wordsworth and his own knowledge of Lake Leman to produce not pseudo mysticism but a simple picture of the scene, and a sense of the beauty, vitality, and joy of Nature [...] everything which participates in the life which has been going on while the hero was in prison, the life which he has learned to do without'.[15] This freedom/oppression dichotomy is never out of consideration when dealing with the water symbols of Byron—this victim of social convention in England and martyr to freedom in Greece. Water's potential for movement seems to set it off from the other phenomena he might have chosen for symbols.

A Hiatus in the Flow: The Lake in the Water-Cycle

Byron's placid, star-reflecting surface of the lake of the third canto is ultimately replaced by the titanic flux of the ocean, as well as by the Velino waterfalls, in the

14 Peter Gunn, *Byron: Seleted Prose* (Harmondsworth: Penguin, 1972) 439.

15 Andrew Rutherford, *Byron: A Critical Study* (Stanford: Standford UP, 1961) 73; see also McGann, *Fiery Dust* 171–72.

fourth canto. Byron neglects a greater dynamism that is to be found in the way poets dwell on the formation of lakes.

When in the first chapter we examined the Storm and Stress ode, Stolberg's 'The Rock Stream: Immortal Youth' / 'Felsenstrom: Unsterblicher Jüngling', we found it divided into inconsistent parts: one part employs the spring-as-falls per se as a continuous process with neither beginning nor end—a symbol for the ideal of eternal youth, whereas the second part uses the image of the stream flowing in an ever-changing process which moves like a piece of music from beginning to end—a symbol for the course of mortal life: 'Why do you hasten down to the green-tinted lake? Don't you feel fine next to a sky that is closer? fine amid the echoing rocks? fine amid the draping scrub oaks? Oh, don't rush so to the green-tinted lake! Now, youth, you are still strong, like a god! Free, like a god!'

Was eilst du hinab
Zum grünlichen See?
Ist dir nicht wohl bei'm näheren Himmel?
Nicht wohl im hallenden Felsen?
Nicht wohl im hangenden Eichengebüsch?

O, eile nicht so
Zum grünlichen See!
Jüngling, du bist noch stark, wie ein Gott!
Frei, wie ein Gott! (ll. 22–30).

The contrast between the stream in its high-mountain realm of freedom, set so near the sky, is juxtaposed with the lake of the flatlands down below; the wild genius existence is set against the placid bourgeois world below. Although not specified, the poet is probably thinking of the utilitarian, farming landscape resembling that of Constable, Austen, and the Goethe of Weimar Classicism.

A comparable theme by which a mountain stream might avoid being a symbol of damming up something—set carefully as it is in its geographic and climatic context with symbolic intent—appears in Goethe's Storm and Stress 'Song in Praise of Mohammed' / 'Mahomets Gesang' (1: 304–05), where the motif of the indomitable seaward drive manifests: the 'Brethren' stream by joining with the 'Prophet' river can save themselves from being lost in the desert or from becoming a marsh or pond: 'a hill dams us up into a pond!'

ein Hügel
Hemmet uns zum Teiche! Bruder [...] (ll. 44–45).

These lines of Goethe, like those by Stolberg before, offer the most perfect embodiment of a theme we have seen in Storm and Stress poetry and in Byron: the high-mountain stream as being emblematic of absolute freedom, liberty, and self-will. The Bacon/Newton-driven science of the time would have told the poet that—in the fullness of time, of the progress of science—all these unpredictable aspects of the stream's flow could be known and predicted. Stolberg and Goethe, however, intuitively anticipated something that scientists over two hundred years later would only dimly begin to understand: there is a sort of absolute liberty in the flow of a stream. It is now called 'randomness'—and it is free of the pathetic fallacy that would term it 'absolute liberty or self-will'.

I am referring to complexity theory. The study of the randomness of the flow of fluids is the focus of considerable complexity research in the physical sciences:

> What are the origins of random behavior? Brownian motion provides a classic example of randomness. A speck of dust observed through a microscope is seen to move in a continuous and erratic juggle. This is owing to the bombardment of the dust particle by the surrounding water molecules that are unseen and exist in great number, the detailed motion of the dust particles is thoroughly unpredictable. Here the web of causal influence among the subunits can become so tangled that the resulting pattern of behavior becomes quite random.[16]

Randomness, then, has replaced determinism. Water in streams, as indeed the entire hydrological-cycle behave the same way; after all, this cycle is, like the Brownian motion above, only water being activated by the application of heat. In the popular presentations of this fluid intricacy made by the complexity scientists themselves, they use the same macro-level water phenomena treated by our poets as a common means of making the concepts clear; and 'the flow of a mountain stream' is one such referent: 'imagine sitting by a mountain stream. The water swirls and splashes as though it had a mind of its own, moving first one way then another'.[17] Even these scientists fall into the rhetorical habit of attributing consciousness; but, of course, it is only as a figure of speech to show how water appears under ordinary scrutiny, while complexity theory posits, instead, a corollary indeterminacy concept, which they refer to as randomness. The insight of the early Goethe, Stolberg, and other Romantics is then truly remarkable. It is even more significant for understanding Romantic culture to observe that they, more often than not, employed the self-willed motion of the stream as a symbol of freedom and of revolt.

Goethe, however, changed his use of the symbol as his world view and lifestyle changed into that of a respectable member of the ruling class, a councilor to the Duke of Weimar. What in the earlier Storm and Stress work, 'Song in Praise of Mohammmed', is a thing to be avoided, becomes something to be desired in the Early Weimar lyric, 'Song of the Spirits' / 'Gesang der Geister' (1: 306–07). This poet implies the determination of the stream/man to move on in the natural course (ll. 18–22). There is something entirely new—unlike the individualism in Stolberg's lyric and in 'Song in Praise of Mohammed', there is a sense in the 'Song of the Spirits' of a higher will governing human fates. Concerning this, Boyd writes the following: 'No emphasis is laid on human volition. Man is now subject to cosmic laws, fulfilling his destiny according to these laws and impelled by his inner motives which conform to a higher will'.[18] The stream of the Weimar classical period begins in the rush and drive of a torrent and ends in the placidity of a lake—just as in

---

16 James P. Crutchfield, J. Doyne Farmer, Norman H. Packard, and Robert S. Shaw, 'Chaos', *Scientific American* Dec. 1986: 45–57; rpt. in *Chaos and Complexity: Scientific Perspectives on Divine Action*, eds. Robert John Russell, Nancy Murphy, and Arthur R. Peacocke (Vatican City: Vatican Observatory; Berkeley: Center for Theology and the Natural Science, 1995) 35–48, qtd. from 35–36.

17 Crutchfield 35, 37. In order to introduce their discussion of the complexity of fluidity, Peter Coveney and Rodger Highfield employ a print, 'The Great Wave off Kanagawa', by a Japanese contemporary of our Romantics—Katsushika Hokusai—*Frontiers of Complexity: The Search for Order in a Chaotic World* (New York: Fawcett, 1995) 43–45.

18 James Boyd, *Notes to Goethe's Poems* (Oxford: Blackwell, 1944) 1: 156.

human life, the passionate drives of youth are tamed into the tranquillity of middle and old age.[19] Here, it does seem that Goethe is employing the lake as a symbol of the poet who, according to Classical esthetics, mirrors reality; however, it is important to notice the unusual and extremely active image by which mimesis is conveyed in the last two lines: 'In its shallow bed, it meanders through the meadow-lined valley, and on the smooth-surfaced lake, all the stars revel in their image'.

Im flachen Bette
Schleicht er das Wiesental hin,
Und in dem glatten See
Weiden ihr Antlitz
Alle Gestirne (ll.23–27).

As the poet reminds us in the concluding lines (ll. 28–35), the water / human existence is subject to fate—here, represented as a thing subject to its stream-bed as it passes through different topography. In this, as in many of the poems treated in this chapter, the stream's flow and its place in the water-cycle are placed in the full complexity of their landscape, of their environment.

Similarly, 'Overwhelming Astonishment' / 'Mächtiges Überraschen' (1: 269) conveys this sense of fateful resignation to a higher shaping destiny. The youthful, impetuous brook of this sonnet 'wanders to the valley with unstoppable strength'. But, here too, self-will is to no avail; for, here Might, in the form of a landslide, stops the stream's youthful forward rush:

But like a demon suddenly tumbles—in great whirlwinds, hillside and forest follow—in order to find Oreas and contentment there, and stems the flow, forms the boundary of the wide basin. The water sprays and recoils in astonishment and retreats, swells uphill, drinking more and more of itself; the drive to unite with Father Ocean is stemmed.

Dämonisch aber stürzt mit einem Male—
  Ihr folgen Berg und Wald in Wirbelwinden—
  Sich Oreas, Behagen dort zu finden,
  Und hemmt den Lauf, begrenzt die weite Schale.
Die Welle sprüht, und staunt zurück und weichet,
  Und schwillt bergan, sich immer selbst zu trinken;
  Gehemmt ist nun zum Vater hin das Streben (ll. 5–11).

By the last stanza, it is clear that the stream's sudden metamorphosis into a lake is a symbol for the change from youth to middle age—'a new life': '[The stream] hesitates and stops, dammed into a lake; stars, mirroring themselves, look at the shimmering water which laps the cliff: [to see] a new life'.

Sie schwankt und ruht, zum See zurückgedeichet;
  Gestirne, spiegelnd sich, beschaun das Blinken
  Des Wellenschlags am Fels, ein neues Leben (ll. 12–14).

As in the 'Song of the Spirits', the lake takes on a smooth, star-reflecting surface after the storm and stress has subsided.

The image of the stream and that of the lake might be taken as symbols for the poet's transition from Storm and Stress to Weimar Classicism. One very good piece

19 Boyd 158.

of evidence for the interpretation of the calm lake as a symbol for middle age lies in the fact that it occurs in the concluding lines of Stolberg's 'The Rock Stream: Eternal Youth'. A mountain lake, the Wallenstädter See,[20] is set in opposition to a mountain falls to symbolize all that is opposed to youth: middle and old age and death or, what was considered worse, capitulation to society—that is to say, a defeat:

> Indeed, the peaceful calm beckons you below—the surging tremble of the silent lake—soon silvered by the swimming moon—soon golden and red in the western glow. Youth, what is silken rest, what is the smile of the friendly moon, the purple and gold of the setting sun, to him, who feels himself enmeshed in bonds of servitude? Now you still wildly flow, as your heart demands! Below it is sometimes the changeful wind that rules, or other times it is the stillness of death in the servile lake! Don't rush so to the green-tinted lake! Now, youth, you are still strong like a god! Free, like a god!

Zwar lächelt dir unten die ruhende Stille,
Die wallende Bebung des schweigenden Sees,
Bald silbern vom schwimmenden Monde,
Bald golden und roth im westlichen Strahl.

O Jüngling, was ist die seidene Ruhe,
Was ist das Lächeln des freundlichen Mondes,
Der Abendsonne Purpur und Gold,
Dem, der in Banden der Knechtschaft sich fühlt?

Noch strömest du wild,
Wie dein Herz gebeut!
Dort unten herrschen oft ändernde Winde,
Oft Stille des Todes im dienstbaren See!

O, eile nicht so
Zum grünlichen See!
Jüngling, noch bist du stark, wie ein Gott!
Frei, wie ein Gott! (ll. 31–46).

Here—as in the two poems by Goethe just examined—one finds the same image of the rush of water being converted into a pond or lake, and we see this image being used allegorically to express rebellious youth being reduced to the staid endeavor of middle and old age. Yet, in Goethe's poems of middle age, this conversion is a desirable event, whereas, it is anything but, in the Stolberg poem. Müller eloquently compares Stolberg's usage of and attitude toward the allegorical use of stream and lake to that of Goethe: Stolberg believes that the stream representing genius declines when it falls into the lake; for the middle-aged Goethe, it does so in order to really live for the first time.[21]

A passage in Byron's *Don Juan* presents itself as an interesting foil for Goethe's use of the dammed-up stream as a symbol of positive order through control. He first presents a river tamed into Denhamesque Neoclassical flow:

---

20 The name of the lake is in Stolberg's letter to Gerhard Anton von Halem, 19 June 1790, in *Briefe*, ed. Jürgen Behrens (Neumünster: Wachholtz, 1966) 262.

21 Richard M. Müller, *Die deutsche Klassik: Wesen und Geschichte im Spiegel des Strommotivs* (Bonn: Bouvier, 1959) 75

Before the mansion lay a lucid lake,
 Broad as transparent, deep, and freshly fed
By a river, which its soften'd way did take
 In currents through the calmer water spread
Around: The wild fowl nestled in the brake
 And sedges, brooding in their liquid bed:
The woods sloped downwards to its brink, and stood
With their green faces fix'd upon the flood (canto 13: stanza 57 [5: 541]).

According to C.M. Bowra, Byron 'catches the charm of an old English country-house in surroundings where nature has been tamed but only to become more truly itself'. Bowra is surely correct in seeing this as Byron's memories of his estate, Newstead Abbey, that he had left behind when he departed for the Continent. 'This', continues Bowra, 'is perhaps as a country gentleman of the Regency might admire it in his own domain, but it has its own charm and beauty and Byron's account reflects his typically English pleasure in nature as it works for the comfort of man'.[22] This, as I think the next stanza makes clear, is a passing bit of nostalgia for the respectable adulthood he might have lived in the eyes of his British coevals—for a placid maturity that might have been Byron's had he loved change and rebellion far less—for a life that might have been more on the heterosexual side of the Kinsey scale.

Byron, unlike Goethe, concentrates on the outflow from the small lake—on the escape of the water from its confinement:

Its outlet dash'd into a steep cascade,
 Sparkling with foam, until gain subsiding,
Its shriller echoes—like an infant made
 Quiet—sank into softer ripples, gliding
Into a rivulet, and thus allay'd,
 Pursued its course, now gleaming, and now hiding
Its windings through the woods; now clear, now blue,
According as the skies their shadows threw (13: 58 [541–42]).

The control is momentary and certainly not absolute. There is certainly no adulthood, but childhood is suggested instead. The images of 'a steep cascade' and the simile of 'like an infant' suggest a somewhat less stormy version of Stolberg's 'The Stream from the Cliff: Immortal Youth'.

Becalmed Waters

Over the next few sections of this chapter, I will be placing Romantic artifacts in a gamut between what might be termed a Romantic-style sea vision and a (Neo)Classical one. While a more Romantic-style sea vision can be found without too many qualifications, I have been hard pressed to find an extensive sea passage which might be taken as a Neoclassical exemplar. A totally tranquil sea has a sinister edge: it is the becalmed sea so feared by sailors during the day of sail-driven ships. In a 24 April 1796 letter, Jefferson seems to imply just such a becalmed-sea metaphor to

22 C.M. Bowra, *The Romantic Imagination* (1949; London: Oxford UP, 1961) 162–63.

distinguish the democrats from what he saw as the sinister reactionary moves of the Federalists—the latter being 'timid men who prefer the calm of despotism to the boisterous sea of liberty' (470).

The becalmed sea is an exceptional form of the receptacle quarter, which must be examined before we can turn to the sea itself; it is the sea's failing to perform its great part in *natura naturans*—its great ebb and flow: we must look at the becalmed sea of the doldrums. This lack of motion was to haunt not only the essentially Romantic minds of Coleridge and of Schubert but also the (Neo)Classical-Romantic minds of Goethe and Beethoven. Rousseau anticipated and codified their dislike of motionless water, when in *Reveries of the Solitary Walker* 'an absolute state of repose is not desirable' and that: 'Without movement life is only a lethargy'. What—in a rowboat on a lake—is distasteful, can—in a sailing ship on the wide ocean—be a nightmare vision of a normal fluidity quieted into sinister stasis.

Sensitivity to this tension between a static and a dynamic sea vision provides a key to understanding Goethe's Weimar Classicism. Too much motion is dangerous chaos, but—just as important—too much stasis is death. Nowhere is this dread of absolute stasis so clear as in Goethe's 'Becalmed Seas' / 'Meeresstille' and its companion poem 'Prosperous Voyage' / 'Glückliche Fahrte' (1: 47–48)—both of which are even more crucial because they also caught the imagination of contemporaneous composers, Beethoven, Schubert, and later Mendelssohn.

The usual translation of the title of this poem (as well as the compositions) based on it is 'Calm Seas and Prosperous Voyage'; however, 'Becalmed Seas' or 'Doldrums Seas' is more accurate for the title of this frightening vision of stasis: 'A profound stillness pervades the water, motionless rests the sea, and the worried captain anxiously scrutinizes the smooth surface of the water all around. Not a breeze from any direction! The horrible deathly calmness! In the vast expanse, not so much as a ripple stirs'.

> Tiefe Stille herrscht im Wasser,
> Ohne Regung ruht das Meer,
> Und bekümmert sieht der Schiffer
> Glatte Fläche ringsumher.
> Keine Luft von keiner Seite!
> Todesstille fürchterlich!
> In der ungeheuern Weite
> Rettet keine Welle sich.

In 1787, Goethe learned the dangers of being becalmed at sea; his recounting of his adventures off Capri aboard a ship that was being driven by a very stiff current, toward the vertically towering cliffs of the southern end of Capri with no wind, are very exciting by any travel-narrative standards (10: 405–10).

The vision is again in keeping with the *liete fine* so characteristic of Goethe's Weimar Classicism: the companion poem brings matters to a happy conclusion—'Glückliche Fahrt': 'The mists dissipates, the heaven is bright, and Aeolus loosens the straining ties. The winds murmur, the captain stirs. Hurry up! Hurry up! The waves part; the distance nears; already I see land!'

> Die Nebel zerreißen,
> Der Himmel ist helle

Und Äolus löset
Das ängstliche Band.
Es säuseln die Winde,
Es rührt sich der Schiffer.
Geschwinde! Geschwinde!
Es teilt sich die Welle,
Es naht sich die Ferne;
Schon seh ich das Land!

This counter-tragic approach to art, stemming as it does from Goethe's world view, is based on a movement from stasis to activity. I view this as the surface sign of a subtext. While it is true that the wind, in this 1795 poem, comes from outside, uncontrolled like the spirit ('Daemon' he termed it) that Goethe felt to drive his own activity, it is part of the dynamic social energy—for lack of a better term driving the French people and their national army—which Goethe witnessed firsthand at the Battle of Valmy (19 September Nachts, *Kampagne in Frankreich 1792, Werke* 9: 727). Like his own Daemonic powers of poetic inspiration, he did not trust the spirit of the French Revolution. But, in the not-too-distant future, he would admire it as it was embodied in the Daemonic energies of Napoleon. A far deeper revolution may also be suggested as another manifestation of energy: the scientific and the technological/industrial revolution.

Beethoven's *Becalmed Seas and Prosperous Voyage / Meeresstille und Glückliche Fahrt* has the same musical-emotive transformation from despondency to joy that he uses in the finale to Goethe's play *Egmont,* that fills all of *Fidelio* and the Third—written for the Consular, not the Imperial Bonaparte—the Fifth, the Ninth, and the last bars of the Sixth Symphony. Both Beethoven and Goethe share preference for the *liete fine* over the unhappy ending—hence, Beethoven's instinctive support for Goethe's counter-tragic, if still unhappy, ending to the Egmont story. They, however, have moved beyond the old Metastasian happy return to status quo; they have little use for the affirmation of the steady state of art and life of Baroque- and Rococo-Classicism. They both trust in the possibility of a change for the better—a central revolutionary impulse—yes, even revolutionary for the darling of the Weimar court and the other darling of the Hapsburg court; it is also crucial that they both had, in varying types, strong feelings of identity with the Napoleon type of enlightened despotism; Beethoven balked not at the Consulship—essentially a dictatorship—but only at the Emperorship; Goethe balked at neither. And at the basis of their belief in a change for the better must be a visceral sympathy for change.

Although an early conservative work, Mendelssohn's *Meeresstille und Glückliche Fahrt* offers in its opening a vision of the becalmed sea that has a darker and richer harmonic feel than Beethoven's. Schubert made the decision to set the 'Becalmed Sea' lyric of Goethe but not the optimistic 'Fortunate Voyage'. Although the reasons for this are not known for certain, this decision seems appropriate; if this is indeed based on pessimism, Schubert demonstrates thereby his closeness to Romantic contemporaries—like, for example, Turner, Coleridge, and Byron.

And in spite of the humor or perhaps in part because of it, Byron's *Don Juan* can be extremely dark. In the second canto, covering Juan's voyage out from Spain, the graphic description of a sea storm and shipwreck is followed by an even more horrifying description of the survivors trapped on a becalmed sea:

The fourth day came, but not a breath of air,
And Ocean slumber'd like an unwean'd child:
The fifth day, and their boat lay floating there,
The sea and sky were blue, and clear, and mild [...] (canto 2: stanza 70 [5: 110–11]).

The horror is made even more clear in a later stanza:

The seventh day, and no wind—the burning sun
Blister'd and scorch'd, and, stagnant on the sea,
They lay like carcases [...] (72).

In another poem, 'Darkness' (4: 40–43), Byron goes even further by presenting a vision of the seas becalmed when the sun hypothetically might no longer shine—that is to say when the great heat and water-vapor machine no longer has the heat of the sun as its great dynamo and comes to a halt:

The rivers, lakes and ocean all stood still,
And nothing stirred within their silent depths;
Ships sailorless lay rotting on the sea,
And their masts fell down piecemeal; as they dropp'd
They slept on the abyss without a surge—
The waves were dead; the tides were in their grave,
The moon their mistress had expired before;
The winds were withered in the stagnant air,
And the clouds perish'd [...] (ll. 73–81).

This terrifying vision of stasis was written in 1816, when the political stagnation of the Restoration and the Metternichian order would have really begun to seem oppressively fixed. It should be noted that other quarters of the water-cycle are included in what is Byron's most comprehensive use of that cosmic symbol.

The doldrums-becalmed sea offers the special horror in Coleridge's *The Rime of the Ancient Mariner*; all the epic machinery of ghosts and nature sprites that fill the poem pale by comparison:

Down dropt the breeze, the sails dropt down,
'Twas sad as sad could be;
And we did speak only to break
The silence of the sea! (ll. 107–10).

Both the (Neo)Classical-Romantic and the more Romantic artists demonstrate a horror of the becalmed sea. In general, the (Neo)Classical-Romantics feel the need to have the return of the breeze and a resumption of the voyage. Coleridge's very Romantic vision is no exception in this regard; but his protagonist's return sea journey is won only after great grief, and few modern readers of the poem consider the voyage fortunate. Further, Coleridge's High Romantic presentation of the supernatural has more in common with Novalis and Eichendorff than with the (Neo)Classical-Realist-Romantic conception of Goethe, who is most often interested in how natural phenomena really behaves.

In the cases just examined, the becalmed sea is an obvious threat to the pursuit of the voyage; the threat of being imprisoned in becalmed seas, left to starve, is patently clear. In the case of the next passage to be considered, however, the threat of the sea is subtler and more insidious.

Germaine de Staël mentions nature quite often in the abstract, but she rarely offers a concrete vision of nature; however, when she does, the significance is unmistakable. The emotional climax of Staël's *Corinne* is set in the Bay of Naples; Corinne has just given a poetic improvisation on Cape Miseno that will alienate her beloved's (Lord Nelvil's) affections permanently. The trip back over the calm waters of the Bay of Naples is described: 'The entire company embarked for the return to Naples. The warm, calm evening made everyone keenly aware of the pleasure of being at sea'.

> Toute la société s'embarqua pour retourner à Naples; et la chaleur et le calme, qui régnaient alors, faisaient goûter vivement le plaisir d'être sur la mer.

Staël makes a reference to 'Der Fischer', one of the many water pieces composed by Goethe that we do not have space to consider in this study: 'In a delightful romance, Goethe painted the fondness people feel for the water in hot weather. A river nymph praises the charm of her waves to a fisherman; she invites him to refresh himself and, seduced little by little, he plunges in at last'.

> Goëthe a peint, dans une délicieuse romance, ce penchant que l'on éprouve pour les eaux, au milieu de la chaleur. La nymphe du fleuve vante au pêcheur le charme de ses flots: elle l'invite à s'y rafraîchir, et, séduit par degrés, enfin il s'y précipite.

Now moving more associationally and lyrically than logically, Staël expresses the lurking fear of the sea, as she now focuses on its waves: 'This magic power of the wave bears some resemblance to the serpent's gaze, which attracts through fear'.

> Cette puissance magique de l'onde ressemble, en quelque manière, au regard du serpent qui attire en affrayant.

Even Byron may have nothing that so perfectly combines the fascination and terror of the sea as does this smile of the hypnotizing gaze of the serpent.

Then, the narrator's mind follows the wave as it approaches the shore, investing it with psychological significance: 'The wave, which rises in the distance and grows bigger as it rapidly approaches the shore, seems to correspond to the heart's secret desire, which begins gently and becomes irresistible'.

> La vague, qui s'élève de loin et se grossit par degrés, et se hâte en approchant du rivage, semble correspondre avec un désir secret du cœur, qui commence doucement et devient irrésistible (1: 777).

Staël's metaphor is most striking for offering a fairly accurate description of wave action, which demonstrates close empirical observation. Staël joins Byron and Turner among those who are both fearless and still pessimistic in the face of change. This passage expressing Staël's fascination for the sea predates the sea passages by Goethe and by Byron, to be considered next.

Goethe and the (Neo)Classical-Romantic Sea Itself

Motion, then, is essential to a positive sea vision. Just as the phenomenon of a torrential stream's being tamed into a placid lake fascinated Goethe, so too did the idea of bringing under control the most uncontrollable elemental force. As with

Goethe's river symbol in *Elective Affinities,* his other Weimar works precariously pose human strivings toward permanence at the edge of the chaos of mutability. Only the most uncontrolled flux could set into sufficient relief its controlled opposite; the tension between two unreconciled opposites is demanded in order to set Faust's chosen task in proper relief and thereby clarify its magnitude. The following epigram sums up the great challenge that the sea offers land-dwelling man: 'The sea always flows; the land can never contain it'.

> Das Meer flutet immer,
> Das Land behält es nimmer (*West-östlicher Divan* 3: 334).

To tame the sea is an impossible task, a Faustian struggle never to be completely won; thus, it is not surprising that Faust's endeavor entails taming the sea by means of dikes. It is this mighty surging and resurgence of the sea that Faust notices above all as Mephistopheles' magic bears him back from the Helena episode in Greece, the 'Classical-Romantic Phantasmagoria', in act 4. This sea passage was among the last composed in the great work: 1831.

In *Faust, Part Two* (5), Goethe draws on his own observations as a practicing scientist when he has his persona describe the wave motion upon the ocean shore: 'My eye was drawn to the high seas: it swelled upwards, towering itself on itself; it then subsided and poured down its waves, in order to storm the breadth of the flat shore'.

> FAUST. Mein Auge war aufs hohe Meer gezogen:
> Es schwoll empor, sich in sich selbst zu türmen,
> Dann ließ es nach und schüttete die Wogen,
> Des flachen Ufers Breite zu bestürmen (ll. 10, 198–01).

Immediately after and in contrast to this precise and objective description, the persona gives vent to the feelings aggravated in him by the uncontrolled flux of this changeful mass of water, which might be seen to anthropomorphically possess the passionate blood rather than the persona: 'And that aggravated me, seeing how willfulness puts the free spirit, who values all the laws, into a state of emotional unrest by means of a passionately excited blood'.

> Und das verdroß mich, wie der Übermut
> Den freien Geist, der alle Rechte schätzt,
> Durch leidenschaftlich aufgeregtes Blut
> Ins Mißbehagen des Gefühls versetzt (ll. 10202–05).

Here Goethe has a water phenomenon represent passions out of control; for, on at least one level, the sea represents the persona/poet wishing to bring the dæmonic forces within him under control; this link is supported by the ambiguity of attribution: precisely who possesses the passionately excited blood? Is it the ocean waves, the revolutionary, or perhaps the Byronic drives of society, or that rebellious side of Faust (Goethe, himself?) that must be dialectically resisted by the other, moderating side?

Faust's revolutionary drive is a drive toward order. Orderly change was Goethe's wish: the quasi-democratic stages of the French Revolution were too disorderly—even at its very start when it attracted many of the English poets—to ever draw his sympathies. *Hermann and Dorothea* (3: 163–243), for example, contains a horrible vision of the revolutionary rabble out of control: 'Quickly peaceful farm implements were now transformed into weapons; pitchfork and scythe were dripping with blood'.

Schnell verwandelte sich des Feldbaus friedliche Rüstung
Nun in Wehre; da troff von Blute Gabel und Sense (6: 72–73).

Only when France passed into the autocratic phase of its political metamorphosis, under Napoleon, did the ideals of the new France catch anything like Goethe's sympathies. Napoleon must have seemed a paradox of order and chaos; he was clearly in control, but who controlled him. To the Old Order, he was the agent of chaos—if a law, then one only unto himself. Any control, moreover, that Napoleon represented is teetering in a delicate equilibrium between two chaoses. Goethe embodied his attitude toward Napoleon in a short poem:

On Judgment Day, before the throne of God, the hero Napoleon stood at last. The Devil—who was a wonderfully despicable being—had a weighty brief against the same and his kindred; Satan began to read. God the Father, or God the Son—one or the other—spoke from the throne—maybe it was even the Holy Ghost who spoke the most: 'Don't be redundant before the divine ear! You speak like a German professor; we know it all—cut it short! On Judgment Day, it's only a fart. If you dare attack him, then go ahead and drag him off to Hell'.

Am jüngsten Tag, vor Gottes Thron,
Stand endlich Held Napoleon.
Der Teufel hielt ein großes Register
Gegen denselben und seine Geschwister,
War ein wundersam verruchtes Wesen;
Satan fing an, abzulesen.
Gott Vater, oder Gott der Sohn,
Einer von beiden sprach vom Thron,
Wenn nicht etwa gar der heilige Geist
Das Wort genommen allermeist:
'Du sprichst wie die deutschen Professoren,
Wir wissen alles, mach es kurz!
Am jüngsten Tag ist's nur ein [...] [Furz].
Getraust du dich, ihn anzugreifen,
So magst du ihn nach der Hölle schleifen' (2: 407–08).

Both Napoleon and Faust achieve their 'salvation' as competent men who against all the odds strive to achieve organized transformations of society; they can act ruthlessly and violently but always with a sense of an order in means and ends. Guiding purpose is set on one side of the tension against change; and, by this negotiation between the two, chaos is avoided.

The complexity goes even deeper: there is yet another dialectic—or rather balancing of antinomies—at work in this passage and, indeed, throughout most of *Faust*: this excerpted portion is a dialogue between Faust and the devil—or, much more precisely, a monologue by the one, intruded upon by the latter. Faust first makes continued empirical, scientific observations to check out his hypothesis; second, the movement of tide and waves further angers him because of the—to use

Stuart Atkins' phrase—'perpetual repetition'[23] of this movement: 'I thought it was a coincidence, I made more careful observations: the wave paused and then rolled backwards, drawing away from the proudly won goal; the time arrives—it replays the game'.

Ich hielts für Zufall, schärfte meinen Blick:
Die Woge stand und rollte dann zurück,
Entfernte sich vom stolz erreichten Ziel;
Die Stunde kommt, sie wiederholt das Spiel (ll. 10206–09).

All this is pure Baconian phenomenology and, for Goethe, only one pole of what science should be. In an aside to the audience, Mephistopheles reacts to the prosaic factuality of Faust's observations, which, of course, are not only essential elements of science but also can in the eyes of the layperson seem like the explanation of the obvious. The devil intends to undercut Faust's conception; however, he rather contributes to it by underlining the age-old process of the sea—one which has gone on for countless millennia: 'This is nothing new to me; I've known about it for a hundred-thousand years'.

MEPHISTOPHELES *ad spectatores.*
Da ist für mich nichts Neues zu erfahren;
Das kenn ich schon seit hunderttausend Jahren (ll. 10210–11).

This eternal recurrence—nihilistic when characterized by the devil—haunts Faust during the entire scene. After all, the verbal dialectic represents the thematic one of the scene: if Goethe advocates the creation of a Utopia, order in a new land, then the Devil must be seen as the advocate of the Ocean and of the chaos which it, in turn, represents. In short, dialectic between Devil and Visionary corresponds to that between land and ocean; the creation of Utopia, as well as art, works on the edge of failure, of destruction, of chaos. For Goethe, both forces—positive and negative—are needed.

And Goethe brings to bear considerable poetic art to create the aural side of the ebb and flow of the waves. By stringing together verbs in an extended conjunctive phrase ('schwillts *und* wächst *und* rollt *und* überzieht'), by phrases implying repetition ('*Well* auf *Well*e'), and by a recurrent alliteration of various combinations of *s, z, l, t,* and *w* sounds, the persona emphasizes this repetition, which is particularly odious to him because it 'accomplishes nothing': FAUST, *continuing passionately.* 'It steals up, at thousands-upon-thousands of shores—sterile itself, dispensing sterility: now it swells and waxes and rolls and covers the abhorrent area of that desolate stretch. There, full of power, wave on wave dominates, withdraws in upon itself, and nothing has been accomplished!'

FAUST *leidenschaftlich fortfahrend.*
Sie schleicht heran, an abertausend Enden,
Unfruchtbar selbst, Unfruchtbarkeit zu spenden;
Nun schwillts und wächst und rollt und überzieht
Der wüsten Strecke widerlich Gebiet.

23 Stuart Atkins, *Goethe's 'Faust': A Literary Analysis* (Cambridge: Harvard UP, 1958) 237.

> Da herrschet Well auf Welle kraftbegeistet,
> Zieht sich zurück, und es ist nichts geleistet! (ll. 10212–17).

The stage directions strategically repeat the word 'passionately' already found with ambiguous application in line 10204: Faust's (Goethe's and every human's) psychological duality is at issue here. Paradoxically, Faust is '*leidenschaftlich*' committed to controlling the waves that behave '*leidenschaftlich*': this is a key to Goethe's deep honesty and cutting irony because persona and phenomena are one and the same.

One would, then, clearly take a too-limited view of Goethe's lines by seeing them dominated by a distaste for the sea; they show his great effort to demonstrate the symbolic connection between his persona and the sea—especially, here: 'This could drive me to despair: the purposeless power of untamed elements!'

> Was zur Verzweiflung mich beängstigen könnte:
> Zwecklose Kraft unbändiger Elemente! (ll. 10218–19).

If one reads 'emotions', 'intellect', 'talents,' and 'poetic genius' in place of the German word 'Elemente'; or better yet, if one sees in the external elements analogues for the internal ones, then the following lines touch very closely Goethe's concern with self-discipline and self-restraint. And, of course, there is a social message too in this, which although not entirely illiberal, is certainly not revolutionary and involving political things 'untamed' or 'unrestrained' / 'unbändiger'.

Goethe's persona has, thus, clearly set forth his ideals: to circumscribe the profligate power of the sea. The task, however, is superhuman in that even if achieved, it must constantly be kept up past the course of one human life, and it must be passed on from generation to generation. All this is what Faust means when he says that, in the conception of such a task, his mind has gone beyond its mortal limitations: 'My spirit dares to transcend itself: here I aspire to struggle, this I aspire to overcome!'

> Da wagt mein Geist, sich selbst zu überfliegen:
> Hier möcht ich kämpfen, dies möcht ich besiegen! (ll. 10220–21).

There is, then, something clearly paradoxical in Faust's drive toward his goal.

An impossible Faustian struggle must begin and be sustained with a conviction that it just might be possible after all. The persona roots this conviction in scientific fact: 'And it is possible! Surge, as it may, it presses past every hill; however boisterously it may stir, a slight height of land proudly looms to thwart it, a slight depth powerfully attracts it'.

> Und es ist möglich! Flutend, wie sie sei,
> An jedem Hügel schmiegt sie sich vorbei;
> Sie mag sich noch so übermütig regen,
> Geringe Höhe ragt ihr stolz entgegen,
> Geringe Tiefe zieht sie mächtig an (ll. 10222–26).

About these lines Atkins writes, 'If, in explaining his project, Faust repeatedly emphasizes the obvious physical principle that water only runs downhill [...] it is because he plans to achieve [his goal] without recourse to magic, not like some successful King Canute but simply by application of the natural laws of physics'.[24] The actual King Canute was different than Faust: he knew he was unable to turn back the sea; he was putting on a demonstration for his courtiers who seemed to think otherwise. Goethe sees clearly that a succession of generations, like the Dutch, can indeed hold back the sea. This sense of being a part of an apostolic succession of creators is one of the great themes of many Neoclassical-Realistic-Romantics. And for all Mephistopheles' magic, Goethe's vision here is, at its philosophic ground, realistic.

The lines, however, that follow show that such a naturalistic reading is only correct up to a point, for Faust also needs the Devil's supernatural powers: 'In my mind, I rapidly made plan after plan: earn for yourself the rare enjoyment of locking the imperious sea out from the beach—of shrinking the borders of the watery expanse, force it further back upon itself! Already I have been able to work it out step by step; it is my wish: dare to promote it!'

> Da faßt ich schnell im Geiste Plan auf Plan:
> Erlange dir das köstliche Genießen,
> Das herrische Meer vom Ufer auszuschließen,
> Der feuchten Breite Grenzen zu verengen
> Und weit hinein sie in sich selbst zu drängen!
> Schon Schritt für Schritt wußt ich mirs zu erörtern;
> Das ist mein Wunsch: den wage zu befördern! (ll. 10227–33).

All Faust's scientific observations and all his 'Plan auf Plan' made 'Schritt für Schritt' are not enough to conquer the eternal change implicit in 'Well auf Welle'. And there is an even more interesting possibility: Mephistopheles' supernatural powers only constitute a symbol for the awe-inspiring possibilities inherent in human science and technology—but such an argument would go beyond the scope of this study.[25] Whatever the supposed nature of the devil's powers may be, in the end, they certainly do not accomplish the task, and Faust dies unsatisfied, leaving his stable and permanent utopia unachieved—and, yet, not undreamed; the conception, if not the completion, in the end, will prove to be his salvation. Faust's (Goethe's) *Weltanschauung* accepts change and places it in the permanent state of *becoming*; the impossible goal and the unending processes directed toward it are all-important; the more remote the goal, the greater the concept, and the greater the mind who conceived it; for what force on earth could offer a greater challenge than the sea?

Concerning Goethe's treatment of the sea—including that in *Faust*—Paul Van Tieghem writes that it 'counts little in the kind of works Goethe wrote, and his attitude toward the sea seems above all philosophical and scientific in its origin'.[26]

---

24 Atkins 238.

25 This problem of the nature of Mephistopheles' help is treated by various scholars. See, for example, Konrad Burdach, 'Das religiöse Problem in Goethes *Faust*', *Euphorion*, 33 (1932): 3–83; Gustav Würtenberg, *Goethes 'Faust' heute: Das Endes des faustischen Menschen* (Bonn: Hümmeler, 1949).

26 The sea 'ne compte pas pour beaucoup dans une oeuvre comme la sienne, et son sentiment de la mer paraît surtout d'origine philosophique et naturaliste' ('Les Débuts' 442); this passage appears as follows in Paul Van Tieghem, *Le Préromantisme*: 'Tout cela ne compte pas beaucoup dans une oèuvre aussi vaste que celle de Goethe, et ses impressions de la mer sont

Van Tieghem makes the common mistake of seeing science and poetry as separate antimonies; he thereby fails to see that Goethe (like his persona Faust) is passionately poetic in his science and that the poet uses the sea as a symbol of an elemental force that challenges Faust to tame it by striving and will. Melchior comes closer to the poetic, but also scientific, truth when he writes that Goethe's entire scheme of salvation ('Erlösungspläne') for his Faust depends on the sea.[27] Faust's task of taming the sea represents his attempt to control and direct his own elemental genius and drive. Atkins points up a considerable internal, formal reason for supporting what seems to be the considerable emphasis Goethe gives to the sea and the task it involves for Faust: 'The high importance of Faust's new seriousness of purpose is underscored by his shift to longer, five-stress lines'.[28]

Just as the waterfall passage, opening part two, marked a turning point from Faust's more common concerns with more common people to an encounter with the entire classical ethos, poesis, and mythos, so, too, does this sea passage mark another, perhaps more important, turning point for Faust: his days of study, of play, and above all, of womanizing are over. He turns now from the experiential to the active—first, in an emperor's service, then, directing the creation of a new land, by walling out the sea with dikes. Giving purpose to this element so symbolic of unrelenting change stands as the ultimate cosmic challenge for Faust. Eventually, during act 5, in Goethe's salvation plan for his Faust, after the bitter experiences and setbacks of his personal effort, and just before he dies, Faust has his utopian vision in which he speaks of his goal, now not merely personal, but rather as a 'Gemeindrang' (l. 11572) of generation after generation. This sociopolitical concept of the individual's place in the larger cycle or upward spiraling of collective endeavor is only touched on in act 5 of *Faust*; however, it is a key point of agreement between the dramatic epic and its dialectic counterpart, the *Wilhelm Meister* novel series—where the political points, only touched on in *Faust,* are elaborately developed; for our purposes, it is only necessary to make it clear how broadly significant the sea symbol is in *Faust*.

The stakes, however, for taming the sea still remain personal and psychological in act 4: by directing himself outward in act 5, he effects inward changes. It must be remembered that his desire to tame the sea symbolizes his desire to tame and direct his own self; he seeks to mold and shape the fluid element for utilitarian purposes in the same way that the persona in 'Song and Form' sought to mold water from the Euphrates for poetic purposes. As in the case of this lyric, it is the ultimate cosmic challenge for Faust to give permanent bounds to this element so symbolic of change; in short, Goethe's persona seeks to turn the fluid into what L.A. Willoughby will term a 'static, plastic, finite, a finished product'[29]—significant words that this scholar used to describe the products of Goethe's (as well as Schiller's) Weimar-Classical style. And yet this is not a Neoclassicism that Boileau or Pope would have

---

surtout, soit celles de l'artiste, soit celles du philosophe ou du naturaliste' (210).

27 Felix Melchior, *Heinrich Heines Verhältnis zu Lord Byron* (Berlin: Felber, 1903) 99.

28 Atkins 237.

29 L.A. Willoughby, *The Romantic Movement in Germany* (1930, rpt.; New York: Russell, 1966) 1.

understood: the static is not the end but merely one antinomy in an ongoing dialectical tension. The essence of Faust's sea vision is the yearning to bring under human control the most uncontrollable of earthly forces. This act entails the reconciliation of opposites, of polarities that Elizabeth M. Williamson and Willoughby perceive as the essential unifying strain in Goethe's entire corpus.[30] Indeed, the more dynamic the clash and dramatic the contrast, the better. Only the untamed flux would set into sufficient relief its opposite; the tension between the two antinomies is vital to clarify the magnitude of Faust's aspirations.

Faust's view that it is possible to rescue this area of low lands from the ocean clashes with the ever-negative view of Mephistopheles that it is not; this clash culminates right before the final visionary moment—the utopian epiphany of the dying Faust, who responds to the sound of workers' shovels. His supernatural proletarians provided by the devil are digging his grave; however, being blinded, he thinks they are excavating drainage ditches: 'How the clanging of shovels delight me! It is the masses, those who labor under my feudal power, uniting land again with land, imposing a limit for the billows, confining the sea in rigid bonds'.

> Wie das Geklirr den Spaten mich ergötzt!
> Es ist die Menge, die mir frönet,
> Die Erde mit sich selbst versöhnet,
> Den Wellen ihre Grenze setzt,
> Das Meer mit strengem Band umzieht (ll. 11539–43).

Ironically, of course, the grave marks the 'limit' of human existence in the way that ditches and dikes mark the finitude between the place of man and the wild ocean; however, Faust's dream of controlling human energy, and in its turn, to control the elemental goes beyond his life, as it goes beyond each generation. There is an apostolic succession of human effort passing from generation to generation, and this energy—collective down through time—transcends the momentary disaster of a breakthrough of the sea: within the month, the dikes are rebuilt better than before; within the decade, the flooded area is desalinated enough to become pasture.

The devil, however, is a slave to Christian eschatology, which always sees disaster coming at the end of any process. So he represents the other side of the ocean-border tension in the most negative terms, where the Christian 'us' is not disguised by the reference to a pagan deity (notice that even he has been doctrinally demonized): 'And yet, with all your dykes and levees, your striving serves none but us; indeed, you're in the process of preparing a grand feast for the water-devil, Neptune. In every respect you are doomed; the elements have sworn to our cause; the end of the struggle will be annihilation'.

> Du bist doch nur für uns bemüht
> Mit deinen Dämmen, deinen Buhnen;
> Denn du bereitest schon Neptunen,
> Dem Wasserteufel, großen Schmaus.
> In jeder Art seid ihr verloren:

30 Elizabeth M. Williamson and L.A. Willoughby, *Goethe: Poet and Thinker* (London: Arnold, 1962) 113–14, 188–96, et passim.

Die Elemente sind mit uns verschworen,
Und auf Vernichtung läufts hinaus (ll. 11544–50).

And of course, there is something sinister in all this, beyond the feudalism, devil, magic, and Christianity, beyond too even the technology and science they also represent, they signify the new force of capitalism and—if truth is told—of its not-too-distant cousin, the corporate (so-called socialist) state: 'Use every means you can to produce masses and masses of workers! Use pleasure and discipline to spur them on! Pay them, entice them, force them!'

Wie es auch möglich sei,
Arbeiter schaffe Meng auf Menge!
Ermuntere durch Genuß und Strenge!
Bezahle, locke, presse bei! (ll. 11545–50).

There is a further tension, indeed, contradiction that—as these lines make clear—Goethe understands: by attempting to create a certain stasis in his fictional Netherlands, Faust must submit to the forces of change that change further. It is this change that Goethe feared by temperament and that Byron welcomed. Lord Byron would certainly never want to confine anything 'with rigid bonds'—let alone the ocean.

There is a tension between the disorder of the sea and the realm of stability that Faust wishes to impose at its edges. Goethe, however, seeks out these seemingly self-contradictory symbols. The symbols convey a sense of being precariously balanced. Goethe designed a garden monument that consists of a stone ball apparently perched on top of a cube; illusionistically ready to roll off at the slightest jar, the ball's instability offers the appropriate symbolism for what Goethe entitled *Altar of Good Fortune* (1777).[31] It is this sense of the ball's resting precariously—of its balancing act between edges of disorder—that makes Goethe's monument and his seaside Utopian fiction into Neoclassic-Romantic symbols. Here again, my research data are seen in light of Lansing's complexity praxis and Foucault's anti-binary heuristic forces qualification of the Neoclassical/Romantic dichotomy: it exists, but only as a sliding scale. In such an intellect as that of Goethe—one which is both Romantic and (Neo)Classical—the opposing elements of each style pull in opposite ways to produce a highly expressive instrument.

Byron and the Romantic Sea Itself

The Faustian urge to control the sea's daemonic force sets the (Neo)Classical-Romantic vision apart from that of Byron, who celebrates the fact that the sea is beyond human control. More, I contend, is at stake than merely the treatment of one symbol: I term Goethe's use (Neo)Classical-Romantic and Byron's High Romantic. Further, if we consider the latter's lake passage treated earlier in this chapter, then we might call the great lover of the verse style of Pope a Romantic-Classicist. It is precisely the sort of symbolic analysis that I just undertook above with a passage

31 Hugh Honour, *Neo-classicism* (Harmondsworth: Penguin, 1968) 129–30.

from Goethe and will undertake below with one from Byron, that can bring one closer to a sensitive and valid use of period and style terms.

Byron's *Childe Harold* was written in 1818 and antedates the older poet's lines, which were composed—according to Ernst Beutler—in 1831.[32] It would be of considerable interest if one were able to prove conclusively that the English work influenced the later German one. As E.M. Butler, however, has observed that this 'most essentially Byronic of all the works' was either 'ignored' or 'by-passed' or 'escaped Goethe's notice entirely'. If he had read it, he might well have recoiled from it, moreover, as he did from *Don Juan*.[33] Just as with influences, simplistic parallels are not key here; once again, it is the contrasts between the passages that prove the more revealing. One partial contrast is that of genre: while it is true that both draw on conventions and receive great narrative energy from being quasi-epics, nevertheless, Goethe has cast his work in dramatic-dialogue form. This difference is important because his Mephistopheles can bring in a considerable element of the Byronic to offset the profoundly Goethean; this is true less of the passage we just examined, but later in *Faust, Part Two,* the devil celebrates, in a pirate song, the sea as a free and a thoughtless place for escape from moral and political restrains: 'The open sea frees the mind; who knows the name of prudence there!'

Das freie Meer befreit den Geist;
Wer weiss da, was Besinnen heisst! (II. 11177–78.).

This passage bears an uncanny resemblance to lines from the 1814 pirate song in Byron's *The Corsair* (3: 148–214):

'O'er the glad waters of the dark blue sea,
Our thoughts as boundless, and our souls as free'[...] (II. 1–2).

---

32 Ernst Beutler, 'Einführung', in vol. 5 of Goethe, *Gedenkausbage der Werke, Briefe, und Gespräche,* eds. Beutler et al. (Zürich: Artemis, 1949) 671–752.

33 E.M. Butler, *Byron and Goethe: Analysis of a Passion* (London: Bowes, 1956) 191–96; Butler's study is essentially an influence study and does not compare the two passages—nor to my knowledge do any of the following studies which have proved to be of the most general sort of use. While not explicit in the ocean passage treated in the text, this theme of escape to the sea is an important theme which permeates Byron's entire corpus. Yax gives this theme a more extensive treatment than I am able to do in these pages: 'Nearly all Byron's heroes', states Yax, 'explicitly prefer a life on the ocean to an existence in the city where natural spirits are imprisoned by social institutions' (25). Concerning the sea as a milieu for moral conflicts within the personality of the Byronic hero, see Bernard Blackstone, 'Guilt and Retribution in Byron's Sea Poems', *A Review of English Literature* 2 (1961): 58–69; rpt. in West, *Byron* 31–41. On the same subject also see the following: W.H. Auden, *The Enchaféd Flood, or The Romantic Iconography of the Sea* (New York: Vintage-Random, 1950) 19–20; Gaston Bachelard, *L'Eau et les rêves: Essai sur l'imagination de la matière* (1942; rpt. Paris: Corti, 1970) 228–29; Kenneth Clark, *Civilization: A Personal View* (New York: Harper, 1969) 293. Martin Opp makes the interesting point that Byron demonstrates in his works a particularly intense and personal relationship with the sea that sets him off from the more generalized 'sympathy for the sea' of the other English Romantics, in *Die See im der Dichtung der englischen Romantiker: Ein Beitrag zur Geschichte der Naturauffassung in der englischen Literatur* (Weida in Thür.: Thomas, 1913) 40–41.

Byron's passage predates Goethe's lines by at least a decade; also *The Corsair* would have been a poem that Goethe would probably have read. This Byronic illusion, like the many found in *Faust, Part Two,* is certainly intentional; it, like the others, is set in a world view antithetical to that of Goethe.

Piave adapted Byron's poem into a libretto for Verdi's *Il Corsaro*. There has been a tendency to undervalue the literary qualities of Piave; however, I think he often shows an ability to translate the original poets and adapt them to a new form with a real fidelity for the spirit of the original. This is nowhere more true than in his chorus for Pirates, which opens the opera: 'As freely as the winds wing their way over the vast plains of the sea, the bold corsairs rush to find battle and booty'.

> Come liberi volano i venti
> Per le immense pianure de' mari,
> Cosi corron gli arditi corsari
> Pugna e preda sull'onde a cercar.

Verdi's broad, Dionysiac setting of these lines exactly catches the wildness of the sea and these free spirits who sail on her surface.

Nor were Goethe, Piave, and Verdi the only beneficiaries of the opening lines of Byron—nor of the larger world view behind them. Jane Austen—who as I have said before rarely used natural scenery unless it reflected in some very significant manner on the narrative—decided to set the climax of the first volume of her *Persuasion* at Lyme Regis, with its technologically impressive application of wave hydraulics to harbor construction—the Cobb. (I place a great deal of stress upon the aesthetic importance of the volume structuring of the novel. We must remember that the volume presentation is a matter both of the presentation of the novel-as-produce for the consumer and its literary form.) It is by the sea that there takes place the fortuitous encounter with Anne Elliot's cousin and heir to her father's estate and Baronetcy, a person who will have crucial importance in the final volume of the novel. In addition, the breeze of the ocean (and surely, viewing it, too) has had a significant effect upon Anne's physiognomy or, what is related but more important, her psychological well-being, which soon reflects itself outwardly:

> When they came to the steps, leading upwards from the beach, a gentleman at the same moment preparing to come down, politely drew back, and stopped to give them way. They ascended and passed him; and as they passed, Anne's face caught his eye, and he looked at her with a degree of earnest admiration, which she could not be insensible of. She was looking remarkably well; her very regular, very pretty features, having the bloom and freshness of youth restored by the fine wind which had been blowing on her complexion, and by the animation of eye which it had also produced. It was evident that the gentleman, (completely a gentleman in manner) admired her exceedingly. Captain Wentworth looked round at her instantly in a way which shewed his noticing of it. He gave her a momentary glance,—a glance of brightness, which seemed to say, 'That man is struck with you,—and even I, at this moment, see something like Anne Elliot again' (*Works* 5: 104).

One cannot help seeing in Wentworth's alert response, quick perception of the situation, and eye semiotics something Byronic—the Byronic image that was in great part based on *The Corsair*.

According to my reading of the text, Jane Austen alludes to *The Corsair* in the course of Anne's discussing Romantic poetry with another sea captain—Captain

Benwick. During their discussions that concentrate on the poetry of Sir Walter Scott and that of Lord Byron, Anne Elliot qualifies her admiration for their Romantic poetry by saying that it needs to be supplemented with prose (*Works* 100–01). Here, Austen indulges in the 'critique on Walter Scott' that she seems to wish she had added to *Pride and Prejudice* for 'a contrast' with its 'playfulness' (*Letters* 203). She certainly admires the talents of Scott, both as a poet and a novelist, when she even writes of him in an ironically grudging mode of admiration (*Letters* 277). Her letters, however, are no help in measuring the sincerity of her admiration for Byron; we know she read *The Corsair,* but since it is mentioned in the same breath with a mended petticoat, it is hard to put any interpretation on this (*Letters* 257).

In *Persuasion,* she makes a direct reference to *The Corsair* during the groups 'adieus to the Cobb':

> Anne found Captain Benwick again drawing near her. Lord Byron's 'dark blue seas [sic]' could not fail of being brought forward by their present view, and she gladly gave him all her attention as long as attention was possible. It was soon drawn per force another way (*Works* 5: 109).

Then comes the climax of the first volume of the novel when Louisa takes a foolhardy leap off 'the high part of the new Cobb'. This is an event that seems to lock Wentworth into a loveless match with Louisa; but also gives him yet another insight concerning Anne's character. It might also be that Louisa's impetuous act is, in some part, due to a sea change—a spirit of reckless abandon. John Halperin states that 'It is no coincidence that after a general conversation about Scott and Byron—Byron's 'dark blue seas' have just been mentioned—Louisa, in an excess of romantic sentiment at Lyme, jumps. [...] Here is the fruit of excessive romanticism'.[34] This is an apt reading of the reference to *The Corsair* in terms of Louisa's jump and excessive romanticism. This is, however, only one half of the dialectic between what we might loosely term the Bryronic and the Goethean tensions within the novel. The sea and the reference to Byron also have a larger significance within the narrative of the life and love of not Louisa, but of Anne Elliot: the sea, its winds, and the talk of Romantic poetry give her the moment of reaffirmation before she faces the full force of the other side of the tension—the chilly Neoclassical architecture and decorum of Bath. Just as the 'dark blue seas' offer a locus wide enough for the Byronic corsair, so too does the sea a place of self-liberation and self-affirmation for the much wiser Anne Elliot, with her qualified, more prosaic Neoclassical-Romanticism.

Leaving style aside, Byron's attitude to the sea is, compared to Austen and Goethe, extreme and high romantic delight in oceanic flux. The persona of this passage delights in contemplating the uncontrolled flux of the ocean: Byron's famous *Childe Harold* sea passage (stanzas 179–84) from the fourth canto (2: 120–86). The persona is not aggravated by the uninhibited and unlimited action of

34 John Halperin, *The Life of Jane Austen* (Baltimore: Johns Hopkins UP, 1984) 303. Jane Millgate finds some possible reactive or dialectic influences that the novels, not the poetry, of Scott may have had upon the narrative strategies of *Persuasion* ('Prudential Lovers and Lost Heirs: *Persuasion* and the Presence of Scott', *Jane Austen's Business: Her World and Her Profession,* eds. Juliet McMaster and Bruce Stovel [Houndmills: Macmillan; New York: St. Martin's, 1996] 109–23, see esp. 113–15).

the ocean waves. The opposite is true: the persona-narrator delights in contemplating the uncontrolled flux of the ocean. Just as the persona of Byron's 1816 Fragment bids the river to flow, so too does the one here bid the ocean:

> Roll on, thou deep and dark blue Ocean—roll!

Rather than desiring, as Faust does, to set bounds to the ocean, Byron's persona takes delight in contemplating that

> Ten thousand fleets sweep over thee in vain;
> Man marks the earth with ruin—his control
> Stops with the shore [...].

In these lines Byron establishes an important dichotomy—one between the land, the realm of humankind, the high seas, and the realm of the almighty God. As already indicated in this study, it is the seaside of the antinomy in which the poet's persona takes great delight. In Byron's conception, humans are great spoilers of the land, but not of the sea.

Mankind as ravager of land, but not the sea appears again in the lines following. Also, here, a simile is drawn between man and a drop; then, the drop is compared with the ocean—the smallest body of water relative to the largest:

> —upon the watery plain
> The wrecks are all thy deed, nor doth remain
> A shadow of man's ravage, save his own,
> When, for a moment, like a drop of rain,
> He sinks into thy depths with bubbling groan,
> Without a grave—unknell'd, uncoffin'd, and unknown (179).

As with the rivers in 'Mont Blanc' and 'Kubla Khan', the sea is a foil for the paltry things of man. The persona delights in the fact that the sea is a law unto itself and a realm in which the human creature is out of his element:

> His steps are not upon thy paths,—thy fields
> Are not a spoil for him,—thou dost arise
> And shake him from thee; the vile strength he wields
> For earth's destruction thou dost all despise,
> Spurning him from thy bosom to the skies,
> And send'st him, shivering in thy playful spray
> And howling, to his Gods, where haply lies
> His petty hope in some near port or bay,
> And dashest him again to Earth:—there let him lay (180).

Byron seems to take sadistic delight in seeing humans suffer from the cruelly playful ebb and flow of the sea.

Not just in this stanza, but in the one that follows the tone becomes so contemptuous as to be downright misanthropic. A 'vile strength', 'petty hope', 'vain title', and 'toys'—all these belong to what is termed a 'clay creator'—a base god:

> The armaments which thunderstrike the walls
> Of rock-built cities, bidding nations quake,
> And monarchs tremble in their capitals,
> The oak leviathans, whose huge ribs make
> Their clay creator the vain title take

Of Lord of thee, and arbiter of war—
These are thy toys, and, as the snowy flake,
They melt into thy yeast of waves, which mar
Alike the Armada's pride or spoils of Trafalgar (181).

Thus, the persona takes particular delight in the fact that the ocean is a law unto itself, or unto God, and much of the fascination of the persona for the ocean is conveyed by the many 'o' sounds and other rounded vowel sounds. G. Wilson Knight has commented on Byron's passage as follows: 'No English poet has written so finely of the sea [...] in the rolling volumes—got by 'o' sounds'.[35] It is this assonance that, in great part, creates the feeling of wild abandon with which Byron's persona addresses the ocean and contemplates the larger implication of its flux and reflux. It is precisely this use of assonance that marks a key technical, perhaps even an ideological, contrast of Byron's passage with Goethe's: Byron may be showing his sympathy to the action of the waves with the softer effect of assonance; whereas Goethe draws upon the harsher effects created with alliteration.

Byron's persona expounds upon how empires constantly change, but one thing does remain constant: they retain their power by holding sway, if always nominally, over the sea:

Thy shores are empires, changed in all save thee—
Assyria, Greece, Rome, Carthage, what are they?
Thy waters washed their power while they were free,
And many a tyrant since; their shores obey
The stranger, slave, or savage; their decay
Has dried up realms to desarts [...].

Jerome McGann pinpoints the great importance of this stanza to understand *Childe Harold's Pilgrimage* and the poet's cyclical world view: 'This idea of an unlimited life flow, a continuum in which Nature and History unfold themselves in successive birth and death cycles, is most forcefully presented in the address to the Ocean at the poem's conclusion, Stanza 182 is a particularly functional addition to this famous series of stanzas'.[36] Unlike the life cycles of these empires, the sea is 'unchangeable' in the sense that its vast bulk of water and its processes seem changeless to human perceptions; however, paradoxically, this essential changelessness is closer to the actuality of the ocean in it appearance as an untamed mass of ('wild') flux and reflux:

—not so thou,
Unchangeable save to thy wild waves' play [...].

Here, again the perception of the poet sees deeper into the reality of things than his contemporaneous science: the waves do, indeed, move with the randomness and

35 G. Wilson Knight finds this phenomenon throughout all of Byron's sea poems (*The Burning Oracle* [New York: Oxford UP, 1939] 197–215; rpt. in *Byron: A Collection of Critical Essays*, ed. Paul West, Twentieth Century Views [Englewood Cliffs: Prentice, 1963] 15–41, qtd. from 29).

36 McGann, *Fiery Dust* 135.

incomprehensible arbitrariness of play as complexity theory now tells us. Fluidity is, so far, even beyond modern computers:

> [S]imulating turbulent fluid flow in a realistic way is still largely beyond the capabilities of today's computers. Improving our understanding of turbulence remains an outstanding fundamental problem in modern science. The principal hurdle we must overcome is modeling the immense complexity of turbulent structures, where the behavior of fluid at the tiniest of length scale has a dramatic effect on behavior over larger distance.[37]

This deep structure of the ocean wave fluidity, which was intuited by Byron, came to embody an almost nihilistic urge to break free from the emphasis on controls and on the bonds of temporality.

Unlike human history, the sea is beyond time. 'The political manifestations of human bondage and freedom', writes Jerome McGann, 'continually rise and fall, but the Ocean herself remains outside this temporal flux and perfectly superior to it. She appears as a kind of infinitely objective consciousness, able to perceive the ebb and flow of human history with self-possessed equanimity'.[38] Since the beginning of time, this great expanse of elemental power—or, for Harold, its creator—has existed:

> Time writes no wrinkle on thine azure brow—
> Such as creation's dawn beheld, thou rollest now (182).

Whatever the true nature of his religious beliefs, Byron intends his persona to give voice to this sense of the fragility of human constructions by the sea—emblem for eternal process, for God. I think it is, for him, a qualification of the Napoleonic, but also English, aspirations of Empire.

Generally speaking, Constable is the very last painter of his age whose works I would consult to find a parallel for those of Byron; however, I think that Constable—whose later paintings sometimes suggest allegorical significance—might have seized on the fragility of one of the signs of empire, the newly erected marvel of contemporaneous steel construction, the Chain Pier, extending out from the beach at Brighton. If this were an example of Constable's allegory, then it was prophetic because, the pier was eventual destroyed by a storm before the century closed. Constable—if not without grudging admiration for 'its long & elegant strides into the sea a full 1/4 of a mile'—describes it as 'unnatural'. In addition to the oil and watercolor sketches which he executed *in situ* there, he painted the large-scale work *Chain Pier, Brighton*. Although it included topographical details such as buildings, vacationers, and fishing paraphernalia, Constable wrote, in the same letter quoted above, that the landscape subject really contained 'nothing [...] for a painter but the breakers—& sky'.[39] In the work itself, the human construction—especially the pier—is made to seem paltry and insubstantial before the flux of breakers and clouds. In other words, although he saw himself as 'a painter of fact' in the broad sense, he

37 Peter Coven and Roger Highfield, *Frontiers of Complexity: The Search for Order in a Chaotic World* (New York: Fawcett, 1995) 67.

38 McGann, *Fiery Dust* 136.

39 Qtd. in Leslie Parris, Ian Fleming-Williams, and Conal Shields, *Constable: Paintings, Watercolours, and Drawings* (London: Tate Gallery, 1979) 148–49; concerning the destruction of the pier, see 148; concerning his latter allegories see also Reynolds, *Constable* 102.

thought beyond mere topography and into the more profound workings and changes of nature—for him—of God. This aspect of his art is even more apparent in his *Brighton Beach, 'Squal[l]y Day,'* one of his many Brighton oil sketches. In this work, all but the elements of the locus are reduced to a minimum and one is surprised to see a wide expanse of water and clouds dominating the small, insignificant human figures and boats.

It is interesting how close in iconography—leaving aside symbolic meaning—this sketch by Constable is to Friedrich's *The Monk by the Sea / Der Mönch am Meer.*[40] In both paintings the large expanses of sand, sea, and sky cover the canvases and dwarf the human figures (two in Constable, one in Friedrich). Friedrich's *Sea Piece / Seestück* comes closer in its iconography to Constable's *Brighton Beach,* particularly since ships are shown on the sea in both, whereas none are shown in *The Monk by the Sea*. Recently x-rays have been used to prove that the sea in the latter originally contained ships. The artist apparently decided to pare down his subject matter by painting over them, 'leaving the monk on the brink of an abyss unprecedented in the history of painting'.[41] The elimination of the ships lends a greater intensity to the work and makes it much less a painting of an actual seascape, as are the *Sea Piece* and *Brighton Beach,* and makes it more a symbol of man's insignificance compared to the vast infinity of nature and its deity. In *The Monk by the Sea,* the stormy sea and its white caps have been carefully rendered with chiaroscuro juxtapositioning of white on dark blue. Friedrich's expressive facture turns the threatening sea into a symbol of the infinite power of the Protestant God made manifest on earth.

Like Friedrich, Byron looks beyond the superficial flux of the ocean when he has his persona envision the 'Almighty', symbolized in part by the cyclic (therefore, permanent process) change of sea moods:

> Thou glorious mirror, where the Almighty's form
> Glasses itself in tempests; in all time,
> Calm or convuls'd—in breeze, or gale, or storm,
> Icing the pole, or in the torrid clime
> Dark-heaving; boundless, endless, and sublime—
> The image of Eternity—the throne
> Of the Invisible [...].

Here, in this 'image of Eternity' one can find the emblem of the real creator, as opposed to the paltry 'clay' one; for, in the Pre-Socratic view promoted by Denis Diderot and Erasmus Darwin, the sea was to be seen as the *Ur*-generator of new lifeforms:

> even from out thy slime
> The monsters of the deep are made—each Zone
> Obeys thee; thou goest forth, dread, fathomless, alone (183).

40 For a general survey of Friedrich's treatment of the sea, consult Eberhard Ruhmer, *Caspar David Friedrich Meereslandschaften* (Munich: Piper, 1978).

41 Robert Rosenblum, *Modern Painting and the Northern Romantic Tradition: Friedrich to Rothko* (New York: Harper, 1975) 13.

For Byron, the sea is essentially God's place, not mankind's: the Byronic hero is exempted from this eternal law of exclusion, one may presume, by some sort of divine dispensation. It may well be that this hero and the sea share some quintessence—something which Goethe would have called the daemonic, the elemental.

One is indeed struck by the absolute and incontrovertible connection among the persona, Harold; the poet, Byron; and the image, the ocean. There is manifestly no antinomian conflicts among them. Indeed, considering Byron's great love for swimming (his famous swim across the Hellespont is well known), it is not surprising that he makes the reaction between the ocean and the speaker so personal as to seem autobiographical. In the following stanza (184), he first feels a mere part of the great ebb and flow—'like thy bubbles', he says to the ocean; then, he feels himself a scion of that ocean; finally, he sees it as an untamed horse—that is, as an objectification of his lifelong defiance of danger:

And I have loved thee, Ocean! and my joy
Of youthful sports was on thy breast to be
Borne, like thy bubbles, onward: from a boy
I wantoned with thy breakers—they to me
Were a delight; and if the freshening sea
Made them a terror—'twas a pleasing fear,
For I was as it were a child of thee,
And trusted to thy billows far and near,
And laid my hand upon thy mane—as I do here (184).

Here, more than just Harold/Byron as a child or adolescent swimmer is at issue; this stanza, the last of the series, offers a celebration of the freedom allowed upon the great sea. All in all, Bryon's stanza 184 reveals a very personal expression of what has been up to this point in the stanza series a generalized vision of the ocean. In a certain sense, the untamable spirit of the high sea and that of the persona/poet are one. Jerome McGann writes of this stanza in particular, 'The Ocean becomes, in fact, the poem's crowning image of the daimonic and eternal source of all life and energy, and the poet presents himself to us as the child of this God and the sharer of his powerful creativity'.[42]

Whether through direct, indirect, or parallel poetic interests, the various ideas and themes running through this ocean passage from *Childe Harold* recur in the works of European poets who chronologically follow Byron and who are, like him, often lumped under the term *Romantic*. The wide influence of Byron's ocean passage will be discovered running throughout the sections that follow.

This wide influence of Byron's ocean passage under consideration can be suggested by examining representative passages from these post-Byronic poets. Beginning with the opening line, 'Roll on, thou deep and dark blue Ocean—roll!', the entire passage is permeated with a fascination for the powerful flux of the ocean. Rather than bewailing the fact that the sea is a realm beyond human 'control', he shows the tools for this control being reduced to 'ruin' and celebrates the sea as an inviolate sanctum and power upon the earth. He continues in this vein of *Weltschmerz* and of misanthropy by taking delight in the sea as a foil compared to which

42 McGann, *Fiery Dust* 136.

the works of humans seem mere 'toys'. In this way, humans are set in opposition to nature. The sea is a thing of freedom and escape—albeit dangerous—from human society. The sea in its exhaustive gamut of moods is the 'image of Eternity' and 'the Almighty's form'—whether reflecting the heavens or being stirred up into a frenzy by storm clouds.

Although transformed by a four-square Neoclassicism which still permeated incipient French Romanticism prior to 1830, Byron's ocean as the 'mirror' of divinity finds an echo in the following two stanzas of Lamartine's lyric, 'Adieux à la Mer' (171–74) from *Nouvelles méditations poétiques*: 'God, who decorated the world with your gracious [watery] element, so that everything responds, made the heavens for shining on the waves, the waves to reflect the heavens'.

Le Dieu qui décora le monde
De ton élément gracieux,
Afin qui'ici tout se réponde,
Fit les cieux pour briller sur l'onde,
L'onde pour réfléchir les cieux (ll. 26–30).

What theism there is in Byron's poem appears almost accidental; however, it is the guiding principle of Lamartine's poetic vision: 'Sublime image of infinity, from wave to wave the passionate eye seeks in vain to follow you from shore to shore, the intellect seeks in vain for your shore, as for the shores of eternity'.

De l'infini sublime image,
De flots en flots l'oeil emporté
Te suit en vain de plage en plage,
L'esprit cherche en vain ton rivage,
Comme ceux de l'éternité (ll. 41–45).

Lamartine's theism does indicate an important concern among French Romantics, as does the poetry of Victor Hugo; a matter of emphasis that marks off the English from the French. It is also significant that the sea is not really infinite; which is, of course, only a metonymy for the sea.

The collection *Nouvelles méditations* contains yet another significant sea poem—'Tristesse' / 'Sorrow' (134–35)—significant because it uses similar imagery as Turner's *The Bay of Baiae, with Apollo and the Sybil* and for comparable symbolic purposes. Turner's painting refers to the Sybil sacred to Apollo who is so central to Virgil's book VI of *The Aeneid*; Lamartine centers 'Sorrow' upon a description of the locus of Virgil's tomb. The poet draws upon images found in the earlier collection; for example, the 'laughing slopes' / 'riants coteaux' of 'The Lake' appear again in 'Sorrow'. The body of water is now the Bay of Naples; it is still an expanse of water that provides a place for lovers to listen 'to the gentle noise of the waves or the murmuring wind',

Au doux bruit de la vague ou du vent qui murmure [...] (l.17).

Indeed, any hasty glance at Lamartine's poetry will demonstrate the poet's predilection for the word *murmur*—whether it is the murmur of the waves, wind, or foliage. The persona of 'Sorrow', as the title implies, suffers in a state of melancholy and nostalgia for a happy past that is lost, never to return. Life and death are joined, a final paradox in which the poet wishes 'to die in the place where he has tasted life'

/ 'de mourir aux lieux où j'ai goûté la vie' (l. 36). Just as in the case of Turner's painting, Lamartine takes us deeper into the landscape than just the placid surface created by the limpid air, full light, and blue surface of the bay. The painter created a landscape and ruins undercut at every point by running rivulets and waterfalls; the poet reminds us of the volcanic and insubstantial land of Campania: 'inflamed Vesuvius once again arising from the bosom of the waves':

encore
Le Vésuve enflammé sortant du sein des eaux [...] (ll. 5–6).

The infinity that fascinated Turner, Lamartine, and Byron is the infinity and eternity of change: when viewed with a godlike eye under the aspect of eternity, this recurrent change is permanent recurrence; when viewed with an eye not affected by poetic sublimity, it is mere change.

Sunsets and Mythic Memories over the Waters

Felix Melchior, in his comparative study of Byron and Heine, observes that Heine shared with Byron a love for the sea.[43] Above all, the exhaustive gamut of moods of the sea offered for them symbols for the infinite. I will concentrate on an aspect of Heine's poetry which Melchior does not touch on: the sea as seen in the light of the sun. True, the sea's reflective properties are mentioned in stanza 183 of Byron's *Childe Harold*. Whereas Byron delights in the danger and the excitement of the cataclysmic flux of a stormy sea, Heine shows a predilection for a calmed sea at sunset. Byron concentrates on movement; Heine, on color. Moreover, it is Heine's heavy concentration on colors that sets him off from the eighteenth century and links him to the late nineteenth and early twentieth century. Among his contemporaries, Heine's heavy use of description, the unprecedented employment of color adjectives to render seascapes, finds its best parallel among the Romantic painters; but composers have—in their programmatic overtures, which came to be termed tone poems—done their own tone painting to capture the play of light over the surface of water. This desire to strive to capture in tone, words, or oils the reflecting properties of water can help with another distinction within the culture of the Romantic era—distinguished like all such cultures by a rich variety and complexity.

Heine uses this motif of Homer receiving inspiration from the sea in one of his poems in the cycle entitled *The North Sea / Die Nordsee* (1: 226–65).[44] This is one of many descriptive passages which extensively treat the visual and aural aspects of the sea—particularly the former. Much of this work is a collection of short narratives taken from Norse or Greek mythology or, more often, an eerie blending of the two. This use of mythology is not a mere pseudo-Classical decoration on the surface of Heine's poems; it goes hand-in-hand with the empirically studied phenomena from nature. Indeed, Heine's vision of nature in *The North Sea* is very close to that of Thoreau: 'I do not know where to find in any literature, whether ancient or modern,

---

43 Melchior 106–17, et passim.

44 *Heinrich Heines Sämtliche Werke*, eds. Oskar Walzel, James Fränkel, Ludwig Kräke, and Albert Leitzmann, 10 vols. (Leipzig: Insel, 1911–15).

any adequate account of Nature with which I am acquainted. Mythology comes nearest to it of any' (*Journal* 2: 153). By this I think he means the profoundest of ecologies: that attitude of humans that addresses nature as a 'thou' and treats it with a holy respect. Thoreau is seeking the noumena behind the phenomena. How far Heine followed in this metaphysical conception of the nature of things is dangerous to speculate, as are all aspects of his religious outlook.[45] It, however, can be said that he uses Norse and Greco-Roman mythological references to help his reader access a link with deep ecology, much in the same way that Wordsworth and Goethe feel themselves deeply into nature through a type of pantheism derived however distantly from Spinoza.

Most of Heine's mythic narratives open with short lyric passages, and in turn, many of these are verbal seascapes—usually at sunset. It would require a lengthy study in itself to analyze how the narrative and lyric passages might be said to grow from the 'mood' of these seascape passages. I would like rather to analyze the opening passages per se as lyric outpourings where the persona shows a strong response to and perception of the phenomenon of the sea with the sky and the sun. (Those which concentrate on the sky will be treated in the next chapter.) In 'Poseidon' (1: 193–94), the sunlit seascape has an effect on the persona similar to that in the Stolberg poem, in spite of the fact that Heine's luxuriant language and freer line structure are worlds apart from the earlier poet; it makes him think of Homer:

> The sunbeams played over the expansively-rolling sea; the ship that was to bear me home glimmered far out in the roadstead; however, there were no favoring winds for sailing, and I sat still peacefully on the white dunes over a lonely beach, and I read the song of Odysseus, the old, the eternally youthful song, out of whose pages—filled with the essence of the sea—these rose exultingly to meet me: the breath of the gods, the radiant spring of human history, and the blooming sky of Hellas.

> Die Sonnenlichter spielten
> über das weithinrollende Meer;
> Fern auf der Reede glänzte das Schiff,
> Das mich zur Heimat tragen sollte;
> Aber es fehlte an gutem Fahrwind,
> Und ich saß noch ruhig auf weißer Düne,
> Am einsamen Strand,
> Und ich las das Lied vom Odysseus,
> Das alte, das ewig junge Lied,
> Aus dessen meerdurchrauschten Blättern
> Mir freudig entgegenstieg
> Der Atem der Götter,
> Und der leuchtende Menschenfrühling,
> Und der blühende Himmel von Hellas (ll. 1–14).

These lines introduce a narrative concerning Zeus. These poems begin with a descriptive prologue followed by a mythos section, and this structure is generally associated with Pindar. But this is not the poet referred to in Heine's lyric, or in that of Coleridge or of Stolberg: they, of course, refer to Homer, his *Odessey,* and his ability to conjure up visions of gods and giants. This is what happens in two works

45 Laura Hofrichter, *Heinrich Heine* (Oxford: Clarendon, 1963) 72–74.

by Turner—*Ulysses Deriding Polyphemus* and the preparatory sketch—which suggest not just the myth of Odysseus, but suggest the mythopoeticizing of the land-, sea-, and skyscapes. In Turner's paintings, the clouds and rock formations suggest the giant Polyphemus.

A sunlit sea brings to the poet's mind Greek mythology just as, in the following opening lines of 'Evening Twilight' / 'Abenddämmerung' (1: 188), the sunset on the sea brings to mind the sagas of Northern mythologies:

> On the pale strand of the sea, I sat, alone with my oppressive thoughts. The sun sank deeper and threw red-glowing bands over the water, and the white, wide-rolling waves, urged on by the tide, foamed and rushed forward nearer and nearer—a weird rush of sound, a whispering and piping, a laughing and murmuring, a sighing and whistling—amongst them a singing of a cozy lullaby—I felt as if I was hearing long-forgotten sagas [...].

> Am blassen Meerestrande
> Saß ich gedankenbekümmert und einsam.
> Die Sonne neigte sich tiefer, und warf
> Glührote Streifen auf das Wasser,
> Und die weißen, weiten Wellen,
> Von der Flut gedrängt,
> Schäumten und rauschten näher und näher—
> Ein seltsam Geräusch, ein Flüstern und Pfeifen,
> Ein Lachen und Murmeln, Seufzen und Sausen,
> Dazwischen ein wiegenliedheimliches Singen—
> Mir war, als hört ich verschollne Sagen [...] (ll. 1–11.).

What Paul de Man said of the flow of the river can here be applied to the flux and reflux of the waves: 'the oblivious spell emanates from the repetitive rhythm of the water, which articulates a random noise into a definite pattern'.[46] Three aspects of the sea are important in this passage: (1) the visual description of the flux of the waves as the tide moves in threateningly, which is furthered by the assertive, incessant alliteration, 'the white, wide-rolling waves' / 'die weißen, weiten Wellen'; (2) and—what is most striking—the persona's careful description of the colors of the sinking sun and their reflection in the water, 'the red-glowing bands' / 'Glührote Streifen'; (3) the sound of the sea's flux. As Laura Hofrichter observes, the poet made a radical departure from the highly structured quatrains of his lyrics to the free verse of *The North Sea*; she also quotes the poet as writing: 'Our customary sweetwater readers are not used to this rocking metre; it may make them a little seasick'.[47] Heine's appeal to verbal music is so strong, it is no wonder that his works were favorite choices of his contemporaries among the composers.

The heightened aurality already intrinsic to Heine's poems bring a musical comparison to mind, and none more appropriate than one that also treats saga material, Mendelssohn's *Fingal's Cave / Fingalshöhle,* op. 26, or as it is also titled *The Hebrides Overture / Ouverture zu den Hebriden,* a revolutionary piece of natural tone painting, without which the other great Romantic sea piece—*The Flying Dutchman / Der fliegende Holländer* Overture—might never have come into being. Mendelssohn's

---

46 Paul de Man, 'Shelley Disfigured', *The Rhetoric of Romanticism* (New York: Columbia UP, 1984) 107.

47 Laura Hofrichter, *Heinrich Heine,* trans. Barker Fairley (Oxford: Clarendon, 1963) 42.

shifting tonality with daringly sustained dissonances, the layering of figures of sixteenth-notes, and the emphatic trills all create the sense of the turbulent flux and the shifting, contrary currents of the ocean, but also of the often pessimistic myths of the northern seas.

Mendelssohn and Heine clearly share an interest in using the elements of their respective arts to evoke the sound and concept of change; however, the essential difference is that Heine, when he wrote *The North Sea,* was beginning consciously to attach his fascination for change as structured by the temporal specificity of history. Nigel Reeves, in his *Heinrich Heine: Poetry and Politics,* discusses how important it is to see *The North Sea* as a cycle of poems—an important formal concept for Heine and other High German Romantics.[48] He, therefore, employs the Third Cycle (4: 89–132)—a prose cycle ('key essay')—of *The North Sea* to frame the earlier poetic cycles in a large political context:

> We have seen that Heine's Utopianism and his skepticism both derive essentially from different attitudes to the flux of human history. What is vital is whether change is seen as part of a meaningful historical process or whether on the contrary it serves as proof that all human process is valuable or whether on the contrary it serves as proof that all human endeavour is destined to perish.[49]

Reeves shows, then, that Heine in *The North Sea* cycle was working his way between two extremes: his future commitment to radical politics; skepticism that such politics would result in a 'meaningful historical process'. Heine's fearlessness in the face of radical political change makes him fearless in the face of the flux and reflux of the sea, which as we saw above was celebrated in the formal construction of the verse form and rhythm. There is also a celebration of temporality in the emphasis on times of day. Even more than for Byron, the cycle context paled in importance for Heine before the mutability of the sea. Heine, as a Jew, faced his own oppression much more openly than did the closeted Byron.

The description of the sinking sun and onrushing waves found in the opening lines of Heine's 'Sunset' / 'Untergang der Sonne' (1: 213–14) of the Second Cycle suggests a more optimistic turn from the world view found above in his 'Evening Twilight' of the First Cycle; both descriptions heavily rely upon the suggestion of the color red and the shade white:

> The beautiful sun has sunk peacefully into the sea; the surging waves are already colored by sable night—only the red glow of evening still strews them with golden light, and the roaring force of the tide drives white waves in toward the shore, which joyfully and quickly gambol, like wooly flocks of lambs, which the singing shepherd youth drives toward home at evening' [...].

---

48 Nigel Reeves, *Heinrich Heine: Poetry and Politics,* 2nd ed. (London: Libris, 1994) 118–21. As his book title suggests, Reeves concentrates on Heine's desire 'to restore the lost connection between life and art' (149); Reeves neglects the connection that these two have with the natural environment. Heine, however, did not. Laura Hofrichter, also emphasizes the important connection between the three cycles (*Heinrich Heine,* trans. Barker Fairley [Oxford: Clarendon, 1963] 43–44).

49 Reeves 149; Heine's politics are also treated in Hofrichter 57–74; Jerome J. McGann, *The Romantic Ideology: A Critical Investigation* (Chicago: U of Chicago P, 1983) 31–39, 49–56.

Die schöne Sonne
Ist ruhig hinabgestiegen ins Meer;
Die wogenden Wasser sind schon gefärbt
Von der dunkeln Nacht,
Nur noch die Abendröte
Überstreut sie mit goldnen Lichtern;
Und die rauschende Flutgewalt
Drängt ans Ufer die weißen Wellen,
Die lustig und hastig hüpfen,
Wie wollige Lämmerherden,
Die Abends der singende Hirtenjunge
Nach Hause treibt (ll. 1–12).

Here, in addition to the color adjectives used in 'Evening Twilight', the persona employs the color gold. The persona gives himself over to description, but not entirely without expressive purposes, for in at least two cases, we have seen how the time of day corresponds to the mood of the narrative. Moreover, the poet has consciously shown a predilection for the particular time of day which he desires for his seascapes—usually at sunset and when the tide is ebbing. In spite of all this, the sea passages of Heine closely resemble the descriptive eighteenth-century poetry. In both, there is an objectivity and a dearth of clearly defined signs.

Heine's twilit poems find their most striking parallels with the sea-skyscapes in the so-called *Colour Beginnings* collection of Turner, and, in one of his very rare forays into oil landscapes, Eugène Delacroix's *The Sea from the Heights of Dieppe,* which beautifully evokes twilight reflecting upon the undulating ocean. The artist's facture creates the sense of waves by wide and heavy application of large strokes of very light shades next to very dark, among other respects, and has caused art historians to see in this painting a foreshadowing of the Impressionists.[50] The expanses of sea and sky in the Turner sketches have a greater sense of immensity than the Delacroix painting. The two sketches which I have chosen, while not representative of the whole series, do represent the more abstracted, rarified compositions of the *Colour Beginnings.*[51] They consist solely of the sun, its light, the sea, the sky, its haze, and wisps of clouds; in fact, the sky and sea are almost indistinguishable. In 'Sunrise at Sea', one is distinguished from the other by a subtle change in technique. Whereas the sky was painted with a very watery application, the sea was applied with a drier brush stroke giving it a more highly textured surface; whereas, the sky was applied with long, straight brush strokes, the sea was applied with shorter, curved strokes. The sunlight reflected on the water is rendered, as is the sun, by an absence of paint; it appears that wax has been applied to the paper to repel water and thereby to signify the sun ball and its reflected light on the sea beneath. The artist has textured the sea by applying the paint of the background with considerably more water than that in the foreground. The drier application of

50 Cf. Maurice Sérullaz, *Eugène Delacroix* (New York: Abrams, n.d.) 146–47. See Peter Greenaway, *Flying Out of This World* (Chicago: Chicago UP, 1994) 75–83.

51 Although I drew mainly on my own formal training in watercolors, I found the general discussion in Andrew Wilton's rather extensive 'A Note on Turner's Technique' a useful touchstone in my formal analysis of watercolors in the so-called *Colour Beginnings* set, *Turner in the British Museum: Drawings and Watercolours* (London: British Museum, 1975) 13–26.

the latter increases the chiaroscuro, giving the surface of the water more 'texture'; this is even more true in the immediate foreground, where the waves were painted with a very dry brush. Very similar—in both the elemental simplicity of subject matter and the techniques used to render them—is 'Sunrise'. Indeed, in the latter the simplicity is even greater: it consists of a sun ball, sea, and sky. The cloud formation is a diffuse stratus layer of mist or fog which envelops the whole. This pervasive mist is purplish grey. The water is distinguished from the sky only by a slightly darker shading and gold coloring. Unlike the 'Sunrise at Sea', the water of the sea in 'Sunrise' is inanimate. The reflection of the sun in the water and the sun itself seem to have been achieved by light application of wax, which partially repelled the watercolor when it was applied.

Edward Lockspeiser—in his suggestive, but not very thorough comparative study *Painting and Music*—hits on a really insightful vision of nothingness in Turner's sea paintings and in the elemental passages from Wagner's operas. Lockspeiser finds in water their shared source of inspiration, and their central motif in rendering nihilistic visions of elemental power and cosmic destruction. In the case of Turner, he refers to the so-called *Colour Beginnings*. On Wagner, he writes:

> Turner's pictorial imagination is present surely in the music depicting the depths of the Rhine, at the opening of *Rheingold*; in the raging storm of *The Flying Dutchman*; both in the Overture and in the final act; in the storm invoked by Donner in *Rheingold*; in the flooding waters in *Siegfried* [...].[52]

But for me the music that was eventually to achieve a parallel to Turner's great stretches of nothingness is best to be found in crucial sections of *Tristan und Isolde*. When Bragäne opens the second scene (according to the French counting of scenes), there is a strange series of descending and hollow intervals, usually fifths and fourths, which suggests the sea and horizon mentioned in the stage directions. (The hollow quality created in the intervals, especially the fifths, is due to the fact that in the three-note chord structure that was the norm in Wagner's time there is an interval of a major third filling in the space between the first note and the fifth. A fourth seems similarly incomplete because it 'wants' a major third above, which makes it an inverted triad). Here, a sequence of chords is descending, perhaps, because it is an analogue for the spiritual separation of the lovers—who are seen together on the stage for the first time by the audience.

The transformed now-ascending figure appears at the beginning of act 3, but here it would represent the physical separation of the lovers by the sea space. The harmonic effect is also a bit less hollow than that of act 1 because there are less open fifths. Concerning the act 3 passage, the great Wagnerian scholar Ernest Newman states that the 'slowly-ascending passage in the violin [...] gives an effect of great sea space'.[53] While I agree with Newman, the symbolic ascription to music must be done, of course, with great care; it is important to note that the same motif appears in the contemporary *Wesendonck Songs* / *Wesendonck Lieder* (where there is

---

52 Edward Lockspeiser, *Music and Painting: A Study in Comparative Ideas from Turner to Schoenberg* (New York: Harper, 1973) 65.

53 Ernest Newman, *Stories of the Great Operas and Their Composers* (1928; Garden City: Garden City, 1948) 103.

no mention of the sea), in which it seems to be associated with the emptiness of separation. And this is precisely what the sea seems to signify in the music drama. I think it remains crucial to the understanding of *Tristan and Isolde* that in the two acts (1 and 3) where the lovers spend the greater part of the time divided by psychological or physical barriers, the stage setting is by the sea, which in accordance with Wagner's own directions is to be visible. Both music and text convey the sea as a void (e.g., 'Öd und leer das Meer!' / 'Desolate and void—the sea'). Act 2, however, where Tristan and Isolde unite in love is set is a fertile, verdant, branch-enclosed garden. The dark enclosed garden offers the wholeness, fulfillment, and freedom of love; the sunny, open sea offers separation, void, and entrapment. All metaphoric expectations are, thereby, inverted.

Either by including expanses of sea or sky with human figures, as in the Turner paintings, or by juxtaposing small human figures against the expanses, as in the Constable and Friedrich paintings, the sense of the infinite is made overpowering. It might well be recalled at this point that the sea is a symbol evoked for the contemplation of the infinite, of God, in Byron's sea passage from *Childe Harold,* a sea lyric by Lamartine and—as the next two sections will demonstrate—in a mountain and ocean lyric, as well as Wordsworth's Mt. Snowdon passage (where the unity of things is implied by the ambiguity of the real sea lying beneath a sea of stratus clouds).

Lake and Sea Storms

Sea storms are certainly not new with Romantic culture. What is different is a new general attitude. Before 1760, a sea storm is a matter of traveling inconvenience and even danger. This is, of course, not to say that a Shakespeare could not also endow them with a certain poetic power, as in the opening of *The Tempest* or the midpoint of *The Winter's Tale*; but these are exceptional and cannot be called celebrations of the storms by any means. The Romantic poets, painters, and composers celebrate them. Even the artistic consciousness, which partakes of the (Neo)Classical-Romantic temperament, demonstrates a fascination with the sea storm and gives it a pride of place in their works rarely before seen. Gluck opens his *Iphigénie en Tauride* with an overture suggestive of a storm at sea. Since he composed the work for Paris, it is not surprising that his harmony and orchestration pay homage to Rameau. Gluck, however, makes his own contributions to the sense of the elemental by breaking with the usual operatic form of separating the overture from the operatic action to follow; he does this by omitting the conventional close of the overture with a full and formal chordal cadence, followed by a full measure rest; he moves directly from the storm music of the overture right into the storm music of the opening chorus of Iphigénie's priestesses singing of the storm. This opening is carefully planned by the composer to point forward to the dramatic centers of the text and music: the psychological instability of Orestes, who is maddened by the furies, and that of the barbarian king, who is equally maddened by his superstitious fears. Mozart's most Gluckian work, *Idomeneo,* also centers on several choruses about sea storms.

As we would expect from a Neoclassical poet, Schiller uses the sea storm as an ambivalent symbol of chaos. In his 1795 poem, 'The Walk' / 'Der Spaziergang' (2.1: 308–14), a sea storm represents the French Revolution, seemingly a product of the Age of Reason. Using this poem as his ground, M.H. Abrams writes 'a résumé' of *Natural Supernaturalism*. Although no quotation is offered, it is surely the sea storm that Abrams refers to as signifying the French Revolution and 'the abyss of primordial savagery and destructiveness beneath the civilized surface of the human mind'.[54] There is, certainly, none of the celebration of cosmic change that we will find in the Romantic poets to be considered below in this section; indeed, it is the anchors of stability that receive the approval. The excerpted passage quoted here seems to suggest a false freedom (anarchy) as opposed to ideal freedom (liberty within limits) that Schiller, of course, celebrates elsewhere:

> Freedom is the cry of reason—freedom is the cry of wild desire, lecherously they wrestle you away from sacred nature. Ah! in the tempest the anchors break loose, that warningly held him onto the shore, and the flow tears him along in its flood—into infinity whirls him—the coast soon vanishes before him; the mastless ship rocks high on the mountainous waves [...].

> Freiheit ruft die Vernunft, Freiheit die wilde Begierde,
> Von der heil'gen Natur ringen sie lüstern dich los.
> Ach, da reißen im Sturm die Anker, die an dem Ufer
> Warnend ihn hielten, ihn faßt mächtig der flutende Strom,
> Ins Unendliche reißt er ihn hin, die Küste verschwindet,
> Hoch auf der Fluten Gebirg wiegt sich entmastet der Kahn [...] (ll. 141–46).

The symbol oddly partakes more of natural fury than of human fury. To my mind, this passage shows a crucial contradiction that—with the exception of Goethe, Shelley, and Byron—poets of this age share: the setting up of nature as a moral force beyond humans, and the employing of that nature to express the human qualities. The sea storm represents the human passion that breaks human laws. Schiller demonstrates a certain ambivalence to sea storms: it is clear that he has a marked dislike of storms combined paradoxically with a desire to employ their dramatic power. The exact opposite will years later be true of Byron.

I would like now to consider the Byronic sea storm and—I think we can safely say—its effect upon his Romantic heirs; first, however, we need to examine what went before *Childe Harold* and *Don Juan*.

I can hardly do justice to the shipwreck section of canto two of *Don Juan* nor is that my intention; it is a complex mixture of proto-Realism and Romanticism—with Neoclassical Popean touches here and there. Certainly the pure dythrambic celebration of the ocean found in *Childe Harold* has been tempered by irony, down to the seasickness of the hero during his first hours under sail (stanzas 19–23). The moment of the wreck or rather submerging itself, however, seems Romantic in its effect: the ship

> gave a heel, and then a lurch to port,
> And, going down head foremost—sunk, in short.

54 M.H. Abrams, *Natural Supernaturalism: Tradition and Revolution in Romantic Literature* (New York: Norton, 1971) 453, 456.

Then rose from sea to sky the wild farewell,
 Then shriek'd the timid, and stood still the brave,
Then some leap'd overboard with dreadful yell,
 As eager to anticipate their grave;
And the sea yawn'd around her like a hell,
 And down she suck'd with her the whirling wave,
Like one who grapples with his enemy
And strives to strangle him before he die.

And first one universal shriek there rush'd,
 Louder than the loud ocean, like a crash
Of echoing thunder, and then all was hush'd,
 Save the wild wind and the remorseless dash
Of billows; but at intervals there gush'd,
 Accompanied with a convulsive splash,
A solitary shriek, the bubbling cry
Of some strong swimmer in his agony (canto 2: stanzas 51–53 [5: 105]).

But the same sea that wrecks Don Juan's ship also gives him his first chance at adult—indeed at heroic—action when he, brandishing 'a pair of pistols', holds off sailors from the ship's store of liquor (2: 35), and this same sea 'providentially' spares and casts him, not without reluctance to part with its victim (107–08), upon the shore of an island. The 'goddess' of this isle is kept pristine by solitude created by—again—that very same sea. A note or two of the old Byronic rhapsody of the deep returns when the now-seasoned young Spaniard sails for England across the gale-tossed North Sea:

 High dashed the spray, the bows dipped in the sea,
And sea-sick passengers turned somewhat pale;
 But Juan, seasoned as he well might be
By former voyages, stood to watch the skiffs
Which passed, or catch the first glimpse of the cliffs (10: 64 [10: 456]).

Dangerous though the speculation might be, I cannot help wondering if the further adventures of his hero might have included even more extensive ocean and sea travel had not Byron's heroic death on the northwestern coast of Greece interrupted their narration.

Heine shared with Byron a love/fear for the untamed nature of the sea, for the sea as a symbol of danger—indeed, of anarchy. As with Byron, Heine revels in the cataclysmic flux of the sea, especially in the midst of a tempest, which is the title of a lyric from the *North Sea / Nordsee* series: in 'Storm' / 'Sturm' (173–74), the persona's ship is shown as the plaything of the sea: 'The storm rages, and it whips the waves, and the waves—foaming with rage and bucking—pile themselves up, and the white water-mountains surge vigorously, and the little craft climbs them, perforce with speed, and suddenly shoots downward into the black wide-yawning abyss of the surging water':

Es wütet der Sturm,
Und er peitscht die Wellen
Und die Well'n, wutschäumend und bäumend,
Türmen sich auf, und es wogen lebendig
Die weißen Wasserberge,

Und das Schifflein erklimmt sie,
Hastig mühsam,
Und plötzlich stürzt es hinab
In schwarze, weitgähnende Flutabgründe [...] (ll. 1–9).

The graphic exuberance of the description seems to me to convey a celebration—if fearful one—of the sea storm.

Significant parallels in Spanish poetry to Byron's love of sea storms occur in Bécquer's 'Giant Waves' / 'Olas gigantes' (69–70).[55] This lyric contains the same admiration for the tremendous flux of the 'giant waves'; moreover, like Byron, Bécquer loves the promise of escape offered by the sea: 'Giant waves that break with a roar on deserted and remote shores: swaddled in a sheet of foam, sweep me away with you!'

Olas gigantes que os rompéis bramando
en las playas desiertas y remotas,
envuelto entre la sábana de espumas,
llevadme con vosotras! (ll. 1–4.).

In several of the lines, there is an unusual density of *s* and *o* sounds, perhaps meant by the poet to suggest the roll and hiss of waves over a beach. The waves, along with clouds in ll. 5–12, which will be examined in the next chapter, become a symbol of complete escape, indeed, of death: 'Sweep me away, out of pity, where vertigo will root out memory along with reason. For pity's sake! I am afraid to remain alone with my sorrow!'

Llevadme por piedad a donde el vértigo
con la razón me arranque la memoria.
¡Por piedad! ¡tengo miedo de quedarme
con mi dolor a solas! (ll. 13–16).

Unlike the aged Goethe, Bécquer and Byron, in the works examined, demonstrate a love for the wild anarchy of the ocean waves, and rather than desiring to control the element, they celebrate its uncontrollable nature.

As in the case of poetry, I am devoting little attention to the sea as a place of voyage, storms, and shipwrecks compared to the immense attention these subjects received from the painters of this period. Even Constable painted at least one color sketch on the subject, *Rough Sea, Weymouth*. This painting of the cataclysmic flux of the sea is highly uncharacteristic of his output, which generally concentrates on subtler movements and changes. But the work shows that Constable was not above a contemporaneous fascination for showing ships and boats in danger of being engulfed by mountainous waves. Of course, it was in the paintings of Turner that this subject achieved its culmination. Turner loved (and feared) the sea storm as much as Byron. Indeed, it is probably Turner who intellectually lies closest to Byron: few, if any among their contemporaries, looked quite as unflinchingly into

55 Gustavo Adolfo Bécquer, *Libros de los gorriones*, eds. Guillermo G. Gallent, Rafael de Balbin, and Antonio Roldán (1868; Madrid: Ministerio de Education y Ciencia, 1971). Estaban Pujals fails to mention Bécquer in 'Byron and Spain: Byron's Poetic Vision of Spain', *Byron's Political and Cultural Influence in Nineteenth-Century Europe: A Symposium*, ed. Paul Graham Trueblood (Atlantic Highlands: Humanities, 1981) 160–78.

the face of change, of despair, of nihilism as did these two countrymen. The chaos of the tempest was an object not of horror but of great interest for them both.

The most famous of Turner's many sea storms, and—perhaps—most extreme example among his oils is the *Snowstorm: Steam-boat off a Harbour's Mouth Making Signals in Shallow Water, and Going by the Lead.* Ironically, for all the attention the title gives to the steamboat, it is a vague shape almost engulfed by the seething vortex of clouds, snow, and sea. A critic of the time labeled it 'soapsuds and whitewash'[56] because the artist used an extremely heavy impasto of white paint in rendering the swirling motion of the snow and water.

Although, as has been discussed, the whitecaps in the *Monk by the Sea / Mönch am Meer* are large and threatening even from the viewpoint of the shore, it is the frozen 'waves' of Friedrich's *Sea of Ice / Eismeer* (once wrongly called *The Wreck of the 'Hope' / Die gescheiterte 'Hoffnung'*),[57] which offer the best comparison with the waves in Turner's *Snowstorm,* and also serve as powerful foils to the pathetic things of Man. The ice, partially rendered with earthy browns and tans, looks more like boulders than ice. The sky is uniformly grey—except for a clear spot in the top middle of the canvas; this general uniformity of sky adds to the bleakness of the scene of ice. The broken hulk of a ship lies off to the right of the ice 'mountain' in the center of the canvas; this hulk constitutes the remnants of a ship which has been caught in the grip of an ice sheaf and then smashed by the cataclysmic expansions, contractions, and shifts of the deceptively static-seeming ice. Concerning this work, Robert Rosenblum writes: 'man's ephemeral aspirations are dashed against the eternal omnipotence of nature and her often malevolent forces'. He continues by perceiving in Friedrich and Turner this unifying aspect of their world view:

> In Friedrich, Turner, and their Northern contemporaries, human passions become more and more relegated to the domain of nature, where man acts either as a luckless or evil intruder, to be devoured by avalanches, snowstorms, tempestuous seas, or as a silent, worshipping mediator, to be equally absorbed by nature's quiet, almost supernatural mysteries: immeasurable vistas of water and luminous sky, distant horizons obscured by fog, and even the more commonplace wonders of flowers and trees.[58]

The northern seas, with the threat of shifting ice, of icebergs, and of being frozen-in, contain the symbolic value that Northrop Frye saw in ice—the rigidity of death.

It was certainly a careful choice on the part of Mary Shelley when her narrative conducts Frankenstein to his place of death: the ice-choked northern polar seas, with the strange and sudden rush of water that shatters the ice sheet. Just when Frankenstein is about ready to catch up with his monster, the sea seems to intervene:

---

56 Qtd. in *Turner*, ed. Giuseppe Gatt, New Grosset Art Library (New York: Grosset, 1967) 37.

57 *Friedrich,* Tate Gallery 83. There have been two significant studies on shipwrecks in painting which complement Bernhard Blume's studies: T.S.R. Boase, 'Shipwrecks in English Romantic Painting', *Journal of the Warburg and Courtauld Institutes,* 22 (1959): 332–46; Edward Hüttinger, 'Der Schiffbruch: Deutungen eines Bildmotive im 19. Jahrhunder', *Studien zur Kunst des 19. Jahrhunderts,* 6 (1970): 211–44.

58 Rosenblum 34–36.

But now, when I appeared almost within grasp of my foe, my hopes were suddenly extinguished, and I lost all trace of him more utterly than I had ever done before. A ground sea was heard; the thunder of its progress, as the waters rolled and swelled beneath me, became every moment more ominous and terrific. I pressed on, but in vain. The wind arouse; the sea roared; and, as with the mighty shock of an earthquake, it split, and cracked with a tremendous and overwhelming sound. The work was soon finished; in a few minutes a tumultuous sea rolled between me and my enemy, and I was left drifting on a scattered piece of ice, that was continually lessening, and thus preparing for me a hideous death (201).

He is saved by the captain of a scientific expedition, to whom he relates his story—the moral of which the captain ignores and continues his ruthless pursuit of knowledge and technological control over the elements:

One man's life or death were but a small price to pay for the acquirement of the knowledge which I sought for the dominion I should acquire and transmit over the elemental foes of our race (27).

The strength of Shelley's work is that throughout the novel she employs visions of nature's power to comment on what she considers to be the limitations set on human science and technology. Earlier off Scotland, Frankenstein has suffered a storm, which carries him to Ireland. He lands there only to be blamed for the murder of his friend Clerval, whose corpse was also carried there by the storm, perhaps helped by the monster (165–66). Ultimately, Mary Shelley culminates her narrative amid the mist and ice of the northern seas.[59] This sublime landscape, like the corresponding mountain scenery, is symbolically crucial to the book in that it demonstrates the power of elemental forces over humans—especially, over Frankenstein, the creator of human life. It seems to me that Mary Shelley neither misses the chance to celebrate nature and its God(dess), nor to claim in the ultimate powerlessness of humans.

The northern creators, however, were not alone in these darker visions of the sea—of course, without ice—in human destiny. In one of his most anthologized poems, 'L'Infinito' (37–38)[60]—indeed, it may well be the most widely known of all lyrics of Italian Romanticism—Leopardi compares his meditations on the infinitude of space and time to being shipwrecked and drowned in the sea: 'There amid this immensity my consciousness drowns: and it is sweet to be shipwrecked in this sea'.

Così tra questa
Immensità s'annega il mio:
E il naufragar m'è dolce in questo mare (ll. 13–15).

Leopardi, who rarely employs symbols from the water-cycle, has made a strategic decision in his powerful employment of the sea, in what has turned out to be the most famous lyric of Italian Romanticism. He joins the other Romantic poets and painters in taking delight in pitting pathetic man against the greater powers of nature—usually the sea or other water phenomena.

Particularly in Byron's sea passage in *Childe Harold,* in Turner's *The Snowstorm,* and in Mary Shelley's conclusion to *Frankenstein,* the sea is celebrated as an

---

59 Concerning the arctic region in the novel, see William Veeder, *Mary Shelley & Frankenstein: The Fate of Androgyny* (Chicago: U of Chicago P, 1986) 85–87.

60 *I Canti di Giacomo Leopardi,* ed. Alfredo Straccali (Florence: Sansoni, 1912).

overshadowing menace to the achievements of man; Romantic composers also find the sublimely elemental as a source of musical inspiration and a chance to employ the expanded vocabulary of the harmonic and musical timbre of their instruments.

Perhaps one of the most striking sea storms in Romantic music is that found in Wagner's *The Flying Dutchman / Der Fliegende Holländer*. The overture paints both the sea and the psychology of the characters, particularly that of the Dutchman. The wide void and expanse of the sea may be signified by the constant use of the octave and especially the open fifth, which lacks the completing third at its center for the rounded harmony of the triad. The lower strings persistently play chromatic scales running up and down for a full octave—suggesting, it seems, the heaving of the sea. Rising and falling scale-figures suggesting waves are not Wagner's innovation; however, it seems that the use of the chromatic or half-stepping scale is. The refinement of moving by half steps rather than whole ones adds fluidity to the figure, and it darkens and enriches the harmonic effect to add a menacing quality to the sea. The liquidity of effect is broken abruptly by *fortissimo* diminished-seventh chords, which sounded very dissonant to the early nineteeth-century audiences of the opera and which in context suggest the crashing of the waves against the shore or a ship. In the opening of act 1 where a storm is in progress, piccolos and other woodwinds may be taken as signifying the whistling of the wind.

Impressive as Wagner might be as a composer, I think his greatest gift was as a weaver of symbols and myths into his libretti. In weaving them, he revised them—as he does with the earlier Romantic vocabulary of the sea metaphor. Doing what Marx did to Hegel's metaphors, he stood them on their head—he transmuted them, as well as combined them. The Flying Dutchman of the title, because he cursed God, is made to voyage perpetually around the world until he finds a sacrificial woman who will redeem him with her love. Daland, another captain as well as foil for the Dutchman, gives us the Greco-Roman and Classical matter-of-fact attitude toward the danger and inconveniences of storms. The Flying Dutchman's monologue uses and transforms the Rousseauesque waves, which were symbols of psychological tranquillity, and the Byronic seas, which are often places for lawless freedom; Wagner's Dutchman combines these two qualities in one, and adds his own—eternal imprisonment down to the drying up of the last wave, even to the last drop of water. In other words, Wagner's sense of time is mid-nineteenth century or 'geological' in scope, and the sea is no longer symbolic of the eternal:

> The deadline has passed, and once again seven years have elapsed!—full of weariness, the sea throws me up on land. ... Ah, proud Ocean. In a short time, you will be carrying me again! Your defiance can be conquered, but my torment is eternal. The salvation that I seek on land is never to be found by me! To you, the waters of the world-sea, I will remain faithful, until your last wave breaks and your last drop of moisture evaporates.
>
> Die Frist ist um, und abermals verstrichen
> sind sieben Jahr'. Voll Überdruss wirft mich
> das Meer an's Land. ... Ha! Stolzer Ocean!
> In kurzer Frist sollst du mich wieder tragen!
> Dein Trotz ist beugsam, doch ewig meine Qual!
> —Das Heil, das auf dem Land ich suche, nie
> werd' ich es finden!—Euch, des Weltmeers Fluten,

blieb' ich getreu, bis eure letzte Welle
sich bricht, und euer letzten Nass versiegt.

Thus, the symbols focused on so far in this chapter, in Wagner's hands, are combined and metamorphosized as the ocean becomes a place of eternal pain and seeming perpetual imprisonment. To accompany these lines he has a dark, rocking figure in octaves played quietly by the violas and violoncellos; it becomes louder over the aria that follows; Ernest Newman calls it an 'undulating motif', already played during the sea-evoking sections of the overture.[61]

This employment of the sea in *The Flying Dutchman,* as well as that in the later *Tristan and Isolde,* runs contrary to that of earlier Romantics as a symbol of freedom; Wagner rather uses it as a emblem of entrapment, and this I think corresponds to the whole direction of Wagnerian drama and of Wagnerian theatre: the energies of his entire career were directed to developing the techniques to control the minds of the audience by continuous music, overpowering harmonic, dramatic, and visual effects—even his all-important innovation of darkening the auditorium, so as to ensure that his viewers were focused on the stage. This may find its political correspondence in how Wagner admired the way Bismarck was submitting Germany to the control of the Prussian Kaiser. In his fairly positive biography of Wagner, Derek Watson describes Wagner's reaction to the Franco-Prussian War and Bismarck:

> During the horrors of the Siege of Paris, Wagner gloated over the terrible privations and hardships of the French in a merry 'farce', *Eine Kapitulation,* which he held to be in the style of Aristophanes. [...] In January 1871 he sent Bismarck a poem, *To the German Army before Paris* in which he insults France, extols militarism and hails the new German Reich. A *Kaisermarsch* followed, with a choral finale which Wagner hoped might become the national anthem of the Empire. The music matches the banality of the text.[62]

The Wagner of these years disliked freedom—for others, anyway. When Bismarck and the Kaiser did not respond according to Wagner's cue, he turned against them. It is not by chance, then, that Wagner turns the Romantic symbol of the sea as a place of freedom on its head.

Wagner offers a certain sinister closure on the sea. I, however, have so far not exhausted the richness of the sea symbols used or even generated by poets, painters, and composers between 1760 and 1870.

Ocean, Shore, and Mountain in Wordsworth, Hugo, and Liszt

Complexity—here as before—demands that we not limit our view to the water-cycle per se—but examine the corollary phenomena. It is, as I have already tried to show, precisely by juxtaposing these phenomena with a segment of the cycle that the full rhetoric of cycles and permanence in change, as well as simple change, are represented. Beyond the cycle and its corollaries can often be much philosophical, indeed theological baggage. And this certainly is true in the case of the next two poets to be considered: Wordsworth and Hugo. In works of each, the sea is a central image;

---

61 Ernest Newman, *The Wagner Operas,* 2 vols. (1949; New York: Colophon-Harper, 1981) 32.

62 Derek Watson, *Richard Wagner: A Biography* (1979; McGraw, 1983) 249–50.

but in each, it is significantly distanced by being experienced from a mountain top and through an intervening atmosphere. In each, the meditation on the mountain top creates a vision of God in the mind of the poet—albeit in widely differing versions. Not only do both works have key places in the oeuvre of each poet and in the lyric of High Romanticism in England and France, but, in addition, Hugo's poem inspired a tone poem by the arch-Romantic Liszt.

As a sort of culminating grand finale for *The Prelude* (the 1805 version),[63] Wordsworth describes how he once climbed Mt. Snowdon near the coast of Wales and how he emerged from the clouds and mist into the clear moonlit world above them:

> I look'd about, and lo!
> The Moon stood naked in the Heavens, at height
> Immense above my head, and on the shore
> I found myself of a huge sea of mist,
> Which, meek and silent, rested at my feet:
> A hundred hills their dusky backs upheaved
> All over this still Ocean, and beyond,
> Far, far beyond, the vapours shot themselves,
> In headlands, tongues, and promontory shapes,
> Into the Sea, the real Sea, that seem'd
> To dwindle, and give up its majesty,
> Usurp'd upon as far as sight could reach (13: 40–51).

The scene before the poet encompasses two seas: one of mist above and the other of water beneath. The mist not only hides the sea, but the land as well, hence there is no sense of a division between land and sea. All is unified by the sea of mist and by the moonlight. Here, as in Goethe's 'To the Moon', the moon offers a face of permanence over a realm of seething mutability.

This change is conveyed in the sounds of various types of water phenomena joined together into one voice to complete this unified vision of the poet:

> Meanwhile, the Moon look'd down upon this shew
> In single glory, and we stood, the mist
> Touching our very feet; and from the shore
> At distance not the third part of a mile
> Was a blue chasm; a fracture in the vapour,
> A deep and gloomy breathing-place through which
> Mounted the roar of waters, torrents, streams
> Innumerable, roaring with one voice.
> The universal spectacle throughout
> Was shaped for admiration and delight,
> Grand in itself alone, but in that breach
> Through which the homeless voice of waters rose,

63 My analysis of portions of this complex Mt. Snowdon passage focuses on one limited aspect and makes no claim of doing it full justice; for a more complete analysis see Newton P. Stallknecht, 'Nature and Imagination in Wordsworth's Meditation upon Mt. Snowdon', *PMLA,* 52 (1937): 835–47. I know very well that this passage has been the focus of recent criticism, and I will use some of this in discussing this passage in the next chapter; yet, I feel Stallknecht's reading remains the classic.

That dark deep thoroughfare had Nature lodg'd
The Soul, the Imagination of the whole (ll. 52–65).

This unified vision of nature gives the poet a vision of God immanent in nature, indeed, indistinguishable from it—in other words, a pantheistic, Spinozian deity:

A meditation rose in me that night
Upon the lonely Mountain when the scene
Had pass'd away, and it appear'd to me
The perfect image of a mighty Mind,
Of one that feeds upon infinity,
That is exalted by an under-presence,
The sense of God, or whatsoe'er is dim
Or vast in its own being, above all
One function of such mind had Nature there
Exhibited by putting forth, and that
With circumstance most awful and sublime,
That domination which she oftentimes
Exerts upon the outward face of things,
So moulds them, and endues, abstracts, combines [...] (ll. 66–79).

This 'outward face of things', of never-ending metamorphoses, covers but also suggests a hidden permanent force. Wordsworth expresses this force in quasi-Spinozian terms; Hugo, in a poem that I will consider next, is somewhat more orthodox in Christian terms—for the latter poet, there is no confusion about God's relation to nature.

Whereas Wordsworth's meditation takes place at the real, specified locus of Mt. Snowdon, Hugo's lyric 'What One Hears on the Mountain' / 'Ce qu'on entend sur la montagne' (2: 24–26) in *The Autumn Leaves* has as its locus the top of an indeterminate generalized mountain, abstracted from several seacoast mountains or perhaps one that is purely an *ex nihilo* fabrication of the poet's imagination: 'Have you—calm and silent—ever climbed a mountain, in the presence of heaven? Was it beside the Sund? on the coast of Brittany? Was the ocean at the foot of the mountain? And there—leaning over the waves and above an immense space, calm and silent—have you listened?'

Avez-vous quelquefois, calme et silencieux,
Monté sur la montagne, en présence des cieux?
Était-ce aux bords du Sund? aux côtes de Bretagne?
Aviez-vous l'océan au pied de la montagne?
Et là, penché sur l'onde et sur l'immensité,
Calme et silencieux, avez-vous écouté? (ll. 1–6).

In the second verse-paragraph, the poet sets the poetic experience in the context of a dream and uses an important symbol of a poet's winged thoughts (this symbol of poetic imagination will be analyzed in the next chapter, in a different lyric): 'This is what one hears:—at least in my imagination one day, my mind suspended its flight on a seacoast and, from the summit of a high hill overlooking a tremendous drop, it observed the land on one side and the sea on the other, I listened, I heard: and such a voice never did escape from a mouth or strike an ear'.

Voici ce qu'on entend:—du moins un jour qu'en rêve
Ma pensée abattit son vol sur une grève,
Et, du sommet d'un mont plongeant au gouffre amer,
Vit d'un côté la terre et de l'autre la mer,
J'écoutai, j'entendis, et jamais voix pareille
Ne sortit d'une bouche et n'émut une oreille (ll. 7–12).

Here, we again find an exceptionally strong sense of aurality already detected in the sea poetry from Stolberg to Heine: 'There was at first a sound—vast, immense, confused, more vague than the wind in leafy trees, full of brilliant harmonies, of soft murmurs, sweet like the song of evening, strong like the clash of arms when blind battle engulfs armies and blows furiously upon the trumpets of war'.

Ce fut d'abord un bruit large, immense, confus,
Plus vague que le vent dans les arbres touffus,
Plein d'accords éclatants, de suaves murmures,
Doux comme un chant du soir, fort comme un choc d'armures
Quand la sourde mêlée étreint les escadrons
Et souffle, furieuse, aux bouches des clairons (ll. 13–18).

In the passage extending from lines 19 through 30 (which can better be treated with the cloud quarter and therefore will be saved for chapter four), there is a cycle symbol through which all things seem unified in an 'eternal hymn', a universal 'symphony' which fragments as the lyric develops.

As with Byron's ocean passages, the dichotomy between the sea and the land is central to Hugo's lyric[64]: 'Soon I distinguished—confused and veiled—two voices mixed one with the other in this single voice, the earth and the sea cried toward heaven, who sang all at once the universal chant; and I distinguished them within the profound rumbling, as if they were two contrary currents in the flood'.

Bientôt je distinguai, confuses et voilées,
Deux voix, dans cette voix l'une à l'autre mêlées,
De la terre et des mers s'épanchant jusqu'au ciel,
Qui chantaient à la fois le chant universel;
Et je les distinguai dans la rumeur profonde,
Comme on voit deux courants qui se croisent sous l'onde (ll. 31–36).

The sounds of the sea are joyful; the sounds of the land, mournful: 'One coming from the sea; chant of glory! happy hymn! It was the voice of the waves, who conversed with one another; the other, which arose from the land, where we exist, was full of sadness: it was the murmur of people; and in this grand concert, which chanted day and night, each wave had its voice and each made its sound'.

L'une venait des mers; chant de gloire! hymne heureux!
C'était la voix des flots qui se parlaient entre eux;
L'autre, qui s'élevait de la terre où nous sommes,
Était triste; c'était le murmure des hommes;

---

64 This important dichotomy is discussed not merely as it occurs in this lyric but in the entire poetic *oeuvre* in a section of Pierre Albouy's 'L'Ocean et les hommes', of *La Création mythologique chez Victor Hugo* (Paris: Corti, 1963) 343–67.

Et dans ce grand concert, qui chantait jour et nuit,
Chaque onde avait sa voix et chaque homme son bruit (ll. 37–42).

As with Byron, the sea and its waves are shown to be under the control of God.

Not only do we get a strong sense of aurality and find a colorful description of a sunset, we also get a rich dose of theistic, even Biblical imagery:

> Now, as I have said, the magnificent ocean poured forth a joyous and tranquil voice, it sounded like the sort of harp found in the temples of Zion, and praised the loveliness of creation. Its sound, which the noise and gust of wind carried, mounted incessantly in triumph toward God, and each of the waves, which God alone is able to tame, when the one finished the other stood to sing. Like the great lion which Daniel played host to, the ocean for a few moments lowered its great voice, and I myself thought I could see—toward the sun setting in fire—the hand of God pass under its mane of gold.

Or, comme je l'ai dit, l'océan magnifique
Épandait une voix joyeuse et pacifique,
Chantait comme la harpe aux temples de Sion,
Et louait la beauté de la création.
Sa clameur, qu'emportaient la brise et la rafale,
Incessamment vers Dieu montait plus triomphale,
Et chacun de ses flots que Dieu seul peut dompter,
Quand l'autre avait fini, se levait pour chanter.
Comme ce grand lion dont Daniel fut l'hôte,
L'océan par moments abaissait sa voix haute,
Et moi je croyais voir, vers le couchant en feu,
Sous sa crinière d'or passer la main de Dieu (ll. 43–54).

It is important to point out one interesting difference between the use of an image by Hugo and the use of a similar one by Byron: whereas the latter's persona claims to have 'laid' his 'hand upon' the 'mane' of the sea, Hugo's persona thinks he sees 'the hand of God passed under its mane of Gold'. Each employs a similar image of the sea as a charger: Byron uses the horse symbol to express an unconventional egotism; Hugo employs this same symbol to voice a conventional religious vision, which relies heavily on Biblical imagery.

Then follows a vision of humankind caught in religious strife:

> However, side by side with this impressive fanfare, the other voice, like a cry from a courser who has been frightened, like the rusty hinge of a gate of Hell, like a brazen bow on an iron lyre, it ground: and tears, and cries, the insult, the anathema, refusal of mass and refusal of baptism, and malediction, and blasphemy, and yells, in the cycling waves of human noise, they passed by, like at twilight when you see in the valleys the sable birds of night flying by in a flock. What noise was this whose thousand echoes vibrated? Alas, it was the earth and mankind who wept.

Cependant, à côté de l'auguste fanfare,
L'autre voix, comme un cri de coursier qui s'effare
Comme le gond rouillé d'une porte d'enfer,
Comme l'archet d'airain sur la lyre de fer,
Grinçait; et pleurs, et cris, l'injure, l'anathème,
Refus du viatique et refus du baptême,
Et malédiction, et blasphème, et clameur,
Dans le flot tournoyant de l'humaine rumeur
Passaient, comme le soir on voit dans les vallées

De noirs oiseaux de nuit qui s'en vont par volées
Qu'était-ce que ce bruit dont mille échos vibraient?
Hélas! c'était la terre et l'homme qui pleuraient (ll. 55–66).

The lines contain no proto-ecological vision.

Hugo's persona can perceive no eternal unity. For him, the cosmos is divided into two parts. On the one hand, there is the sea as a symbol for Nature, controlled only by God; on the other, there is the land as a symbol for Humanity, seemingly abandoned by God: 'Brothers! of the two weird voices, ceaselessly reborn, and ceaselessly die away, who the Eternal hears during eternity; the one said: *Nature*! and the other: *Humanity*!'

Frères! de ces deux voix étranges, inouïes,
Sans cesse renaissant, sans cesse évanouies,
Qu'écoute l'Éternel durant l'éternité,
L'une disait: NATURE! et l'autre: HUMANITÉ! (ll. 67–70).

Again, the reader is reminded of a nature/human split. The poet ends the poem in a state of questioning, which remains unresolved:

Thus I meditated; for my faithful soul, alas, had never flown with nobler plumage; in my gloom never had there been such light of day; and I dreamed a long time, contemplating everything, after the dark abyss that the billow hid from me, the other bottomless abyss in my soul opened out. And I asked myself why does one exist—toward what, after all, does all this tend—of what use is the soul—is it better to exist or to live—and why did the Creator, who alone can read his own cryptic text, mix eternally in a fateful marriage the song of nature with the cries of humans?

Alors je méditai; car mon esprit fidèle,
Hélas! n'avait jamais déployé plus grande aile;
Dans mon ombre jamais n'avait lui tant de jour;
Et je rêvai longtemps, contemplant tout à tout,
Après l'abîme obscur que me cachait la lame,
L'autre abîme sans fond qui s'ouvrait dans mon âme.
Et je me demandai pourquoi l'on est ici,
Quel peut être après tout le but de tout ceci,
Que fait l'âme, lequel vaut mieux d'être ou de vivre,
Et pourquoi le Seigneur, qui seul lit à son livre,
Mêle éternellement dans un fatal hymen
Le chant de la nature au cri du genre humain? (ll. 71–82).

The natural and the human are caught in a bad alliance—this is not Wordsworth's vision.

On the other hand, in the works by Hugo and Wordsworth the clouds/waters seen and heard from the mountain tops bring thoughts of God into the minds of the poets. Each poet's concept of God differs considerably in conception: Hugo's conception of God expressed in this passage is of a deity outside and beyond nature—yet, in control of it; this conception seems to derive directly from eighteenth-century deism. Wordsworth's conception of God is partly pantheistic, hence inseparable from nature. Thus, whereas Wordsworth sees only 'The perfect image of a mighty Mind', Hugo closes his poem with a question about evil in God's universe, asking 'why did the Creator—who only can read his cryptic text—eternally counterpoint the song of nature with the cries of humans in the hymn of destiny?'

Both a theistic conception of God and the sea infuse Franz Liszt's tone poem based on Hugo's poem and bearing the same title. The sea appears in a hymn-like motif that becomes especially prominent near the end of the work, in correspondence with the structure of the poem. The religious strife is represented by fairly conventional battle music. In accordance with the musical tradition that extends back through the aural storms of Beethoven, Gluck, and Rameau, tremolos and trills form a key element in the orchestral texture—sometimes even being elevated to the status of themes—such as the sea/God in 'What One Hears on the Mountain'. A motif that also appears in sea or wind passages has been a standard pattern for an allegro or a presto movement at least since the time of Vivaldi in the early eighteenth century: this motif consists of a quick succession of notes (eighths or sixteenths, usually). Thus the same motif can suggest natural elements or not suggest anything at all, but merely be a fast series of notes. It cannot be stressed too much that musical language is much more subtle and variable in meaning. One needs a verbal title or program, as Liszt has provided (like the other composers treated in the study). Since there is such a title relating it to a sea, this motif can be safely taken as a representation of the turbulent movement of wind and sea.

In Liszt's tone poem and Hugo's lyric on which it is based, the land is the habitat of man—controlled and thereby corrupted by him. The sea is not subject to the control of man but only to that of God ('ses flots que Dieu seul peut dompter', as Hugo put it). Thus too, in the sea passage of Byron's *Childe Harold,* the ocean is 'the throne / Of the Invisible' as well as a place apart from human control where 'Ten thousand fleets sweep over thee in vain; / Man marks the earth with ruin, his control / Stops with the shore'. In contrast, in the Mt. Snowdon passage of Wordsworth's *Prelude,* there is no division of land and sea; the moon and the mist make each part of the land-, sky-, and seascape indistinguishable from the other. The mist-covered sky is a 'still ocean' and below it is 'the Sea, the real Sea' and through 'a fracture in the vapour' comes 'the roar of waters, torrents, streams'. In Hugo's poem, there are two separate sources of the sounds—that of man and that of the sea; in Wordsworth's passage, all things roar 'with one voice'. In order to find an equivalent to Hugo's dichotomy in Wordsworth, it is necessary to look beyond the Mt. Snowdon passage and examine the Hellish cityscape of London in book seven (e.g., from the 1850 version):

> Rise up, thou monstrous ant-hill on the plain
> Of a too busy world! Before me flow,
> Thou endless stream of men and moving things! (ll. 147–51).

The political turmoil leading up to the 1830 Revolution had an even more immediate effect on Hugo's poem. To thereby understand this second voice of Hugo's 'What One Hears on the Mountain Top', we need to return to the lyric (touched on in the introduction), 'Fantasy of a Passerby: Speaking of a King' / 'Rêverie d'un Passant: A' Propos d'un roi', also in *The Autumn Leaves* (2: 20–22). Hugo describes an event on the eve of the 1830 revolution. It is a conscious rhetorical strategy on the part of the poet when he draws on water imagery to express his point, by viewing the mob of miserable ones, as a sea, giving out a great roar: 'The court is giving a gala! while all around it—like a ship under which the sea heaves—ceaselessly stir and roar a vast people, whom the glance of a king could never fathom'.

La cour est en gala, pendant qu'au dessous d'elle,
Comme sous le vaisseau l'océan qui chancelle,
Sans cesse remué, gronde un peuple profond
Dont nul regard de roi ne peut sonder le fond.

This is the human voice, not that of the real sea or the voice of God-as-Nature. This voice represents the sad voice of human misery that is heard by the poet and haunts him on the mountain top.

Old Age and Death: The Ocean in the Water-Cycle

Goethe, in his 'Song in Praise of Mohammed' / 'Mahomets Gesang' (1: 304–05) uses the river at its fullest breadth and running through the flatlands—where the river becomes navigable—as an allegorical symbol for full adulthood. The river-bearing boats might be interpreted with some certainty in direct allegorical terms as the shouldering of adult responsibilities: 'This Atlas carries abodes of cedar on his gigantic shoulders; over his head a thousand flags flap in the breeze, testifying to his magnitude'.

Zedernhäuser trägt der Atlas
Auf den Riesenschultern; sausend
Wehen über seinem Haupte
Tausend Flaggen durch die Lüfte,
Zeugen seiner Herrlichkeit (ll. 60–64).

Death—and, for the Prophet and his disciples, afterlife—is allegorically represented by the sea: 'And thus he carries his brothers, his treasures, his children to the waiting Creator tumultuous with joy'.

Und so trägt er seine Brüder,
Seine Schätze, seine Kinder
Dem erwartenden Erzeuger
Freudebrausend an das Herz (ll. 65–68).

Or as has been said earlier in the poem, the prophet-poet flows 'to the ancient father, to the everlasting ocean, who awaits us with outstretched arms' / 'zu deinem alten Vater, / Zu dem ewgen Ozean, / Der mit ausgespannten Armen / Unser wartet' (ll. 36–39).

The Duddon of Wordsworth's sonnet series (3: 244–61) is also navigable, but the traffic is humbler than that of Goethe's river—as is the man it symbolizes:

CONCLUSION
BUT here no cannon thunders in the gale;
Upon the wave no haughty pendants cast
A crimson splendour; lowly is the mast
That rises here, and humbly spread, the sail [...].

The rocks and falls ('vicissitudes') have been left behind:

While, less disturbed than in the narrow Vale
Through which with strange vicissitudes he passed [...].

Now, the impulse is toward the sea where his process of flow comes to an end:

The Wanderer seeks that receptacle vast
Where all his unambitious functions fail.

Thus the relationship between the life of the poet and the 'life' of the river is clearly established. Above, the river is described as a 'Wanderer' bound toward the sea; below, the Poet is like the stream bound toward Sea/Eternity:

And may thy Poet, cloud-born Stream! be free—
The sweets of earth contentedly resigned,
And each tumultuous working left behind
At seemly distance—to advance like Thee;
Prepared, in peace of heart, in calm of mind
And soul, to mingle with Eternity! (sonnet 33).

The apostrophe 'cloud-born Stream' introduces the larger cycle pattern by its formal return to the beginning sonnets of the series, where the apostrophe is 'CHILD of the clouds!' (sonnet 2).

In the rivers mixing with the sea, Wordsworth sees a symbol for the oblivion of death similar to that used by Byron in his 1816 Fragment, where the river flows 'until it glides / Into the number of the nameless tides'. But Wordsworth is not content with a final vision of mere change and death, for he adds another perspective to the symbol of the Duddon River by means of an appended sonnet entitled 'After-thought'.

This final sonnet partakes of the unlabored naturalism of Wordsworth's 1805 *Prelude,* perched as it is between the topographical poetry of the *Descriptive Sketches* and the heavily moralizing poetry of the main section of *Duddon River,* which often reduces the stream image to tendentious allegory. The 'After-thought' is different in that it derives from the actual mechanism of the stream and draws forth an even deeper concept than the poem proper. To call attention to the importance of this poem to the entire series, the poet had it printed with emphasis (italics). In this sonnet, Wordsworth states that the river—which was previously a symbol of transience, growth, and death—is rather eternal and fixed; for the river which always changes and 'passes away' remains the same in form and function:

AFTER-THOUGHT

I THOUGHT of Thee, my partner and my guide,
As being past away:—Vain sympathies!
For, backward, Duddon! as I cast my eyes,
I see what was, and is, and will abide;
Still glides the Stream, and shall for ever glide;
The Form remains, the Function never dies [...].

Thus the river cannot die like mortal man:

While we, the brave, the mighty, and the wise,
We Men, who in our morn of youth defied
The elements, must vanish;—be it so!

But in the final section of the sonnet, the life of man, like that of the river, comes in for its own 'after-thought'. Man's mortality may be only at the surface of things like the beginning-to-end flow of the river; man's immortality—especially a poet's—may be the earthly legacy he leaves behind or the transcendent hopes of his Christian faith:

Enough, if something from our hands have power
To live, and act, and serve the future hour;
And if, as toward the silent tomb we go,
Through love, through hope, and faith's transcendent dower,
We feel that we are greater than we know (sonnet 34).

Thus, this 'After-thought' is not, as the title indicates, merely tacked on to the series; it is not an inconsistent usage of the image as it is in Stolberg's 'The Stream from the Cliff: Immortal Youth'; rather, this final sonnet in the series is an example of Wordsworth's drawing back and making further reflections concerning something which he had previously felt or created—setting the symbol in its full complexity.

By means of a careful reading of the water imagery and its allegorical usage, we can see that *The River Duddon,* taken as a whole with both its 'Conclusion' and 'After-thought' sonnets, is a further example of the poet's fascination with water and its ambiguous nature of seeming at once to change and to remain the same—that is, of water's ability to symbolize the paradox of permanence in change. The 'Function' or process of the stream will continue because, after its water has flowed into the sea, it does not remain there in stasis (symbolically speaking, 'in death'); this water will evaporate, form into clouds, condense, and precipitate as rain or snow. These latter processes form the next quarter of the cycle, to be treated in the chapter that follows.

Wordsworth's 'Conclusion' offers another of those recalcitrant symbols that goes against the intentionality of the poet. Wordsworth here and elsewhere shows no sympathy with reincarnation. Not even at his most heretical moments, Spinozian or Neoplatonic, does he endorse reincarnation. Wordsworth's final sonnet, however, juxtaposes the course of water in its cycle with the destiny of a single human being. The hydrological-cycle returns water to its beginning, poetically the spring or glacier; the individual soul, as the symbolic relation implies, returns to a human body. Wordsworth in the 'After-Thought' shows that he understands the disjunction between the symbol and the symbolized.

As in the 'Song in Praise of Mohammed' and *Duddon River,* the river's flowing into the ocean can be used allegorically to represent man's merging with the Infinite through death. Cole's *Voyage of Life* concludes with a vision of death on the ocean—'Old Age'. Here is the artist's program dealing with the ocean part of the symbolism (the cloud portion will be quoted and discussed later):

> Portentous clouds are brooding over a vast and midnight ocean. A few barren rocks are seen through the gloom—the last shores of the world. These form the mouth of the river; the Boat, shattered by storms, its figures of the Hours broken and drooping, is seen gliding over the deep waters. [...]
>
> The stream of life has now reached the Ocean to which all life is tending. The world to Old Age is destitute of interest. There is no longer any green thing upon it. The broken and drooping figures of the Boat show that time is nearly ended [...] (29).

Clearly, in the river's debouching into the ocean—combined with the more obvious boat, ruined and containing a now aging voyager—we have an image of death.

In 'Winter' of Friedrich's *Lebensalter-Zyklus / Cycle of Life's Ages*, representing old age, two old men in the foreground are digging a grave. The lighting suggests that it is dusk and the moon is out. According to Börsch-Supan, 'the sea in the background, into which the river flows, is [...] a symbol of death'. G.H. von Schubert's contemporaneous essay on this series is helpful in interpreting Friedrich's allegory: 'Speed on then, river, down your way! Where your waves flow into the infinite sea on a far distant shore we have heard of a last place of rest'.[65] The scene rendered here of a churchyard and graves, with a stream near the back flowing out to sea, suggests one of the last sonnets in the Duddon series, where a graveyard is associated with a stream that nears its ending in the sea. The narrator of the poem has been following the stream to its end through a cycle of the day. It is evening, near the end of a long hike, he describes himself as a 'pilgrim' looking forward to his rest at the end of day:

THE KIRK OF ULPHA to the pilgrim's eye
Is welcome as a star, that doth present
Its shining forehead through the peaceful rent
Of a black cloud diffused o'er half the sky [...].

The undertone suggested is the rest at the end of a weary life:

How sweet were leisure! should it yield no more
Than 'mid that wave-washed Churchyard to recline,
From pastoral graves extracting thoughts divine;
Or there to pace, and mark the summits hoar
Of distant moonlit mountains faintly shine,
Soothed by the unseen River's gentle roar (sonnet 31).

The cycles of day, the hike, the river (in one sense), and of life are all reclosed in the penultimate line when the narrator looks back toward the mountains where he began the day-long hike at the source of the Duddon.

Just as Wordsworth, as we have seen earlier, suggests in the 'After-Thought' sonnet the tension in employing the water-cycle as a symbol for human destiny, so too does Friedrich push the hydrological-cycle to it full limits. The sea is used not only as an allegory for death in the *Cycle of Life's Ages*; it is used as an open, multifaceted symbol of resurrection, of the source of new life, and of the link in the life cycle; all of these meanings are suggested in the 'Sea with Rising Sun' / 'Meer mit aufgehender Sonne'. Concerning this picture, Börsch-Supan writes: 'The sea is an image [symbol] of eternity. The sun stands for God, and the foam on the crests of the waves for the impermanence of existence. The clouds, which are formed over the sea through the influence of the sun, are the source of the rain, and the origin of the river' which runs through much of the *Life Cycle*.[66]

The 'Sea with Rising Sun' bears a strong resemblance in choice of subject to 'Sunrise at Sea' and 'Sunrise'. Although the latter two works reveal no detectable allegorical purpose in their images, all three works share a stark simplicity. The sea, sky, and sun ball can be seen as abstract forms or patches of color which stray into

65 *Friedrich,* Tate Gallery 83, 106.

66 *Friedrich,* Tate Gallery 82; see also Platte 13.

the realm of *imago ignota* or abstract shapes and symbols, usually related to the four elements—earth, air, water, and fire (here represented by sunlight). Finally, while the Friedrich panel portrays the unity of the water-cycle by being itself a part of a cycle of life, in Turner's two works the all-enveloping mist creates a unity of sky and sea. The 'Sunrise' in particular embodies the most extreme form of Turner's vision of the all-embracing unity of things and is described by Finberg as follows: 'The bubbling waters are but clouds made captive, the clouds the freed souls of the brooks'.[67] In 'Sunrise' the mist is but the exhalation of the sea, which will eventually draw this water vapor back into itself again. Lord Clark describes this unity in a different way from Finberg by saying that Turner's 'chief aim in life was to see the sun rise above water: he owned a number of houses from which he could see this happening. And he was particularly fascinated by the line where the sky and the sea join each other, that mingling of the elements which seems, by its harmony of tone, to lead to a general reconciliation of opposites'.[68] Indeed, I think, in 'Sunrise' Turner dissolves this 'line where sky and sea join', implying that the two are not really 'opposites' and that they find a common unity in water—whether liquid or vapor.

Friedrich's 'Sea with Rising Sun' and Turner's two color sketches imply the close link between the waters of the sea and of the sky—evaporation. To find in poetry a comparable vision of the world ocean and its place in the cosmic process of the water-cycle, I turn to Goethe's later poetry for *Faust, Part Two*. The passage presented comes amid a whole series of circle/cycle images and symbols and just before Homunculus, a man-created man, dashes himself against a rock to spill out, perhaps for greater regeneration, into the sea. Here, Goethe has Thales—the Ionian Greek philosopher who proposed that water was the first cause and sustainer of things—sing the praises of the ocean as the prime agent of the water-cycle:

> Hail! Hail again! How my joy culminates in being penetrated by beauty and truth: all things have their origin in water! all things are sustained by water! Ocean, may you always favor us with your agency! If you were never to send the clouds, never to give abundant streams, never to change back and forth the directions of rivers, never to completely replenish the streams, what would the mountains, the plains, and the world be like! It's you who preserves what is most freshly vital!

> THALES: Heil! Heil aufs neue!
> Wie ich mich blühend freue,
> Vom Schönen, Wahren durchdrungen:
> Alles ist aus dem Wasser entsprungen!
> Alles wird durch das Wasser erhalten!
> Ozean, gönn uns dein ewiges Walten!
> Wenn du nicht Wolken sendetest,
> Nicht reiche Bäche spendetest,
> Hin und her nicht Flüsse wendetest,
> Die Ströme nicht vollendetest,
> Was wären Gebirge, was Ebnen und Welt!
> Du bists, der das frischeste Leben erhält! (ll. 8432–43).

67 A.J. Finberg, *Turner's Sketches and Drawings* (1960; rpt. New York: Schocken, 1968) 37.

68 Kenneth Clark, *Civilization: A Personal View* (New York: Harper, 1969) 287.

Freed from allegory, the water-cycle is allowed to stand empirically, as a cosmic vision contained in a few lines.

CHAPTER FOUR: LINK

Chapter four offers the link that will give closure to the water-cycle model: the processes of clouds and rain that complete the circuit and that constitute above all others the most impressively Romantic aspect of the water-cycle. The sky or air quarter adds a new dimension to the hydrological-cycle: a new mode of change. The first three of the four quarters have brought to the model only *translational* changes—changes of place. This quarter brings the more fundamental and dramatic mode: the *substantial* changes of evaporation, vaporous condensation, and precipitation. These processes provided the link in the cycle with absolutely certain knowledge of where the water moves rather than with conjecture. What is more essential is that the key scientific works on hydrology and meteorology were being produced after 1760: these works offered yet more changes, which this time might be termed *informational* changes concerning the water-cycle. There were even a few individuals within the realm of the humanities who were aware of this new information; however, these few—including Shelley and Goethe—understood these discoveries to a surprising degree. For the rest, these discoveries were part of—or were fast becoming part of—a less-exact knowledge, a climate of opinion, to use Whitehead's term. Yet, the knowledge closing the link, even at its simplest, was known and used by the few who kept up with developments in science; it is for this reason that this chapter is shorter than the previous two; only the consideration of the important scientific discoveries prevents it from being the shortest of them all. The works this chapter treats, however, are especially remarkable because—to a somewhat greater degree than the other works treated previously—they could only have been first produced by the culture termed Romantic.

Toward the Romantic Link in the Water-Cycle

Scientists, as we saw in the introduction, had already filled in the knowledge gap concerning sources in the hydrological-cycle by the late seventeenth century; they proved that rainfall was sufficient to provide more than enough water for the sources within a river basin, allowing plenty left over to account for run-off. Poets, however, took at least a century to catch up, since ideas need time to spread through the various epistemic strata of a culture. This completion of the cycle and its magnificent employment as a symbolic rhetoric of great complexity is the glory of the (Neo)Classical-Romantic-Realistic creators. Among these were specific painters and poets who avidly read, even researched, in science; the rest seem to have picked up the theory by cultural osmosis from the climate of opinion, which was created by the widely spreading interest in weather. Marjorie Hope Nicolson—a great scholar of the effect of science upon the humanities—sees this trend well underway by the eighteenth century from Thomson and others who formed 'a generation of poets even more conscious of science than their predecessors'.[1]

---

1 Marjorie Hope Nicolson, *Mountain Gloom and Mountain Glory* (1959; New York: Norton, 1963) 330.

Now more than ever, a consideration of Henry Vaughan's waterfall poem will set the forebears off from their (Neo)Classical-Romantic-Realistic successors. There is, also, no clear indication that Vaughan, when composing 'The Water-fall' around 1650, had any direct knowledge of water-cycle research and theory. Alan Rudrum, in notes to the poem, uses the indirect evidence of citing lines from a 1605 poem by Joshua Sylvester to prove that Vaughan knew of the water-cycle with its corollary processes of evaporation and precipitation:

> The purest humour in the Sea, the Sun
> Exhales in the Air, which there resolved, anon
> Returns to water; and descends again
> By sundry ways unto his Mother Main.[2]

These lines, however, indicate only a partial knowledge of the water-cycle that could have been found in Lucretius; the phrase 'by sundry ways' is vague in the extreme. Still it must be said that what these lines do say is correct: they do offer an attempt at a full vision of the process. They further demonstrate that a partial vision of the water-cycle was available for poets.

Vaughan's use of the water-cycle, in 'The Water-fall', differs from that of Sylvester: Vaughan employs the image for symbolic purposes; further, as we will see, Vaughan manipulates scientific knowledge or lack of it for his own purposes. By means of a series of questions, lines 13 through 26 of 'The Water-fall' introduce this cycle by its effect rather than its process. With a certainty that is more religious than scientific, the poet implies that the same water has moved before down this river's channel and into the sea, from where it rises again, and then moves back to the source:

> Dear stream! dear bank, where often I
> Have sate, and pleas'd my pensive eye,
> Why, since each drop of thy quick store
> Runs thither, whence it flow'd before,
> Should poor souls fear a shade or night,
> Who came (sure) from a sea of light?
> Or since those drops are all sent back
> So sure to thee, that none doth lack,
> Why should frail flesh doubt any more
> That what God takes, hee'l not restore?
> 0 useful Element and clear!
> My sacred wash and cleanser here,
> My first consigner unto those
> Fountains of life, where the Lamb goes? (ll. 13–26).

These questions are more than merely rhetorical; indeed, they are exactingly dialogical in that the reader is expected to meet each question with an absolute Christian faith to the effect that no true believer need fear because, first, waters and human souls are part of the eternal workings of God's divine plan, and second, the process of the water is eternal, as is the soul of man. Capping this catechism is a

---

2 Qtd. in *Henry Vaughan: The Complete Poems*, edited by Alan Rudrum (New Haven: Yale UP, 1981) 638, n. 16.

final apostrophe to God. The questions, then, are ritualistic; the answers, presumably those of the poet, are as inevitable as those in a catechism are inevitable.

Either in lieu of certain scientific knowledge or in spite of it, the poet seems unconcerned with the precise material means by which the water moves from the sea to the spring; indeed, he looks upon it as a divine mystery, one that can only be comprehended with faith and divine inspiration. Vaughan employs the hydrological-cycle's ever-replenishing water as a symbol of God, the ultimate source of human immortality; thus, it is the lacuna in the poet's scientific knowledge of the water-cycle that becomes the symbol of the divine mystery:

> What sublime truths, and wholesome themes,
> Lodge in thy mystical, deep streams!
> Such as dull man can never finde
> Unless that Spirit lead his minde,
> Which first upon thy face did move,
> And hatch'd all with his quickning love (ll. 27–32).

Vaughan's focus in lines 13 through 32 has been on theology; one finds, accordingly and appropriately, a lack of precision in the verbal description of external nature. We have, in microcosm, the view that science must necessarily be subsumed under orthodox theology.

The imagery becomes more focused when the poet, in lines 33 through 40 concluding the poem, drops the water-cycle and turns to the turbulent flow of the waterfall, a familiar and—for him, better understood—aspect that bears directly on the favorite Baroque theme of human transience:

> As this loud brook's incessant fall
> In streaming rings restagnates all,
> Which reach by course the bank, and then
> Are no more seen, just so pass men.
> 0 my invisible estate,
> My glorious liberty, still late!
> Thou art the Channel my soul seeks,
> Not this with Cataracts and Creeks (ll. 33–41).

These final lines of the lyric indicate, to a considerable extent, that Vaughan's use of water imagery draws not only from nature but also from scripture—especially from Ecclesiastes 1:7—'All the rivers run into the sea, yet the sea is not full; unto the place where the rivers come thither they return again'. The precise hydrological way that all of this happens has little bearing on Vaughan's poem or his Scriptural source: scientific 'how?' becomes completely submerged in metaphysical 'why?'

It could be that, for Vaughan, faith fills in the gaps of his scientific knowledge; or it might be that, even if the poet had possessed the knowledge, he would not have employed it. His countryman and contemporary, John Milton, might be considered briefly as a parallel in this instance. Milton, during his sojourn in Italy, had met the great pro-Copernican scientist Galileo; moreover, he came to learn more and more about the new heliocentric theory during his later life; he even includes such knowledge in a dialogue between Raphael and Adam in book eight of *Paradise Lost* (especially ll. 120–30); however, the cosmos of this great epic is based solidly on Aristotlean and Ptolemaic astronomical theories. He did so, as Lawrence Babb suggests, primarily because Milton's 'acceptance of the heliocentric hypothesis would mean the rejection of astronomical concepts which had been tested through the centuries by the human reason and which were—in a fashion—compatible with Scripture'.[3] Something quite similar may be true of Vaughan's 'Water-fall': either through a lack of knowledge or—what is more likely—through an unwillingness to employ that knowledge poetically, he has failed to offer the concept of the relation of rain and springs that would incorporate a more, if not entirely, complete vision of the hydrological-cycle.

Yet, even allowing for a certain artistic and metaphysical traditionalism, 'The Water-fall' still shows signs of being touched by those attitudes which, at the time the lyric was composed, were contributing to the creation of modern science. Although Vaughan's knowledge of the water-cycle appears to have been incomplete (like that of his scientific contemporaries), he is drawn to contemplation and speculation on the movements of the waters of the world. This poet's description of water, specifically his use of images, bears comparisons with Ruisdael's rendering of the same element and with that of Rubens. A poet's aural evocation of water, particularly through assonance and alliteration, can be compared to a similar recreation by painters' visual ones, especially through varied brush strokes. While the poet and these painters borrow from art, they all seek to reproduce the actual. They have carefully observed water with eye or ear and have reproduced something that bears comparisons with the waters as found in nature. Like Rubens and Ruisdael, Vaughan partakes—if often tentatively—of the incipient spirit of modern empiricism.

Some painters and poets, however, chose either the new theory or, at least, the correct aspects of the old. We must now recall that the older theory allowed for movement from the sea, to the sky, and finally, through the rain and the run-off water, to provide part of the water for the streams; however, these painters and poets joined the old-school scientists in denying that enough rain water would have fallen to provide the discharge of the springs. They either lacked the knowledge of or ignored the work of those contemporaneous scientists who were providing a new hypothesis with the required empirical measurements to prove that the water from rainfall was not only enough for run-off but also for spring water. And after all, the scientists of the eighteenth and early nineteenth centuries had yet to make their contributions to fill in the gaps on precisely how evaporation and precipitation function within the cycle. It must be admitted that even this partially complete view offers a context in which to place one part of the cycle and makes an appeal to a broader scientific conception. Whether this full appeal is made to the correct, partly correct, or incorrect aspect, it remains a conscious decision on the part of the artist to draw inspiration from the scientific process.

---

3 Lawrence Babb, *The Moral Cosmos of Paradise Lost* (East Lansing: Michigan State UP, 1970) 92. Concerning this, also, Nicolson writes: 'Philosophically Milton denied the idea of infinity, as theologically he denied the God of Plenitude. But aesthetically Milton was [...] responsive [...] to the vastness of a universe in which Infinite God was reflected in Infinite Space' (274).

In the Vienna Kunsthistorisches Museum, there is a painting by Rubens that is known both as *Philemon and Baucis* and *Landscape with Thunderstorm,*[4] in which waterfalls fill the canvas; but what is more, the waterfalls are shown in context with other parts of the cycle. In the upper-left middle of the canvas a huge thunderstorm and downpour is in progress; a mammoth cirrocumulus cloud fills the sky, up and outside the canvas; and large streaks of falling rain are boldly rendered with long vertical and diagonal brushstrokes. The part of the cycle that moves from cloud to rain, from run-off (instead of spring) to stream and waterfalls is shown with empirical or, perhaps, scientific precision. There is no clear indication as to whether or not Rubens knew about the current developments in hydrology; however, since Rubens moved in the most cosmopolitan and intellectual circles in both England and France, he had every opportunity to meet the scientists doing research in this area.

The importance of clouds in Ruisdael's works and the importance of Ruisdael's clouds in art history are set forth by Clark so cogently, in his *Landscape into Art,* that I feel it necessary to quote *in extenso*:

> This feeling he [Ruisdael] expressed, as all the greatest landscape painters have done, through the large disposition of light and dark, so that even before we approach his pictures we feel their dramatic significance: it is only on looking more closely that we find them full of observed facts. Once more light lifts these facts on to a new plane of reality; but here light has a new character. [...] It is in continual movement. Clouds pile up in the sky, shadows sail across the plain.[5]

An examination of Ruisdael's *The Landscape with a Ruined Castle and a Church*[6] bears out Clark's analysis even if some qualifications must be made concerning his application of 'motion' to the static art of painting. First, the shapes of the cumulus and nimbus clouds are carefully rendered with an intense chiaroscuro—conveying a strong sense of corporal extension and opaque mass. Second, the shapes of the clouds, especially the nimbus clouds on the left, are so precariously situated in unstable positions as to make one aware of the transient motions of the clouds, showing that they would in actual nature soon shift to new positions and that, for all their expansiveness, they are things of air and water vapor. Third, the land below is rendered in such a way that the clouds appear to screen out the sun at some points and let it through at others: the artist used this fragmentary lighting (or, at least, a pretense of it) in order to highlight some points of the landscape and darken others. Here, too—perhaps only by association with similar phenomena one has observed in nature—we sense motion in the patches of light, which seem to have been frozen in their rapid passage across the landscape.

Most significant for us, now, is Lord Clark's sweeping statement that Ruisdael's clouds show that the clouds of Constable and his contemporaries have a context—a

---

4 Peter Paul Rubens, *Philemon and Baucis, or Landscape with Thunderstorm,* Kunsthistorisches Museum, Vienna. The Louvre's *Deluge* by Nicolas Poussin offers a few parallels; however, both the clouds and water are rendered schematically.

5 Kenneth Clark, *Landscape into Art* (New York: Harper, 1976) 64.

6 Jacob van Ruisdael, *The Landscape with a Ruined Castle and a Church,* National Gallery, London; illus. in *National Gallery Catalogue* 71 or, for an excellent print, see illus. 35 in *The Treasury of Painting through Five Centuries,* by William Gaunt (New York: Excalibur, 1976).

line of descent, as it were—and should not be viewed as a complete break with or rebellion against the past:

> Constable said that the best lesson on art he ever had was contained in the words, 'Remember light and shadow never stand still'. Nowhere, in the art which preceded him, was this lesson so beautifully demonstrated as in the paintings of Ruisdael, and in fact Ruisdael was to become the model for the whole East Anglian school of landscape painting.[7]

On the very basic level of a fascination for its mutability, Ruisdael and Vaughan can be cited as forerunners of the Romantic water symbol; on the basis of an attraction to its cataclysmic flux, Rubens certainly also can; moreover, both the Baroque and the Romantic artists share considerable elements in their treatment of water—in opposition to the eighteenth-century Neoclassicists that separates them. There is one trend that the Baroque, the Neoclassical, and the Romantic artist all participate in: an opening out spatially of their creative visions. By the eighteenth century, this widening of view and greater complexity of content is well underway, as Nicolson describes: 'The eighteenth-century earth has become a wide, wide world indeed, with extraordinarily extensive panoramas and prospects, in which mountains, plains, rivers, and seas are on a majestic scale and caverns seem actually measureless to man'.[8] Taking an example from Thomson's *Seasons*, Nicolson also sees a corresponding growth in 'the widespread interest of the descriptive poets in volcanic eruptions, earthquakes, and tempests. Storms and other violences of Nature were nothing new in poetry, but no generation before this had treated [them] in such detail'.[9]

In one crucial respect my treatment of seventeenth- and eighteenth-century painting and poetry share one common methodological tool: water symbols of the Baroque can be distinguished from descriptive and topographical images of eighteenth-century Neoclassicism. A contrast of the rendering of skies in Wilson's *The Valley of the Dee* and *Holt Bridge on the River Dee* with that in Ruisdael's painting demonstrates the different intentions between these Neoclassical and Baroque approaches. The skies in Wilson's two paintings are filled with vaguely rendered clouds, a dull generalized lighting with little shading, and a smooth, oily application with no impasto. On the other hand, in Ruisdael's *The Waterfall*, the sky is filled with clouds that are rendered with as much attention as the water. Ruisdael has painted, in the upper left with a dappled effect, what Luke Howard has since taught us to term cirrus clouds. The cumulus and nimbus clouds which dominate the upper right are rendered with an impasto application and an extremely high-contrast chiaroscuro. In short, whereas Ruisdael concentrates on capturing the flux of the water itself, Wilson concentrates on the environs and the human figures. This comparison demonstrates that the chronologically more distant Baroque style of the seventeenth century was more the precursor of Romantic treatments of water phenomena than was the nearer Neoclassism of the eighteenth century.

Neoclassicism, in poetry if not in painting, bears some examples of proto-Romanticism and (proto)environmentalism. The seeds, at least, of a more comprehensive

---

7 Clark, *Landscape into Art* 64.

8 Nicolson 332.

9 Nicolson 339.

vision of clouds—one approximating a scientific conception of them—are apparent in Thomson, although clouds are included by him as an addition to the catalogue of natural phenomena in *The Seasons*. The following cloud passage from *Summer* might be taken as an example:[10]

> As from the face of Heaven the shattered clouds
> Tumultuous rove, the interminable sky
> Sublimer swells, and o'er the world expands
> A purer azure. Nature from the storm
> Shines out afresh; and through the lightened air
> A higher lustre and a clearer calm
> Diffusive tremble; while, as if in sign
> Of danger past, a glittering robe of joy,
> Set off abundant by the yellow ray,
> Invests the fields, yet dropping from distress (ll. 1223–32).

While this passage contains an image use indicative of eighteenth-century descriptive poetry, it also reveals several characteristics of cloud images and symbols contained in a significant body of poems written after 1770. In this passage, the persona uses impressive words to convey the sense of clouds in a culminating moment of change, when a storm is breaking up to reveal the sun, for example—'shattered', 'tumultuous', 'sublimer swells', 'higher lustre', and 'diffusive tremble'. Thus, while heavily descriptive, it is not statically 'pictorial' nor is it completely lacking in a conceptual framework; for the poet has chosen a change-fraught image and, in the voice of his persona, strives to describe it in all its nuances. Finally, the heavy—almost excessive—use of adjectives, adverbs, and verbs denotes a certain enthusiasm for natural phenomena.

This enthusiasm for rendering, in painting, the details of changing natural phenomena first appears in Loutherbourg's landscapes—and more especially in Diderot's art-theoretical works written under the inspiration of them. Indeed, Loutherbourg's greater significance for this study lies in the impetus his early canvases gave to the first great theoretical expression concerning the use of water phenomena in landscape painting, and by extension in poetry, in music: this expression comes in Diderot's writings concerning the canvas which Loutherbourg exhibited at the Salons of 1763, 1765, 1767, 1769, and 1771—especially that of 1765. Generally regarded as a great harbinger of Davidian Neoclassicism, Diderot was no less perceptive as a theorist *ante factum* of Romantic landscape painting—specifically, with regard to water images. One of Diderot's instructions to landscape painters[11], while written with Loutherbourg in mind, bears striking parallels with Turner's seascapes in the *Colour Beginning* series: 'Go out before the return of the sun. View his veiled disk, the obscured borders of his orb, and all the extent of the mass of the rays lost, dispersed, smothered in the immense and profound mist which only receives a faint and roseate tint. Already the nebulous

10 This passage was brought to my attention by J.R. Watson, *Picturesque Landscape and English Romantic Poetry* (London: Hutchinson, 1970) 30. Watson uses it and others to make some important comparisons between Thomson, Constable, and Turner.

11 Denis Diderot, *Salons*, 2nd ed., 4 vols., ed. Jean Seznec (Oxford: Clarendon, 1979).

mass begins to sink under its own weight: it condenses near the ground; it dampens and refreshes it, it drenches it, and the softened glebe sticks to your feet'.

> Devance le retour du soleil. Vois son disque obscurci, les limites de son orbe effacées, et toute la masse de ses rayons perdue, dissipée, étouffée dans l'immense et profond brouillard qui n'en reçoit qu'une teinte foible et rougeâtre. Déjà le volume nébuleux commence à s'affaisser sous son propre poids: il se condense vers la terre; il l'humecte, il la trempe, et la globe amollie va s'attacher à tes pieds.

Only Turner would eventually prove to have the temperament, intellect, and technique to carry out Diderot's strictures on the effect of light in mist and fog.

Friedrich, as well as Wordsworth, was to gloriously realize Diderot's vision concerning 'the ocean of clouds' seen from a mountaintop: 'Turn your observations toward the mountain top[s]. Those that begin to pierce the ocean of clouds. Run; quickly climb upon some high hill; and from there contemplate the surface of this ocean which gently undulates over the earth'.

> Tourne tes regards vers le sommet des montagnes. Les voilà qui commencent à percer l'océan vaporeux. Précipite tes pas; grimpe vîte sur quelque colline élevée; et de-là contemple la surface de cet océan qui ondule mollement au-dessus de la terre [...].

Water phenomena both in the sky and on the earth are assigned by Diderot for the painter's study: 'Take up the brush which you have just soaked in the light, in the water, in the clouds; the diverse phenomena that fills your head only demand—as they fade away—to be preserved upon the canvas'.

> Prends le pinceau que tu viens de tremper dans la lumière, dans les eaux, dans les nuages; les phénomènes divers dont ta tête est remplie, ne demandent qu'à s'en échapper et à s'attacher à la toile (2: 165).

Perhaps, in the realization of these subtler aspects of light effects upon water and in the clouds, the real honor will ultimately belong to Constable. Only his stippling facture in white and yellowish-white was to prove able to evoke or signify in the frozen realm of paint on canvas such evanescent effects as the *philosophe* describes.

Diderot bids the painter to be conscious of all the meteorological changes and corresponding light changes in the sky, leading ultimately to storms, rain, and on the earth below, torrents:

> While you are occupied—during the torrid hours of the day—at painting the freshness of the morning hours, the sky prepares for you new phenomena. The light dims; the clouds are agitated, separate, join, and the storm begins to brew. Go watch the storm originate, break, and finish; and when—in two years from now—I find on display at the Salon the trees broken by the storm, the torrents which it has swollen, and all the scene of its ravages; and then, my friend and myself both leaning on the other for support, with our eyes glued to your painting, we will re-experience the terror of the scene.

> Tandis que tu t'occupes, pendant les heures brûlantes du jour, à peindre la fraîcheur des heures du matin, le ciel te prépare de nouveaux phénomènes. La lumière s'affoiblit; les nuages s'émeuvent, se séparent, s'assemblent, et l'orage s'apprête. Va voir l'orage se former, éclater et finir; et que, dans deux ans d'ici, je retrouve au Salon les arbres qu'il aura brisés, les torrents qu'il aura grossis, tout le spectacle de son ravage; et que, mon ami et moi, l'un contre l'autre appuyés, les yeux attachés sur ton ouvrage, nous en soyons encore effrayés (165–66).

While the sublime aesthetic of the late eighteenth century is appealed to in the last phrase, none of Diderot's contemporaries, even Loutherbourg, were able to fulfill the *philosophe*'s appreciation of cosmic process. Only Constable and, sometimes, Friedrich would be able to capture these nuances of atmospheric transformation, and only Turner would be able to achieve such a comprehensive spectacle of cloud, rain, and flood. But even poets like Byron and Bécquer, as well as composers like Beethoven and Wagner, utilized the water-cycle rhetoric combined with other cycle rhetoric and with the vastly expanding techniques of their arts to achieve storms of the cosmic grandeur envisioned for painting by Diderot.

It seems no accident that the human catalyst of the *Encyclopedia* that helped to spread the scientific method and technological knowledge throughout Europe and North America should draw science and art into a single creative unity in the conception of the workings of the hydrological-cycle. And in this respect it remained the task of two coevals of Constable and Turner to travel even further in presenting a holistic conception of water phenomena: Friedrich and Cole symbolically employed all four of the quarters of the water-cycle in a series of panels to represent the narrative of human existence.

During Jefferson's stay in Paris, he wrote (12 October 1786) a delightful and whimsical dialogue between his head and his heart to a woman friend, Maria Cosway, who seems to have been a painter of some talent, like her husband. His advice to her about choices of subjects is not whimsical and, in fact, resembles that of Diderot very closely:

> Where could they [Mr. and Mrs. Cosway] find such objects as in America for the exercise worthy of immortality to render her pencil immortal. The Falling spring, the Cascade Niagara, the Passage of the Potowmac thro the Blue mountains, the Natural bridge. It is worth a voiage across the Atlantic to see these objets; much more to paint, and make them, and thereby ourselves, known to all ages.

We have already given considerable attention to 'the Cascade Niagara'. More profoundly, however, Jefferson further offers a 'sublime' water-cycle subject—the storm—viewed from his own mountain home, which contains apperceptions comparable to the works of Diderot, Hugo, Turner, Constable, Goethe, and Beethoven:

> And our own dear Monticello, where has nature spread so rich a mantle under the eye? mountains, forests, rocks, rivers. With what majesty do we there ride above the storms! How sublime to look down into the workhouse of nature, to see her clouds, hail, snow, rain, thunder, all fabricated at our feet! (404).

Jefferson's advice to artists, then, and that of Diderot offer a surprisingly comprehensive view of the sky quarter—preliminary to them are the clouds.

The Cloud Itself

Since the processes of evaporation and precipitation are so crucial to understanding the cloud itself and since these processes form an essential link in the water-cycle, extended and significant poetic usage of the cloud is almost impossible for the artist to convey outside the cycle context. The visual and musical arts are not quite so dependent upon the full model; however, they were vastly aided by the additional knowledge. If such a limited and simplistic usage does occur, it is often heavily descriptive—as is Thomson's passage—or rather superficially symbolic—involving clouds as stimuli for idle imaginings.

Sometimes there may be a profound and analogous relationship between the image of the cloud and the state of mind of the persona, but such a correspondence is rarely immediately apparent. As was pointed out in the last chapter, Heine's descriptive sea- and skyscapes of *Die Nordsee* are strongly reminiscent of the eighteenth-century tradition of descriptive poetry. Thus, the opening lines of 'Sunset' / 'Sonnenuntergang' (164–66) make significant use of cloud imagery:

The glowing red sun sinks down into the wide and blustering, silver-grey ocean; creations of air, touched with rosy breath, surge sunward; and opposite, out of the autumnal and twilit veil of clouds, a mournful, deathly pale countenance—the moon—emerges, and behind her, like tiny twinkles of light, the stars—made misty by distance—are shimmering.

Die glühend rote Sonne steigt
Hinab ins weit aufschauernde,
Silbergraue Weltmeer;
Luftgebilde, rosig angehaucht,
Wallen ihr nach; gegenüber,
Aus herbstlich dämmernden Wolkenschleiern,
Ein traurig todblasses Antlitz,
Bricht hervor der Mond,
Und hinter ihm, Lichtfünkchen,
Nebelweit, schimmern die Sterne.

Just as sunlight and moonlight play an important part in works of Thomson and Goethe, here too Heine concentrates, if perhaps to a greater degree, on the color- and shade-creating effect of the sun and moon—for example, 'red' / 'rote', 'silver-grey'/ 'silbergraue', 'rosy' / 'rosig', 'deathly pale' / 'todblasses'. The clouds in the west are lighted by the sun, while the clouds in the east are under the influence of the moon.

In the great Turner legacy left to the state, there was a large collection of watercolors and oils, usually capturing some rare effect of light on a land-, sea-, or skyscape and somewhat misleadingly termed *Colour Beginnings* by art historians. I have chosen two of Turner's watercolors, which in turn come from a 'sunset sequence', where he shows the changes of cloud forms and light. In the first, the red-orange of the sunset glow has been rendered with a wet wash, while the cirrus clouds have been achieved by dry brush stipples and slashes. In the second, the sky area has changed from red-orange to a silver-grey. Although such direct comparisons are dangerous because one artist is working with color and the other with color-adjectives, the sunset (with moonset) of Turner's sketch closely parallels the opening lines of Heine's 'Sunset'.

Heine's description of the sunset strikes the tone for the story of discord among the Gods, in particular, between the husband, Sol, and Wife, Luna: 'But evil tongues were hissing discord, and the noble, radiant wife and husband separated in enmity'.

Doch böse Zungen zischelten Zwiespalt,
Und es trennte sich feindlich
Das hohe, leuchtende Eh'paar (ll. 16–18).

The persona feels some reconciliation to his mortality in knowing that the gods have an immortality filled with strife: 'And the poor gods—high in the heaven—wander, agonizing and hopeless on infinite paths, and cannot die, and behind them they drag their radiant misery. But I, the human, the lowly planted, death-blessed—I no longer complain'.

Und die armen Götter, oben am Himmel
Wandeln sie, qualvoll,
Trostlos unendliche Bahnen,
Und können nicht sterben,
Und schleppen mit sich
Ihr strahlendes Elend.

Ich aber, der Mensch,
Der niedrig gepflanzte, der Tod-beglückte,
Ich klage nicht länger (II. 53–55).

For the persona, earthly change is better than a permanence of heavenly discord and, worse, eternal ennui. For the persona as for the poet behind him, change is a thing not to be dreaded, but rather to be celebrated. Mortals are fortunate enough to experience the greatest change of all: death. It is no accident that Heine wrote this lyric on the brink of a major turning point when he would devote himself to a vision of political change. In light of this poem, one might consider for a moment Kenneth Clark's statement that Turner's clouds 'had a symbolic meaning. In his work clouds the colour of blood become symbols of destruction. He identified skies of peace and skies of discord'.[12] Although of course with Heine, we are dealing with adjectives and not with paints, it might still be similarly observed that he also portrays a sky of discord.

The symbolic meaning of Turner's clouds at sunset is more safely interpreted in *Ulysses Deriding Polyphemus* than in the sketches simply because, as in the case of Heine's 'Sunset', there is an accompanying myth. The orange-red and yellow sunset, which dominates the right of the canvas, is an appropriate symbol for the myth in that it foreshadows the trials and tribulations of the hero as well as the death of his crew. The monster's revenge is to be effected by the sea storm which is seen where the sea meets the sky. This storm—which, as one knows from Homer's version and that of Ovid, is being brewed by the father of Polyphemus, Poseidon—is rendered by blackish blue and purple.

Far from perceiving and describing natural phenomena passively, the persona of Lenau's 'To the Clouds' / 'An die Wolke' (1: 67–68) addresses a cloud and demands that it act in sympathy with—that is in a state analogous to—his state of mind: 'Don't speed so quickly over this still heath; don't pass so quickly by my profound sorrow, you clouds above, pause before my woe'.

Zieh nicht so schnell vorüber
An dieser stillen Heide,
Zieh nicht so scheu vorüber
An meinem tiefen Leide,

12 Clark, *Civilization* 287.

Du Wolke in der Höh,
Steh still bei meinem Weh! (ll. 1–6).

Toward the end of the lyric, however, Lenau breaks the subtle web of the analogue between his persona's state of mind and the cloud by calling upon the latter to turn into a storm cloud and to be an agent of revenge on his behalf against the falseness of women.

In his chapter on clouds in *Air and Dreams / L'Air et les songes,* Bachelard discusses a passage from another lyric by Lenau, 'The Tavern on the Heath' / 'Die Heideschenke' (1: 68–73): 'The clouds are the herds—clouds assembled in a winding gallop, while the wind, an excellent horseman, presses them on by cracking "the whip of lightning"'.[13] The stanzas he refers to are ten and eleven: 'The clouds seemed like horses to me, rapidly mixing themselves together and galloping through the echoing quarters like thunder. The storm seemed a valiant groom, singing his cheerful little song, swinging the lightning's whip so that the herd moves properly'.

Die Wolken schienen Rosse mir,
Die eilend sich vermengten,
Des Himmels hallendes Revier
Im Donnerlauf durchsprengten.

Der Sturm ein wackrer Rosseknecht,
Sein muntres Liedel singend,
Daß sich die Herde tummle recht,
Des Blitzes Geißel schwingend (ll. 37–44).

In this passage, the persona actively uses the external, natural phenomena of clouds as material for his imaginative vision. It is significant that this political conservative, having matured in the midst of Metternichian hegemony, must have seen the flux and reflux of a cloudy sky in terms of a controlling order.

An important aspect of the clouds which makes them analogous to horses is their quality of movement and change, as well as their sense of progress in one direction toward and even arrival at an endpoint. Bachelard continues: 'One can say that the contemplation of clouds places us before a world where there is as much form as movement; the movement creates the forms, the forms are in motion, and the motion always unbuilds the forms. It is a universe of forms in continuous transformation'.[14]

In the opening lines of the sonnet 'Fancy in Nubibus, or The Poet in the Clouds', Coleridge describes the clouds as material for the poet/persona's imagination:

O! it is pleasant, with a heart at ease,
Just after sunset, or by moonlight skies,

---

13 My translation of: 'Les nuages sont des troupeaux, nuages assemblés dans un galop tournant, tandis que le vent, bon écuyer, les presse en faisant claquer "le fouet de l'eclair"' (Gaston Bachelard, *L'Air et les songes: Essai sur l'imagination du mouvement* [Paris: Corti, 1943] 222).

14 My translation of: 'On pourrait dire que la contemplation des nuages nous met devant un monde où il y a autant de formes que de mouvements; les mouvements y donnent des formes, les formes sont en mouvement, et le mouvement toujours les déforme. C'est un univers de formes en continuelle transformation' (Bachelard 222).

To make the shifting clouds be what you please,
Or let the easily persuaded eyes
Own each quaint likeness issuing from the mould
Of a friend's fancy; or with head bent low
And cheek aslant see rivers flow of gold
'Twixt crimson banks; and then, a traveller, go
From mount to mount through Cloudland, gorgeous land!

Here, too, the clouds are seen in the moonlight or at sunset; also, considerable attention is given to the description; but the most significant aspect of this passage is that the mind of the persona makes the clouds represent what he wills. But even more important is the fact that the clouds are 'shifting'; for, the constantly changing forms offer a richly varied show, which by its varied movement excites the imagination.

Another example of the use of clouds for fanciful imaginings occurs in a section of Keats' *Endymion*:

Although, before the crystal heavens darken,
I watch and dote upon the silver lakes
Pictur'd in western cloudiness, that takes
The semblance of gold rocks and bright gold sands,
Islands, and creeks, and amber-fretted strands
With horses prancing o'er them, palaces
And towers of amethyst,—would I so teaze
My pleasant days, because I could not mount
Into those regions? (1: ll. 739–47).

This fantasizing of things or creatures into the clouds which takes place in this passage by Keats, as in Lenau's 'Heideschenke' and Coleridge's 'Nibibus', appears in Turner's rendering of his *Polyphemus*: the painter has rendered the giant in such a way as to create ambiguity concerning whether it is a cloud or giant. In his preparatory sketch, there is less ambiguity. The final version teases the viewer—by the title and the rendering of an odd shape on a mountain—to ask questions about what exactly is perceived. The viewer's imagination is thereby activated and challenged. Is it a real giant at which the Homeric Greek shakes his fist? Or is it an oddly shaped cloud on which the pagan hero and indeed his poet's dreams are based? I believe it is the latter: in this work, Turner creates by means of this cloud/giant a commentary upon the mythopoetic process among the pagans.

For poetry and painting, the cloud suggests objects, animals, and people. In Romantic music, the cloud per se, apart from the storm or the cycle, would have to wait for Debussy to achieve its aural apotheosis.

Cataloging the Clouds: Howard, Goethe, and Constable

Strikingly similar in some aspects to the cloud reveries we have just examined are those of the *Howards Ehrengedächtnis / Commemoration of Howard* by Goethe. Like these latter works, the clouds of this lyric cycle are made to represent other things that they only vaguely resemble. It, however, must further be pointed out that these poems offer a paradoxical treatment of natural phenomena; for, in addition to being the objects of poetic fancy, the clouds in these poems are also the objects—or perhaps one should say the results—of empirical observation in accordance with contemporaneous science. Written in honor of, and in accordance with, the theories of the English meteorologist Howard, each of the lyrics is centered on one of Howard's cloud categories. Thus, as Bachelard states, 'It seems that Goethe has desired to found poetic imagination solidly upon empirical knowledge'.[15] In this series of poems, description and imagination, impression and perception, scientific objectivity and poetic subjectivity are all melded into a unified whole.

It is crucial to examine carefully the theories of the dedicatee Howard in order to understand, not just Goethe's great commemorative-poem series but all the works to be treated in the remainder of this chapter. The basic classification of his system was presented in the introduction, which I will summarize:

A. The simple types are:
   1. the high atmosphere *cirrus* clouds, broken clouds, racks;
   2. the mid-level *cumulus* clouds, the fat-bellied cloud;
   3. the lower-level *status,* covering a vast area.
B. The intermediate modifications, which lie at the boundary of the simple types, are:
   4. *cirro-cumulus*; possibly on its way to building up into a nimbus;
   5. *cirro-stratus,* clouds in a turbulent atmosphere, often in an updraft over a mountain.
C. The mixed or compound modifications are:
   6. *Cumulo-stratus*; possibly on its way to building up into a nimbus;
   7. *Cumulo-cirro-stratus,* or *nimbus,* the thundercloud.

Throughout this chapter, I will try to classify the painted or described clouds of poets and painters according to the Howard categories. It will be found that they are very true to types, as well as functions.

As one would expect, given the phenomenon, Howard's emphasis is upon change. Concerning cirrus clouds he writes, 'Their duration is uncertain, varying from a few minutes after the first appearance to an extent of many hours. It is long when they appear alone and at great heights, and shorter when they are formed lower and in the vicinity of other clouds' (6). Howard links this change to other variables, like temperature, found within the cycle of a normal-weather day: 'Their appearance, increase, and disappearance, in fair weather, are often periodical, and keep pace with the temperature of the day. Thus they will begin to form some hours after sun-rise, arrive at their maximum in the hottest part of the afternoon, then go on diminishing, and totally disperse about sun-set'. He further points out that the change is even more drastic when the weather is unsettled: 'in changeable weather they partake of the vicissitudes of the atmosphere; sometimes evaporating almost as soon as formed, at others suddenly forming and as quickly passing to the compound modifications' (7).

The first of a series of lyrics composed in honor of Howard was originally published at the end of Goethe's scientific article 'Cloud Formations According to Howard' / 'Wolkengestalt nach Howard' (17: 647–69) in 1820 and published as a

15 My translation of: 'Il semble que Goethe ait désiré mettre des connaissances objectives à la base même de ces images poétiques' (Bachelard 222).

separate poem in 1822 entitled 'Atmosphäre' (1: 524). In 1820 and again in 1823, Goethe published a series of articles on Howard and a report on the poet-scientist's experimental applications of his classification system: one offers a series of readings at Karlsbad/Karlovy Vary during 1820; another, at Marienbad/Mariánské Lázne during 1823 (17: 647–69). These reports on observations often include the barometric readings, detailed descriptions of the weather and, of course, the clouds typed according to the Howard terminology. In another of the scientific journals Goethe edited (*Zur Morphologie,* 1822), he published a short biographical sketch of the English meteorologist and especially his wide-reaching impact on his own thoughts: 'To what degree I was drawn to Howard's method of cloud classification, to what extent I must have desired to see the formless formed [...] can be seen in my entire work in science and art. I sought to immerse myself in this set of principles, applying them as well as I could at home and away, in every season and under considerably different barometric conditions'.

> Wie sehr mich die Howarisch Wolkenbestimmung angezogen, wie sehr mir die Formung des Formlosen [...] folgt aus meinem ganzen Bestreben in Wissenschaft und Kunst; ich suchte mich von dieser Lehre zu durchdringen, befleißigte mich einer Anwendung derselben zu Hause wie auf Reisen, in jeder Jahreszeit und auf bedeutend verschiedenen Barometerhöhen [...].[16]

I think it is absolutely significant that Goethe feels called upon to draw attention to the classification system and what it classifies; this to us today might seem rather strange; however, if we consider that up until the Romantic period the great prototype for all scientific classifying is that model of the static world view of the Enlightenment: Linneus's botanical typing system, with every plant in its compartmentalized species- and genus-group. Goethe, in the same work, goes on to make clear the importance of visual representation in meteorology by stating that he has asked the artists for help in making illustrations of cloud formations. In the frozen forms of visual arts, the whole feat of categorizing the transitory into discrete units becomes even more paradoxical. Howard, however, desires the same thing. What Howard's classification attempts to organize is that most changeable thing—the weather; accordingly, his classification takes on the incredible feat of seeking to fix and mark-off from one another a series of transitory events.

In the same scientific journal where his prose observations on Howard first appeared, Goethe also published a series of short lyrics on the subject. This cycle, already mentioned, is *Commemoration of Howard* (1: 524–26); it is this latter version which will be examined because its meaning is clarified by some very useful quotation marks, which distinguish two personae. In the first stanza, what seems to be one of the two speakers sets forth the problem: despair in understanding the great world and its workings: '"The world—it is so large and broad—the sky, as well, so noble and wide—all this must I grasp with my eyes although it transcends the power of the mind"'.

> 'Die Welt, sie ist so groß und breit,
> Der Himmel auch so hehr und weit;

16 This, along with other meteorological material, is not contained in the Artimis edition (the most inclusive of the twentieth-century editions) but is in the so-called Weimar edition: *Goethes Werke,* 143 vols. (Weimar: Böhlaus, 1896) 76: 43.

Ich muß das alles mit Augen fassen,
Will sich aber nicht recht denken lassen'.

The problem of the immensity and complexity of the cosmos posed in the first stanza is answered in the second: understanding of that great order comes only through the process by which the scientist, first, collates and describes the multiplicity of phenomena and, then, intuits their unity. The solution offered is, in short, the scientific method of Howard and of Goethe: 'In order to find yourself within the Infinite, you must differentiate, then synthesize; therefore, my winged song thanks the one who classified clouds'.

Dich im Unendlichen zu finden,
Mußt unterscheiden und dann verbinden;
Drum danket mein beflügelt Lied
Dem Manne, der Wolken unterschied.

If one remembers from the introduction that Howard's system of classification was based on the concept that each form was on its way to becoming another form, one sees that the meteorologist has done all in one packet exactly what the poet calls for—'differentiate, then synthesize'. Beyond the seven labels of cirrus, cumulus, or so on, Howard's larger categories or groupings are based on distinguishing one moment of metaphorposis from another: 'simple modifications', 'intermediate modifications', and 'compound modifications'. This fluid grouping and its transformational terminology was pleasing to Goethe, who—like Foucault in our century—called into question the static, discrete classification systems, like that of Linnaeus.

In the opening section of *Commemoration of Howard,* the clouds are described collectively as a Hindu deity: for a devout Spinozean pantheist, this could only be a poetic avatar of the transcendent or hence the eternal nature of the process behind the clouds. Moreover, the persona—whom, with relative safety, we can closely identify with Goethe himself—emphasizes the unity of opposites which this process creates. This unity of opposites, which is expressed through the oxymoron of lightness//heaviness, consists also of the seemingly opposed actions of doing//undoing, of forming//unforming, and of drawing-together//dispersing: 'When Kamarupa's godhead, lofty and noble, swaying through the air wanders lightly and heavily—gathers and scatters the folds of his veil—[he] takes delight in the mutability of phenomena, now holds himself stiff, then dwindles like a dream; we, at that moment, are full of awe and hardly believe our eyes'.

Wenn Gottheit *Kamarupa,* hoch und hehr,
Durch Lüfte schwankend wandelt leicht und schwer,
Des Schleiers Falten sammelt, sie zerstreut,
Am Wechsel der Gestalten sich erfreut,
Jetzt starr sich hält, dann schwindet wie ein Traum,
Da staunen wir und traun dem Auge kaum.

The Hindu mythological reference seems to be to Kama, the God of Love—sometimes a bringer-together and other times a splitter-apart; obviously he is also a fitting emblem for the processes of clouds.

Then, however, the persona's imagination takes over and employs animals as analogies for the cloud's shapes in order to express the naive, childlike fascination with forms and, above all, the *variety* of forms which is the creative basis for the scientific process of perceiving and distinguishing various forms of clouds:

> Now the internal power of creation boldly arouses itself, molding the indeterminate into the determined: here looms a lion, there surges an elephant, a camel's neck changed into a dragon, a multitude starts moving but fails to triumph since its power breaks upon steep crags; the truest cloud-messenger vanishes, before he reaches the distance to where we long to travel.

> Nun regt sich kühn des eignen Bildens Kraft,
> Die Unbestimmtes zu Bestimmtem schafft:
> Da droht ein Leu, dort wogt ein Elefant,
> Kameles Hals, zum Drachen umgewandt,
> Ein Heer zieht an, doch triumphiert es nicht,
> Da es die Macht am steilen Felsen bricht;
> Der treuste Wolkenbote selbst zerstiebt,
> Eh er die Fern erreicht, wohin man liebt.

In these last lines, we have two themes that we shall encounter in other poems treated in this chapter: the cloud as a messenger and as a thing moving freely 'to far places where one would like to go'.

In the penultimate line of the final stanza, the metamorphosis of the cloud is carefully, if concisely reviewed—thereby reminding us that Howard's 'precision' of distinguishing forms is imposed upon a process which unites all these forms in one gradual continuum:

> But with a pure mind, he, Howard, gives us the most glorious prize of a new discipline: what cannot be held, cannot be obtained, he touches it; he is the first to take hold of it; he makes the imprecise precise, confines it, names it exactly!—let the honor go to you [Howard]!—whenever a cloud-fragment rises, builds itself up, scatters, falls, may the world gratefully remember you.

> Er aber, Howard, gibt mit reinem Sinn
> Uns neuer Lehre herrlichsten Gewinn:
> Was sich nicht halten, nicht erreichen läßt,
> Er faßt es an, er hält zuerst es fest;
> Bestimmt das Unbestimmte, schränkt es ein,
> Benennt es treffend!—Sei die Ehre dein!—
> Wie Streife steigt, sich ballt, zerflattert, fällt,
> Erinnre dankbar deiner sich die Welt.

This celebration of the scientist introduces a series of sections treating each of Howard's cloud types, which Goethe also connects with the underlying process of the cycle.

The first of these is the evaporation from the surface of sea or lake and, from this moisture, the creation of the low-altitude clouds: 'When from the still, flat mirror of water a mist begins to raise its level carpet, the moon—joining the undulation of this apparition—seems like a ghost fabricating ghosts, then, oh Nature, we are all, we must admit, revived and refreshed children!'

> *Stratus*
> Wenn von dem stillen Wasserspiegel-Plan
> Ein Nebel hebt den flachen Teppich an,
> Der Mond, dem Wallen des Erscheins [sic] vereint,

Als ein Gespenst Gespenster bildend scheint,
Dann sind wir alle, das gestehn wir nur,
Erquickt', erfreute Kinder, o Natur!

In the last line, Goethe finds the cosmic inclusiveness of the water-cycle too confining: drawing in nature, including humankind. The clouds, new-born from the collection of the recently evaporated mist, become emblematic of all children created from the ceaseless processing and reprocessing of *Dea sive Natura*.

These stratus clouds are formed from the mist rising from mountains—or those already created are added to: 'Then upon the mountain [the mist] rises, collecting together fragment after cloud-fragment, so that all about the middle strata is wrapped in gloom—equally favorable for falling as rain or rising skyward'.

Dann hebt sichs wohl am Berge, sammelnd breit
An Streife Streifen, so umdüsterts weit
Die Mittelhöhe, beidem gleich geneigt,
Obs fallend wässert oder luftig steigt.

In the last line above, *either* there is a smaller water-cycle of locally generated precipitation (known to people in the area of large lakes as 'lake effect')—such a short-circuiting of the full cycle is indeed closer to the actual workings of the real hydrologic cycle (see scientific summary in introduction) than is the ideal poetic apperception of a complete one—*or* there is a first step in a more expansive water-cycle—which seems to be indicated by the link Goethe establishes between the next two lyrics.

Moisture that rises locally makes up the stratus cloud and becomes collected into a cumulus at a higher level: 'And when straight from the upper atmosphere, the appropriate substance is called, the cloud rises up, piled up most splendidly, announcing, strongly built, powerfully made, and, what you fear and well may live to observe, when it looms above, it shakes down below'.

Cumulus

Und wenn darauf zu höhrer Atmosphäre
Der tüchtige Gehalt berufen wäre,
Steht Wolke hoch, zum herrlichsten geballt,
Verkündet, festgebildet, Machtgewalt,
Und, was ihr fürchtet und auch wohl erlebt,
Wie's oben drohet, so es unten bebt (ll. 33–38).

The moisture of a cumulus cloud rises to become a cirrus: 'The noble impulse mounts ever higher! Release is a heavenly, light compulsion. A tangled pile, it breaks up into fluffy bits—like tiny tripping lambs—lightly combed in clusters, so finally that which easily developed down below will flow quietly into the lap and hand of the Father above'.

Cirrus

Doch immer höher steigt der edle Drang!
Erlösung ist ein himmlisch leichter Zwang.
Ein Aufgehäuftes, flockig löst sichs auf,
Wie Schäflein trippelnd, leicht gekämmt zuhauf.
So fließt zuletzt, was unten leicht entstand,
Dem *Vater* oben still in Schoß und Hand.

Do not be fooled by the simile to lambs or the God-reference: what Goethe is describing, underneath epic simile and religion, is good science. Evaporated moisture, as it moves higher, becomes more diffuse and, at high attitudes, becomes the cirrus clouds—small and dispersed into small units.

Then, moisture—even that which has been in the higher cirrus clouds—can form into the nimbus or thundercloud, and the moisture is pulled down by the earth's gravity, thereby completing the evaporation / cloud-formation / precipitation portion of the cycle: 'Now let that which was piled up on high also fall down, pulled by terrestrial forces, exploding in furious thunderstorms, like an army fanning out and scattering!—'

Nimbus
Nun laßt auch niederwärts, durch Erdgewalt
Herabgezogen, was sich hoch geballt,
In Donnerwettern wütend sich ergehn,
Heerscharen gleich entrollen und verwehn!—

In the lines which follow, the earth's passivity in the process of evaporation and activity in precipitation are paralleled with the scientist/poet's passivity of perceiving phenomena and his activity in describing it: 'The Earth's active/passive destiny! Yet with the image you look up: the Word falls down, for it describes [...]'.

Der Erde tätig-leidendes Geschick!
Doch mit dem Bilde hebet euren Blick:
Die Rede geht herab, denn sie beschreibt [...].

But observation and even description are partial and temporal, whereas the spirit that motivates the process is eternal. The scientist/poet becomes a part of this spirit through his aspiration to understand: 'The Spirit strives upwards, where it will remain eternally'.

Der Geist will aufwärts, wo er ewig bleibt (ll. 52).

Goethe carried this interest into the visual realm. Following the guidelines of Howard's comparative morphology of clouds, the poet did cloud sketches.

Goethe was not the only creative mind of the time to come under the influence of Howard: the English scientist's countryman John Constable also did. Of all the parts of the cycle, clouds in the paintings of Constable, as well as several of his contemporaries, have received the most attention from art historians.[17] And this is significant for, among other reasons, clouds are important keys in evaluating to what extent Constable was a topographical painter, a scientific painter of factual landscapes, and a symbolic painter of expressive landscapes. First, it should be mentioned that clouds would only be a transient part of a particular locus; therefore, by definition, Constable's great concern with clouds would be one aspect of his art which lies outside the topographical tradition and can more properly be considered under what Clark

17 In addition to the works cited elsewhere in this chapter, there is E.H. Gombrich's discussion of Constable's clouds in *Art and Illusion: A Study in the Psychology of Pictorial Representation*. Rev. ed. (Princeton: Princeton UP, 1961) 174–78. Also, see Louis Hawes, 'Constable's Sky Sketches', *Journal of the Warburg and Courtauld Institutes* 32 (1969): 344–65.

calls the landscape of fact. Second, Constable's careful attention to clouds and the processes which make them is sufficiently methodical to be termed scientific. Indeed, this cloud work is scientific to such an extent that he recorded all the data relevant to the conditions favorable to the particular category of cloud: the date; time of day; wind and other weather conditions evident when he made sketches of clouds. Above all, he has himself linked his work to the scientific theory as well as the praxis (minus taking barometric pressures) of Howard. Yet, thirdly, Constable's clouds are not the result of pure objectivity. We have the author's own testimony concerning the expressive capabilities of landscape, or, rather, skyscapes: he makes it clear that he considers the sky 'the chief *Organ of sentiment*' of landscape.[18] Finally, his conscious choice of types of clouds was made for reasons which seem to combine scientific interest with a desire to give expression to a world view. William Vaughan offers an important caution for our overemphasis upon the scientific side of Constable's clouds: 'if Constable's concern is for accurate observation, he is not interested in classification. He compiled no inventory of clouds'; further, Vaughan reminds us that Constable's 'studies were largely devoted to those kinds of cumulus associated with changeable weather. He sought only to gain a close knowledge of those transient moments that attracted him'.[19] These transient clouds offered the artist a symbol with which to embody the evanescent 'chiar'suro' which he perceived in nature.

It may be, however, that Vaughan errs in the opposite direction of emphasizing the symbolic over the scientific aspects. Constable is, after all, doing some actual classification when he seeks to coordinate the clouds of actual skies with those in his canvas—and to synchronize them with the season, time, and meteorological conditions. While we can agree this is not 'hard' science, to repeat, it has an empirical and a theoretical depth that still makes it 'scientific'. From these empirical cloud studies, I have chosen an 1822 *Study of Cirrus Clouds*. In this sketch the clouds are rendered with thick daubs of white—shaded from pure to grey—over a red-primed background. This sketch bears an almost illegible inscription, 'Cirrus', which seems to indicate that Constable was familiar with Howard's cloud classifications; it thereby lends further evidence to the hypothesis that Constable had a strong scientific bent.

Clouds and, particularly, their coordination with time of day and environmental conditions all constitute an important part of the subject matter of his major canvases. The work that we know as *The Hay Wain*—but which was significantly entitled by Constable himself *Landscape: Noon*—offers the clearest but by no means the only example of this science-based coordination. 'The Hay Wain', writes Kenneth Clark, 'has a serene fullness of nature which is eternally refreshing. Few pictures express more completely the confidence in nature which was the chief faith of the nineteenth century'.[20] The clouds are rendered with an oily brush in this work. The artist's facture, especially in the application of the thin streaks of pure-white paint, indicates light. The heavier application is the white paint at the top of

18 Qtd. in *Constable,* Tate Gallery 127. Italics are in original.

19 William Vaughan, *Romantic Art* (New York: Oxford UP, 1978) 203.

20 Kenneth Clark, *John Constable: 'The Hay Wain'* (London: Lund, 1944) 16.

the clouds, lit thereby to suggest the time of day—noon. This combination of intense chiaroscuro and impasto effects gives a sense of mass, particularly to the cumulo-stratus clouds over the top of the house and the trees at the left of the canvas. The more lighted clouds—both simple cumulus and cirro-cumulus—hang over the field; this part of the sky has clear patches which allow for the field's being more brightly lit than the rest of the scene. As in the Ruisdael landscapes, the clouds allow for the dramatic variety of lighting to take place below. And yet the shading and the texture of the clouds make them interesting in themselves; it is this emphasis on the image and the mimetic that makes *Hay Wain* the essentially Neoclassical-Realistic-Romantic work that it is. Clark indicates that Constable preferred the more Romantic canvases—'The Leaping Horse, Hadleigh Castle, and above all Salisbury with a Rainbow'.[21]

For what might be termed Constable's fully symbolic and entirely Romantic usage of clouds, I will turn to one of the greatest paintings of his last years: *The Stonehenge*. Here the ruined monument made by man is dominated by what Louis Hawes calls an 'overpowering cataclysmic sky'. Hawes points out how the painter of fact yields to the painter of symbols: 'Here Constable strives for utmost expressive intensity rather than maximum meteorological truth'.[22] It is a matter both of emphasis and expressive use, but all this is not meant to imply that the less scientific phenomena are the more Romantic they are. The science of hydrology, especially in its holistic and theoretical dimensions, offered the Romantic creative minds a powerful means of embodying the very act of their intellect—one that sought unity in a vortex of change and diversity.

While not cataloging clouds quite so precisely as Constable, Turner gave considerable attention to carefully observing and to factually rendering clouds. Gerald Wilkinson shows how the young artist would make rapid series of sketches to 'keep pace with the shifting clouds' over the locus of Durham Cathedral.[23] This fascination with cloud observation and sketching was for Turner lifelong; for, around 1820, the now middle-aged artist executed a series of sixty-one sky studies. Concerning the techniques used in rendering them, Wilkinson writes: 'The brushwork is extremely versatile and economical—as it must be to record such ephemeral subjects. Sometimes, with only a blue and a neutral grey, he exploits a wide range of textures, from the wettest, softest blots for cumulus to the driest of scumbles for light flecks of cirrus'.[24]

This attention to detail of Constable matches Goethe's desire to emphasize the process of clouds over any static category. He suggests this scientific/artistic process in his 'Carefully Take Note' / 'Wohl zu merken' of 1822 (1: 526), also printed with an article, 'Goethe's Tribute to Howard' / 'Goethe zu Howards Ehren' (17: 672–73). To understand is, for him, to distinguish separate stages or forms within the entire process, then, and only then, to see how they are part of a continuous whole: 'And

21 Clark, *Hay Wain* 16.

22 Louis Hawes, *Constable's 'Stonehenge'* (London: HM's Stationery Office, 1975) 1.

23 Gerald Wilkinson, *Turner's Early Sketch Books, 1789–1802* (London: Barrie, 1972) 111.

24 Gerald Wilkinson, *Turner's Colour Sketches, 1820–34* (London: Barrie 1975) 20–21. Also, see Hawes, 'Sky Sketches' 355–58.

when we have completed the classification, then we must again breathe life back into the separated items and take delight in the continuity of all living things'.

Und wenn wir unterschieden haben,
Dann müssen wir lebendige Gaben
Dem Abgesonderten wieder verleihn
Und uns eines Folge-Lebens erfreun.

What is transitory must be understood before the creative intellect can mold and combine an apperception of the whole: 'Thus, when the painter, or the poet, is familiar with Howard's categorization, during early morning, in late evening, carefully examines the atmosphere, he can evaluate its nature; meanwhile the airy world endows him with the transformation of consciousness, the peace of mind, that he may grasp, feel, and embody the categorization in his art'.

So, wenn der Maler, der Poet,
Mit Howards Sondrung wohl vertraut,
Des Morgens früh, am Abend spät
Die Atmosphäre prüfend schaut,

Da läßt er den Charakter gelten;
Doch ihm erteilen luftige Welten
Das Übergängliche, das Milde,
Daß er es fasse, fühle, bilde.

In addition, the reader of these lines should take note of the emphasis Goethe places on painters' and poets' needing a solid grounding in scientific classification and experimental methods. Of even greater significance to this study is the need to observe that in the last three lines above, the transitoriness and liberality of the cloud become a correspondence for the tentative, and yet expansive manner, by which hypotheses are formulated and tested, and by which scientific method proceeds. This use of the cloud to symbolize the poet/scientist is comparable to the use of the cloud to represent the poet as prophet found in the poems of Hugo and Shelley, which are to be treated in the last section.

There is an extraordinary passage (composed in 1827) that opens act 4 of *Faust, Part Two* and that contains, in what can only be called a cosmic combination, many of the images from the Howard cycle. A cloud floats down upon a mountain peak, it opens, and Faust steps forth. It is from the vantage point of this water-vapor zeppelin that Faust has his great vision of taming the ocean and creating his Holland-like land. This is not a cloud messenger, but cloud transport: 'As my eyes see the deepest solitude below, I step with care onto the margin of these peaks and send away the cloud flight-machine that during sunlit days gently carried me across the land and sea'.

Der Einsamkeiten tiefste schauend unter meinem Fuß,
Betret ich wohlbedächtig dieser Gipfel Saum,
Entlassend meiner Wolke Tragewerk, die mich sanft
An klaren Tagen über Land und Meer geführt (ll. 10039–42).

In the context of *Faust Two* where so much of the work is about the power of illusions and their eternal dialectic with actualities, Faust's fantastic reverie over the clouds takes on extra meaning. The cloud and the human imagination acting upon it summon up the power of the Mothers—collectively the Eternal Feminine offering mythic visions of Greek goddesses and legendary women—Helen above all:

> It slowly draws away from me without dispersing. The greater part, a massive curling thing, is pressing eastward, where the eye must strive to follow: although its changing billows, as they move, divide, it seems to shape a figure.—Yes, my eyes perceive it right!—I see, stretched out in sun-gilt splendor couched, a giant, yet still a godlike woman's form. In its majestic loveliness it hovers there within my sight, resembling Juno, Leda, Helen! Already it moves on! Like distant icy mountain masses piled high upon each other, there in the east it stays, a dazzling symbol of the significance of these fleeting days.

> Sie löst sich langsam, nicht zerstiebend, von mir ab.
> Nach Osten strebt die Masse mit geballtem Zug;
> Ihr strebt das Auge staunend in Bewundrung nach.
> Sie teilt sich wandelnd, wogenhaft, veränderlich;
> Doch will sichs modeln.—Ja, das Auge trügt mich nicht!
> Auf sonnbeglänzten Pfühlen herrlich hingestreckt,
> Zwar riesenhaft, ein göttergleiches Fraungebild,
> Ich sehs! Junonen ähnlich, Ledan, Helenen,
> Wie majestätisch-lieblich mirs im Auge schwankt!
> Ach! schon verrückt sichs! Formlos-breit und aufgetürmt
> Ruht es in Osten, fernen Eisgebirgen gleich,
> Und spiegelt blendend flüchtger Tage großen Sinn (ll. 10043–54).

There is immense German-Idealistic perceptual egocentrism, which is reminiscent of Goethe's Jena philosopher acquaintances, in this apperception, where the will of Faust's consciousness transforms scientific phenomena into a highly personalized remembrance of his wandering in the geography and mythology of Greece.

From a cumulus cloud, the speaker turns to a cirrus. A small wisp of a cloud becomes the symbol of the youthful spirit within his soul:

> Yet one bright tenuous streak of mist still hovers near and cheers me with its cool caress on heart and brow. Now, lightly it rises, hesitates, goes higher still, and draws together.—Am I deceived by a mirage of what, when young, I valued most, but lost long since? From the depths of my heart, youth's first treasures well up; it shows me Aurora's love, a bit of verve—that swiftly felt, first, scarcely comprehended glimpse, which—had it been firmly held—would outshine all other treasures. Like inward beauty of the soul, the lovely form intensifies, rises, not dissolving, to the ether, and draws away with it that which is best deep within myself.

> Doch mir umschwebt ein zarter, lichter Nebelstreif
> Noch Brust und Stirn, erheiternd, kühl und schmeichelhaft.
> Nun steigt es leicht und zaudernd hoch und höher auf,
> Fügt sich zusammen.—Täuscht mich ein entzückend Bild
> Als jugenderstes, längstentbehrtes höchstes Gut?
> Des tiefsten Herzens frühste Schätze quellen auf:
> Aurorens Liebe, leichten Schwung bezeichnets mir,
> Den schnellempfundnen, ersten, kaum verstandnen Blick,
> Der, festgehalten, überglänzte jeden Schatz.
> Wie Seelenschönheit steigert sich die holde Form,
> Löst sich nicht auf, erhebt sich in den Äther hin
> Und zieht das Beste meines Innern mit sich fort (ll. 10055–66).

In these lines, we find the quintessential image of weather for a Neoclassical-Romantic vision. The sense of or 'feel' for cosmic order, however, can best be heightened by local disorder—loved by the extreme Romantics for its own and by the Neoclassical-Romantics for the polarity it offered to the essentially placid ground of their existence and creation.

The Storm Cloud

Poets, painters, and musicians fascinated with dynamic change turned easily to the vortex of storm clouds. Goethe's *Werther,* the masterwork of the Storm and Stress movement, has thunderstorms at key moments.

A thunderstorm appears on the horizon at the very moment when Werther is about to meet, for the first time, the woman whom his own neurotic personality will turn into his nemesis; the approaching storm, in fact, makes itself felt right when the carriage of his party turns into the entrance drive of Lotte's home:

> The sun was still a quarter of an hour from the mountains when we started to drive off at the gate of the courtyard. It was very sultry, and the women voiced fears of the thunderstorm that seemed to be brewing on the horizon, where humid whitish-grey clouds were gathering. I set their minds at rest by pretending expertise in meteorology, though soon I was myself starting to suspect that our pleasure-party would be dealt a blow.
>
> Die Sonne war noch eine Viertelstunde vom Gebirge, als wir von dem Hoftore anfuhren. Es war sehr schwül, und die Frauenzimmer äußerten ihre Besorgnis wegen eines Gewitters, das sich in weißgrauen dumpfichten Wölkchen rings am Horizonte zusammenzuziehen schien. Ich täuschte ihre Furcht mit anmaßlicher Wetterkunde, ob mir gleich selbst zu ahnen anfing, unsere Lustbarkeit werde einen Stoß leiden (4. 397).

Later at the party, when Werther is dancing with Lotte, the stormy weather of lightning and thunder appears in a human context. In order to express Werther's new-found feelings of physically and latent sexuality for Lotte, Goethe employs a lightning metaphor: 'I was no longer a mortal. Holding the most loveable creature in my arms and flying about with her like lightning [literally: weather], so that I forgot my surroundings, and—Wilhelm, to be honest, I swore that a girl I loved and had a claim on should never waltz with anyone but me, not even if it cost me my life. You take my meaning!'

> Ich war kein Mensch mehr. Das liebenswürdigste Geschöpf in den Armen zu haben und mit ihr herumzufliegen wie Wetter, daß alles rings umher verging, und—Wilhelm, um ehrlich zu sein, tat ich aber doch den Schwur, daß ein Mädchen, das ich liebte, auf das ich Ansprüche hätte, mir nie mit einem andern walzen sollte als mit mir, und wenn ich drüber zugrunde gehen müßte. Du verstehst mich! (402).

Metaphor quickly becomes reality.

Goethe continues to have weather symbols reveal the psychology of Werther and Lotte. The actual, natural lightning soon intrudes upon the party, symbolically proving Werther's predictions wrong: 'The dance had not yet finished when the lightning, which we had seen flaring on the horizon for some time and which I had pronounced to be merely sheet lightning, began to flash more violently, and the thunder drowned out the music'.

> Der Tanz war noch nicht zu Ende, als die Blitze, die wir schon lange am Horizonte leuchten gesehn, und die ich immer für Wetterkühlen ausgegeben hatte, viel stärker zu werden anfingen und der Donner die Musik überstimmte (402–03).

This thunderstorm demonstrates the courage and emotional strength of Lotte, just as the marauding French soldiers allow Dorothea to demonstrate hers. Lotte has everyone engage in a game to get their minds off the thunder. She later admits she was afraid, after the thunder is over. The passing of the storm brings Lotte and Werther together in a common love of the leading German poet of the older generation: 'The thunder was passing by and a delightful rain was falling on the land, filling the warm air with the most refreshing fragrance. She stood there resting on her elbows, gazing deep into the country about us; she looked up to the sky; I saw her eyes filled with tears; she laid her hand on mine and said, 'Klopstock!''

> Es donnerte abseitwärts, und der herrliche Regen säuselte auf das Land, und der erquickendste Wohlgeruch stieg in aller Fülle einer warmen Luft zu uns auf. Sie stand, auf ihren Ellenbogen gestützt, ihr Blick durchdrang die Gegend, sie sah gen Himmel und auf mich, ich sah ihr Auge tränenvoll, sie legte ihre Hand auf die meinige und sagte—Klopstock! (404).

Klopstock was one of the older exemplars of the Storm and Stress poets. Werther, moreover, as we have seen, is psychologically motivated and expresses his mental states through literature, from Homer to Ossian. The reference, then, would have been taken by the contemporaneous reader as signifying a profound sympathy between the two. So, Werther's love for Lotte is conceived in the midst of a thunderstorm.

One wonders if Scott was under the influence of this omen-like appearance of a thunderstorm when he penned a similar occurrence into his *Bride of Lammermoor*. The outcast Master of Ravenswood joins a hunting party of his clan-enemies, as a distant thunderstorm is approaching. The masked patriarch of this opposing clan, the Lord Keeper Sir William Ashton, introduces his daughter Lucy, also masked, to Ravenswood:

> At this moment the cloud which had long lowered above the height on which Wolf's Crag is situated, and which now, as it advanced, spread itself in darker and denser folds both over land and sea, hiding the distant objects and obscuring those which were nearer, turning the sea to a leaden complexion, and the heath to a darker brown, began now, by one or two distant peals, to announce the thunders with which it was fraught; while two flashes of lightning, following each other very closely, showed in the distance the grey turrets of Wolf's Crag, and, more nearly, the rolling billows of the ocean, crested suddenly with red and dazzling light (114).

The Lord Keeper and Lucy accept the offer of hospitality from Ravenswood to go to Wolf's Crag, the last part of his family's estate that has not be taken over by the Keeper. When they enter, the lightning illuminates the Ravenswood arms:

> —the old arms and scutcheons which hung on the walls of the apartment, were for an instant distinctly visible to the Keeper by a strong red brilliant glare of light. Its disappearance was almost instantly followed by a burst of thunder, for the storm-cloud was very near the castle; and the peal was so sudden and dreadful, that the old tower rocked to its foundation, and every inmate concluded it was falling upon them (125).

Keeper Ashton's playing with the affections of Ravenswood and Lucy—he dabbles with the possibility of a marriage alliance between the two—causes the end of one family and the slow decline and fall of the other. And it is the thundercloud that announces it, never to recur in the novel again; here the parallel of Scott's with Goethe's novel ends.

The weather imagery of *Werther* returns again and culminates in a storm scene in which the cloud motif is subtly linked back to earlier moments in the novel to signify an ultimate and radical psychological transformation of the protagonist. The cloud is used as an analogue for the state of mind of the persona; two important examples of this usage occur in the two river passages, which were introduced in chapter two. In Werther's happier days, there had been 'the gentle stream' / 'der sanfte Fluß' which had reflected 'the charming clouds' / 'die lieben Wolken'; these, in turn, are gently wafted along by an evening breeze from heaven (430–31); whereas, over the same, but now storm-swollen, stream there are dark clouds which become an extension of the persona's disturbed consciousness: 'Oh Wilhelm, how happy I would have renounced life in order to rend the clouds with the stormy wind [...]' / 'O Wilhelm, wie gern hätte ich mein Menschsein drum gegeben, mit jenem Sturmwinde die Wolken zu zerreißen [...]' (483). These two types of clouds—placid and stormy—correspond respectively to the happy and the unhappy era of the persona's life; nevertheless, these passages still remain descriptive, to a great extent, and one might just barely term these clouds symbols in the full sense of the term, as are those in poems by Goethe, Shelley, and Hugo—all to be treated in the next section.

The storm symbol is central to Bécquer's 'Giant Waves' / 'Olas gigantes'. In the previous chapter, the sea symbol of this poem was examined; now, the cloud symbol needs to be looked at, as well as the related image of the storm. In stanza two, the poet desires to be borne off by a stormy wind: 'Hurricane gusts tearing the dead leaves from the mountain forest: pulled away into the blind vortex, sweep me away with you!'

> Ráfagas de huracán que arrebatáis
> del alto bosque las marchitas hojas,
> arrastrado en el ciego torbellino,
> llevadme con vosotras! (ll. 5–8).

The third stanza offers the same symbol but in a more tangible form—the clouds of the storm. These clouds are somewhat elaborately described as having their edges lit by lightning: 'Storm clouds split by the lightning bolt and whose bloody fringes erupt in fire: carried away in the dark mist, sweep me away with you!'

> Nubes de tempestad que rompe el rayo
> y en fuego encienden las sangrientas orlas,
> arrebatado entre la niebla oscura,
> llevadme con vosotras! (ll. 9–12).

This is a powerful and very gloomy vision of cataclysmic change; but most horrific of all is the way the poet welcomes it, celebrates it.

In order to find in painting a comparable pessimism embodied in a cosmic storm, we must appeal to Turner and take from a choice of many canvases—*Snowstorm: Hannibal and His Army Crossing the Alps*. Concerning this painting, there is sufficient information to connect Turner the painter of symbols with the painter of facts: there is eyewitness evidence indicating that the sky or snowstorm of this work was based on sketches made from nature in the same way that his sketches of the sunset were made. Although the sketches do not seem to be extant, there is firsthand evidence from a friend of Turner. This eyewitness describes how Turner made hasty sketches of the storm: 'There was the storm rolling and sweeping [...] over the Yorkshire hills. Presently the storm passed, and he [Turner] finished. "There", he said "[...] in two years you will see this again, and call it Hannibal crossing the Alps"'.[25] This quote shows how closely connected the natural phenomena of clouds combined with sunset or storms are to the mythic or the historic events, which they accompany in Turner's works—except that as with Claude Lorrain the latter are secondary to the former. One is not merely symbolic or an analogue of the other; they all seem to have been concurrently apperceived. Too often critics have seen an inconsistency between the private Turner of the *Colour Beginnings* and the public history paintings; but the rendering of actual phenomenon of the one plays a crucial role in the background of the latter. Indeed, the real theme of the picture, I think, is the vortex of the snow storm; it is, in the words of John Canaday, 'the cosmos as a spectacle of the elements in chaos—with air, fire, and water fusing and spinning in a vortex that might be the primordial womb from which the solid forms of earth were brought forth'.[26]

Impasto application was used on the clouds, especially for the white, sunlit rims where the heaviest application of paint was made. The sky areas on the sides (especially on the left) have been rendered with a thick impasto of white and tan over the base of ultramarine blue. The sky in the central area of the canvas is rendered with a very oily application of the paints—hence the smooth texture. This is the area where the clouds give forth into snow, seen as dark grey streaks. The swirling cloud formation with the equally swirling precipitation emanating from it overshadow the heroic, historic action below. Turner's use of the storm, in conjunction with other Alpine imagery, to convey the insignificance of man bears comparison with the following lines from Percy Shelley's 'Mont Blanc': 'The race / Of man flies far in dread; his work and dwelling / Vanish, like smoke before the tempest's stream, / And their place is not known' (ll. 117–20).

In her great expressive marshalling of natural symbols to follow the tragic destiny of the inventor and his monster, Mary Shelley preludes the bridal night of Frankenstein and his bride, who has been threatened by the monster, with an approaching mountain storm:

> The wind, which had fallen in the south, now rose with great violence in the west. The moon had reached her summit in the heavens and was beginning to descend; the clouds swept across it swifter than the flight of the vulture and dimmed her rays, while the lake reflected the scene of the busy heavens, rendered still busier by the restless waves that were beginning to rise. Suddenly a heavy storm of rain descended (188).

Preluding as it does the fatal appearance of the monster in the bridal chamber, one might dismiss this as merely an Early Romantic, 'Gothic' storm, were it not part of a network of nature symbolism that, as I have tried to show over the past chapters, runs through the entire novel. The storm here seems to signify the indignation of

25 Qtd. in Jack Lindsay, *Turner: His Life and Work* (New York: Harper, 1966) 117.

26 John Canaday, *The Lives of the Painters* (New York: Norton, 1969) 738.

violated nature and that of the atavistic Shelley: having had a child 'unnaturally'. Frankenstein must be prevented from natural conception. The monster prevents this by murdering Frankenstein's wife, with woman as the innocent victim of nature and a woman's (Mary Shelley's) 'justice'. The storm is the overture, as it were, to retribution. I suspect that today we cannot—even with the best will in the world—evoke the power a storm appearing in a work of literature would have caused in the minds of a Romantic reading public; it may be only in the musical storms, as it were, that we can dimly sense what they might have felt.

Storms provide an important key to understanding that most misunderstood of musical forms in Romantic culture—program music. The appeal of composers to the elements in music is not a radical innovation of Romantic culture; we have already seen it was present in Rameau and Gluck; however, the way that they use these aural evocations of streams and clouds in the massive framework of their musical structure is very radical. Beethoven quite consciously places a brook in the second movement of *The Pastoral* which symmetrically balances with the thunderstorm of the penultimate movement of the same symphony. He creates the sense of the wind and rain by tremolos in the upper stings broken by almost random-seeming short rhythmic figures. The lower strings meanwhile reiterate short scale passages in thirty-second notes; these scale passages, after being repeated six to twelve times, ascend in upward-moving sequences. Thunder seems represented by fortissimo chords by the whole of the orchestra.

Rossini's *William Tell / Guillaume Tell* contains several storm passages; the most famous is the overture to the same opera which, in reality, is a separate programmatic overture, like those of Mendelssohn, which of course do not precede operas. Rossini creates the sensation of the rushing swirl of a storm clouds, rain, and wind with much the same techniques as Beethoven uses, with one important exception; he has the upper strings, rather than the lower, play the thirty-second scale figures; the lower strings play the tremolo, with the exception of the double basses, which play reiterated notes in eight-note patterns. The Paris *Opéra* orchestral forces—the largest and most varied in the world at the time—allowed Rossini to add other colors to his already amply self-enriched palette: the shrill interjections of the piccolo add to the pandemonium; the extra power of the trombones, the bass drum, and cymbals add force to the fortissimi crashes. Rossini's storm crescendos relate to his other crescendos: they usually suggest psychological stress (as in the *stretta* of the finale to act one of *The Barber of Seville / Il Barbiere di Siviglia*). In the case of *William Tell,* the storms throughout the opera correspond to the great emotional tension felt by the oppressed people of the Swiss Cantons and their sympathizers under Hapsburg domination.

Rossini was one of the few composers about whom the middle-aged Richard Wagner had positive things to say. Building on what Rossini—along with Gluck and Beethoven—had done, Wagner's *Flying Dutchman* opens with a storm at sea and ends with the self-sacrificial drowning of the heroine, the sinking of the Dutchman's ship, and the reception of the two lovers into heaven. The *Ring* Cycle, as we have already seen, opens with the suggestion of the creation of life out of the water of the Rhine and ends with the destruction of the world of man by the flooding of the same river. But if we look at just the first opera of the Ring—*Rhinegold*—we have a work that not only begins with a tonal vision of the Rhine but ends with the conjuring up of the clouds by Donner, the Storm God, for a thunderstorm—the Romantic musical climax par excellence. The swirl of clouds and crosswinds is notched up to the highest degree possible. Wagner divides the first violins into six sections; he does the same to the second violins; and the violas into three sections. They play scales and arpeggio figures that run in opposite directions from one another; some of the divisions play sustained notes, along with the woodwinds.

Although the section that follows treats poetical and fine-art works that contain the water-cycle sequentially, such a treatment in music would need to be accompanied with a verbal text, stage visuals, and take a large space of time to be suggested at all—the space best offered by a Beethoven symphony, by Rossinian grand opera, or by Wagnerian cosmic music dramas. In these, the water-cycle—which seems to lack any real direct connection to scientific knowledge—appears in a fragmentary form. Yet, the very fact that the water quarters appear so centrally in such contexts demonstrates a Romantic culture fascinated by water phenomena, weather, and the elemental in general. Wagner's *Rhinegold* offers a surprisingly comprehensive vision of the water-cycle—opening as it does in the depths of a great river and closing with the thunderstorm; a similar pattern being implicit in the opera closing the cycle—*Die Götterdammerung,* which ends with a flooding by the river.

The Great Link: The Cloud and Rain in the Water-Cycle

To qualify as a cycle, the water-cycle needs closure. Only a scientific theory of rain employing processes of evaporation and precipitation could close definitively the final gap and fuse the hydrological-cycle. As was discussed in the introduction, James Hutton was most responsible for closing this gap in the cycle with his theory of rain. He states in one of his 1784 papers read at the Academy of Science of Edinburgh that 'RAIN is the distillation of water, which had been first dissolved in the atmosphere, and then condensed from that state of vapour or solution. It is the explanation of this condensation that must form the theory of rain' (47). Condensation—which is a process graduated along a gamut of various states of the vapor—causes clouds as well as rain:

> WATER, indeed, is condensed in a cloud equally as in rain, and yet clouds may subsist without rain. But, it is evident, that, without condensation of aqueous vapour in the atmosphere, no rain could be produced; and that, however different causes may influence water condensed in the atmosphere, and operate variously, in either retaining it longer in a suspended state, or bringing it sooner to the ground, the condensation of the water is properly the cause of rain (48).

The cycling of weather from cloudy to partly cloudy to sunny is generally caused by this condensation into rain: 'RAIN having fallen in a place, in consequence of the mixture in the atmosphere above, this will naturally be followed by clearness in the sky and sunshine, which is so necessary for warming the surface of the earth, and for giving health and vigour to growing plants' (50).

Shelley, Hugo, Blake, Friedrich, Cole, and others discussed in this section used not merely a single natural phenomenon but rather a full-blown scientific theory as a symbol. Underlying the works of these painters and poets is the same theoretical structure that underpins Goethe's 'Song in Praise of Mohammed' and Wordsworth's *River Duddon.* In short, all of the works considered in this section offer the final connecting link in the water-cycle, in part or as a whole, which offers a symbol for the life of a poet or prophet and, by extension, for the life of mankind in general. This comprehensive symbol allows those writers to accept the inevitability of change and multiplicity while at the same time allowing for unity and continuity. Similar concerns have been discerned among painters of the same time-frame as these poets.

Shelley's poetic use of clouds, along with evaporation and precipitation—in short, clouds in the context of the water-cycle—is extensive and varied and undergoes a development throughout his poetic works as complex as Goethe's usage of spring/fountain and stream symbols. An early 1816 poem by Shelley, 'Mutability' (1: 203), opens with the image of the fading light on the clouds as objectifying the abstract notion which is the theme of the poem; in addition, a relationship by means of a simile is established between human beings and clouds as both being subject to the laws of change:

We are as clouds that veil the midnight moon;
How restlessly they speed, and gleam, and quiver,
Streaking the darkness radiantly!—yet soon
Night closes round, and they are lost for ever [...] (ll. 1–2).

During the next year, Shelley wrote an eight-line fragment (3: 167) which gives expression to the themes touched on in poems treated earlier in this chapter, like Goethe's 'Howards Ehrengedächtnis': 'the truest cloud-messenger vanishes, before he reaches the desired distance'. / 'Der treuste Wolkenbote selbst zerstiebt, / Eh er die Fern erreicht, wohin man liebt'. In Shelley's poem, the cloud objectifies the persona's desire for the change of scenery, the freedom from the mundane, and the dream of faraway places:

O THAT a chariot of cloud were mine!
Of cloud which the wild tempest weaves in air,
When the moon over the ocean's line
Is spreading the locks of her bright grey hair.
O that a chariot of cloud were mine!
I would sail on the waves of the billowy wind
To the mountain peak and the rocky lake,
And the [...].

A similar use of the cloud symbol aimed at crystallizing the persona's desire for escape was seen in Bécquer's 'Olas gigantes'. Both poets put the transient cloud to similar expressive purposes. Although it must be noted that the yearning of Shelley's persona does not share the morbidity of Bécquer's persona, who out of despair yearns for escape, perhaps into death. This and other of Shelley's poet/cloud-speakers want to sail over the world, sampling the diversity of life and landscape that spreads out over the earth. As in Plato's *Phaedrus,* the chariot symbol represents a ground of being on which learning or experiential development can take place.

Nowhere in the works of Shelley, or in Romantic poetry for that matter, is the cloud theme handled with the complexity and subtlety as in an 1820 lyric 'The Cloud' (2: 299–301). The meteorological aspects of 'The Cloud' have been considered to some degree, and one aspect has been treated exhaustively: Shelley's ideas on electricity in clouds, ideas which he derived from the works of Humphry

Davy and Adam Walker.[27] Both the water-cycle and the larger world view it embodies must be considered to fully comprehend the rich, profound beauty of 'The Cloud'.

Beyond the scientific considerations, J.A. Notopoulos, in *The Platonism of Shelley,* comes closest to offering a sensitive reading of the poem; however, even he gives it scant, albeit perceptive, mention. For him, 'the cloud' is 'a symbol of the change which cannot die'—in other words, of the permanence in change.[28] This poem is not only about a cloud: the entire water-cycle is involved in the opening portion of the poem and must be taken into account. This poem, to my mind, demonstrates Shelley's ability to transform the latest in contemporaneous scientific observations and theories into flexible metaphors, freighted with the greatest possible significance. In speaking of the poet's *Prometheus Unbound* (with which 'The Cloud' was published), Marjorie Hope Nicolson writes that Shelley's 'science does not intrude—as in some of the eighteenth-century poets—into the poetry. It is an integral part of it'.[29] Shelley proceeds by tracing the cloud through various metamorphoses:

> I BRING fresh showers for the thirsting flowers,
> From the seas and the streams;
> I bear light shade for the leaves when laid
> In their noon-day dreams (ll. 1–4).

The cloud of the opening two lines, which brings rain, changes into that of the next two lines, which floats across the sky on a sunny day. But an even larger process is outlined in lines one and two, which contain the essence of the water-cycle—the movement from evaporation to precipitation. The cloud is the agent for transporting water to where there is none.

In considering the opening lines, it is necessary to view them in several important contexts: (1) the cloud is an archetypical symbol for a messenger; (2) the cloud is traditionally a symbol employed in religious art to represent the prophets[30]; (3) in the preface to the edition of poetry in which 'The Cloud' was first published, Shelley uses the cloud/mind symbol in a passage to be quoted later; (4) in Plato's *Ion,*

---

27 See for example: Desmond King-Hele, *Shelley: His Thought and Works,* 2nd ed. (Teaneck: Fairleigh Dickinson UP, 1971) 219–27; Timothy Webb, *Shelley: A Voice Not Understood* (Atlantic Highlands: Humanities, 1977) 245–51.

28 Notopoulos, *The Platonism of Shelley* (Durham: Duke UP, 1949) 270. Strangely enough, David Perkins omits 'The Cloud' from his discussion of Shelley's use of the water-cycle; he rather bases his argument entirely upon the fragments; to my mind, these fragments are only useful as tools to illumine this poem; see his *The Quest for Permanence: The Symbolism of Wordsworth, Shelley, and Keats* (Cambridge: Harvard UP, 1959) 118–21.

29 Nicolson 381.

30 J.E. Cirlot writes that one important 'aspect of clouds reveals their family connection with fertility-symbolism and their analogous relationship with all that is destined to bring fecundity. Hence the fact that ancient Christian symbolism interprets the cloud as [being] synonymous with the prophet, since prophecies are an occult source of fertilization, celestial in origin' ('Clouds', in his *A Dictionary of Symbols* [New York: Philosophical Library, 1962] 48); similarly Bachelard writes, 'Le nuage est pris aussi comme un messager', *L'Air et les songes* 218.

translated by Shelley (7: 238), the poet is seen as a mobile being who ministers to the wants of the common man, chained as he is in metaphysical stasis:

> Thus *the composers of lyrical poetry* create those admired songs of theirs in a state of divine insanity, like the Corybantes, who lose all control over their reason in the enthusiasm of *the sacred dance*; and during this supernatural possession, are excited to the rhythm and *harmony* which they communicate to men. Like the Bacchantes, who, when possessed by the God draw honey and milk *from the rivers*, in which, when they come to their senses, they find nothing but simple *water*. For *the souls of the poets*, as the poets tell us, *have this peculiar ministration in the world*. They tell us that these souls, *flying like bees from flower to flower*, and wandering over *the gardens and the meadows* and the honey flowing *fountains* of the Muses, *return to us laden with the sweetness of melody*; and arrayed as they are in *the plumes* of rapid imagination, they speak truth. *For a poet is indeed a thing ethereally light, winged, and sacred.*

Many of the related symbols in this passage are to be found in the poem (my emphasis indicates phrases with similar phrases in ll. 1, 2, 5 as well as 8, 47, and 54 of the poem). (5) Shelley more overtly uses the cloud as a symbol for human intelligence in 'The Soaring Mind', another 1817 fragment (3: 167) of six lines; he sees, in the movements of the cloud and in the lightning motif, qualities which are analogous to the creative mind of the poet:

> Behold sweet mine; once more descend
> The shadows of my soul upon mankind,
> For to those hearts with whom they never blend,
> Thoughts are but shadows which the flashing mind
> From the swift clouds which track its flight of fire,
> Casts on the gloomy world it leaves behind.

There is an interesting relationship between the world and the cloud/mind: the shadow cast by the latter on the former is fleeting, but there is no reason to believe that the process of the mind/cloud is transitory like the world it leaves behind. When one reads the above lines in these contexts, it becomes clear that the cloud is a metaphor for what Shelley, in his *Defense of Poetry*, sees as the prophet-poet (7: 112); it is this figure who, to use the language of the poem, brings the 'fresh showers' of intellectual prophecies in the form of poetic pleasures to his readers ('the thirsting flowers') from out of a collective repository of poetic conventions, of philosophical and scientific concepts, of observations from nature, and even, from the realm of Platonic ideals ('the seas').[31] Since 'flowers' connote ultimate fruition, it would seem that Shelley is suggesting, through his persona, the intellectual and social fruition resulting from the works of the prophet-poet. Just as the cloud must transport water from 'the seas and the streams' to immobile, earthbound flowers, so too does the prophet-poet act as an intermediary between the ideal and everyman.

The lines, 'I bear light shade for the leaves when laid / In their noon-day dreams', suggest another purpose of the clouds that we take for granted. The meteorologist Howard, however, does not take this aspect for granted: 'Independently of the beauty and magnificence it adds to the face of nature, the cumulus serves to screen the earth from the direct rays of the sun, by its multiplied reflections to diffuse, and, as it were, economize the light, and also to convey the product of evaporation to a

31 Notopoulos, *Platonism* 6.

distance from the place of its origin' (7–8). Howard, then sees not only the contributions of the cloud to the water-cycling process, but the moderating effect of clouds upon temperature on the earth's surface.

Furthermore, for Shelley's part, the gentler offices are not enough: a poet must also be a satirist (as Shelley himself is in *Queen Mab* and *Peter Bell the Third*) and a protester of social ills (as he is in *The Mask of Anarchy* and 'Song to the Men of England'):

> I wield the flail of the lashing hail,
> And whiten the green plains under,
> And then I dissolve it in rain,
> And laugh as I pass in thunder (ll. 9–12).

These images even taken on a most fundamental level—the combination of destructiveness and creativeness in the manifestations of the cloud—would appeal to Shelley's revolutionary spirit.[32] The poem, seemingly simplistic in its metaphor, well disguises its sociopolitical signification. It, for example, may be admitted that the stanza that follows suggests an alternative, even paradoxical aloofness from the fray below on the part of the poet:

> I sift the snow on the mountains below,
> And their great pines groan aghast;
> And all the night 'tis my pillow white,
> While I sleep in the arms of the blast (ll. 13–16).

Action must be combined with contemplation. The cloud always remains above the fray. Thunder, hail, and snow constitute what might be termed an intermediary between the cloud and what is below—all corresponding to the part poetry plays for poet-as-prophet.

The supreme intermediary, as well as symbol for the revolutionary spirit itself, is the lightning which emanates from the cloud and the electricity which—according to Shelley's science—is the creative impulse in the cloud:

> Sublime on the towers of my skiey bowers,
> Lightning my pilot sits [...] (ll. 17–30).

In these lines, the lightning symbolizes human thoughts, as shown by extrinsic evidence. The use of lightning as a more straightforward symbol for thought is implied in 'The Soaring Mind': 'Thoughts are but shadows which the flashing mind / From the swift clouds which track its flight of fire, / Casts on the gloomy world it leaves behind'. These lines contain an interesting relationship between the world and the lightning/mind which is important for understanding 'The Cloud': the shadow cast by the latter on the earth is fleeting, but there is no reason to believe that the perpetual process of lightning/mind is transitory like the world it affects from a distance and leaves behind.

---

32 Two recent studies pursuing Shelley's figure of the poet and his revolutionary vision in minor as well as the major works—completely omits 'The Cloud' from any discussion: Timothy Clark, *Embodying Revolution: The Figure of the Poet in Shelley* (Oxford: Clarendon, 1989); David Duff, *Romance and Revolution: Shelley and the Politics of a Genre* (Cambridge: Cambridge UP, 1994).

But 'The Cloud' offers a symbol that is far more intricate, in that the cloud like the mind has two parts: the congregation of water vapors of the cloud is similar to the synthesis of thoughts by reason; the electricity that pilots the cloud is the imagination. In other words, the cloud is to the electricity what, to adapt Shelley's words in *A Defense of Poetry,* 'Reason is to the imagination' (7: 109).

An even more immediate and compelling comparison is to be found in Shelley's preface to an edition of poems in which 'The Cloud' first appeared. In this preface, hc writes,

> The great writers of our own age are, we have reason to suppose, the companions and forerunners of some unimagined change in our social condition, or the opinions which cement it. The cloud of mind is discharging its collected lightning, and the equilibrium between institutions and opinions is now restoring, or is about to be restored (2: 173).

Although Timothy Clark seems to disregard 'The Cloud', he does offer a profound gloss upon this passage from *Defense*: 'The contemporary state of manners and opinions was, in Shelley's understanding, one of crisis and latent revolution. [...] The extreme instability was due to the universal sense of want arising from the massive discrepancy between the state of knowledge and opinion within the public mind and the oppressive reality of political institutions'.[33] I agree, but only if one (as I think Clark does) interprets 'political institutions' in the widest sense because, among other aspects, Shelley's prophet-poet also takes on the religious superstructure of his society. I think it is by following a similar associational line of thinking that Lamartine, in discussing religious revelation, also draws upon the thunder and lightning image. His 'The Revolutions' / 'Les Révolutions' (510–19) conceives that along with everything else religion evolves—even, in light of its different interpretations, the sacred revelation of the Bible. Each generation reads its structures into the seemingly fixed text: 'Page by page your epochs spell out the Gospel: therein you read but a single word, and you shall read a thousand; therein your more venturesome children will read even more still!'

> Vos siècles page à page épellent l'Évangile:
> Vous n'y lisiez qu'un mot, et vous en lirez mille;
> Vos enfants plus hardis y liront plus avant! (ll. 267–69).

God's revelation to man, for Lamartine, is not something fixed in time but is, rather, a dynamic, winged phenomenon: 'In thunder and lightning your Word soars':

> Dans la foudre et l'éclair votre Verbe aussi vole [...] (l. 273).

Although few late twentieth-century theologians, in the Judeo-Christian mainstream at least, would be upset by this notion of revelation, in Lamartine's day it represented a clearly heretical conception of the relationship between man and God, and it is not surprising that at least one of his works (*Jocelyn*) was put by the Church of Rome on the Index.

What Lamartine—like Shelley in *Defense of Poetry*—employs as a rhetorical metaphor (if certainly one that expresses a larger world view behind it) will in Shelley's 'The Cloud' be treated much more extensively, even with a show of

---

33 Clark 157.

contemporaneous scientific knowledge. It is, of course, true that a modern meteorologist would take exception with Shelley's having separated the phenomena of the lightning from that of the thunder, so that he can use the thunder to suggest the passions that must be controlled and directed toward creative ends:

> In a cavern under is fettered the thunder,
> It struggles and howls at fits [...] (ll. 19–20).

When the cloud tells of how it passes under the control of its electric charge (which is *perhaps* being drawn by a opposite charge in the earth, 'the genii'), the rhythm moves smoothly and hypnotically—like some magical incantation:

> Over earth and ocean, with gentle motion,
> The pilot is guiding me,
> Lured by the love of the genii that move
> In the depths of the purple sea;
> Over the rills, and the crags, and the hills,
> Over the lakes and the plains,
> Wherever he dream, under mountain or stream,
> The Spirit he loves remains;
> And I all the while bask in heaven's blue smile,
> Whilst he is dissolving in rains (ll. 21–30).

The cloud persona, here, claims greater longevity than the electricity that dissipates in the thunderstorm.

The poet who therefore masks himself in the persona of the cloud spins out a series of images based on the cloud's processes and on those phenomena which act on it from outside—especially light. A useful tool in understanding the symbolic nature of these light phenomena and their relation to the cloud/poet is Shelley's own dictum: 'Poets, the best of them are a very camoeleonic [sic] race; they take the colour not only of what they feed on, but of the very leaves under which they pass'.[34] Similarly, the cloud/poet reacts to the sun and moonlight. The cloud and the poet share in this creativeness and share eternal illumination/inspiration which derives from the sun and from the moon—representing respectively, given Shelley's Neoplatonism, a primary and a secondary relation to the Ideal. These lines offer yet another cycle to juxtapose with the water-cycle—the cycle of the day, which begins with a sunrise over the cloud in a cirrus ('rack') formation:

> The sanguine sunrise, with his meteor eyes,
> And his burning plumes outspread,
> Leaps on the back of my sailing rack,
> When the morning star shines dead;
> As on the jag of a mountain crag,
> Which an earthquake rocks and swings,
> An eagle alit one moment may sit
> In the light of its golden wings (ll. 31–38).

---

34 Qtd. in Notopoulos, *Platonism* 168; Notopoulos takes this quotation from a letter of 13 July 1921 to John and Maria Gisborne, in which Shelley is referring to the influence of Goethe's *Faust I* upon his own *Adonais*; this letter is printed in *The Complete Works* 10: 283.

The sun lights up the mountains created by the seismic forces that are always at work in the earth and that have pushed them up—reminding us of that other great contribution of Romantic science: the theory of the perpetual, cyclic changes of geological structures described by James Hutton's other great cycle theory. This is the perch of an eagle, who offers a metonomy for the cloud, just as the dove offers a feathery, airborne simile:

> And when sunset may breathe, from the lit sea beneath,
> Its ardours of rest and of love,
> And the crimson pall of eve may fall
> From the depth of heaven above,
> With wings folded I rest, on mine airy nest,
> As still as a brooding dove (ll. 39–44).

Shelley ensures, in every line above, that the day vision with the sun, in the stanza above, and, below, the night vision presided over by the moon are both filled with the sense of metamorphosis and of flight.

Under the moon's orb, the clouds create their own emblem of mutability, and most wonderfully, too, by transforming those sometime emblems of permanence; it might be remembered how the moon, for example was used as such a symbol in Goethe's 'In the Moonlight', or the mountain peaks in Shelley's 'Mont Blanc', the stars in many poems:

> That orbed maiden with white fire laden,
> Whom mortals call the moon,
> Glides glimmering o'er my fleece-like floor,
> By the midnight breezes strewn;
> And wherever the beat of her unseen feet,
> Which only the angels hear,
> May have broken the woof of my tent's thin roof,
> The stars peep behind her and peer;
> And I laugh to see them whirl and flee,
> Like a swarm of golden bees,
> When I widen the rent in my wind-built tent,
> Till the calm rivers, lakes, and seas,
> Like strips of the sky fallen through me on high,
> Are each paved with the moon and these (ll. 45–58).

When performing the gentler offices outlined in the lines above, the cloud takes on the form of the cirrus or rack. It is important to notice not only that the cloud passively receives the light of the heavenly luminaries, but that it seems to act upon them, thereby creating imagery mirrored upon the reflecting bodies of water below. It should, moreover, be observed that the viewpoint of the persona is violated slightly in this portion of the poem when the cloud is allowed to view the visual effects it creates from the vantage point of earthbound recipients of beauty and gifts of its cloud-offices. This is another clue that something more than the description of a cloud is at issue: whereas, the poet is a manipulator of sounds and images, the cloud creates in light and colors.

Above all, in this stanza, there is a sense of play. In a passing treatment of 'The Cloud', Michael Henry Scrivener writes somewhat contradictorily that: 'Although "The Cloud" is apolitical, it is a poem of liberty in the sense that it fully embodies the Utopian concept of play, for which the cloud's processes provide a natural model'. It seems to me that this is certainly 'political'—more profoundly so than a more obvious political poem like 'Men of England'. Scrivener goes on to say that the poem celebrates 'the changes of the water-cycle, from vapor, to cloud, to rain or snow, and back again to vapor'.[35]

In the lines that follow, the cloud's more dynamic processes are set forth. Its capability of transforming all the images of the world corresponding to the imagination of the Poet-Prophet works on the imagery in the mind's eye—images which the senses have filmed and which reason edits or synthesizes. Finally, the imagination of the poet is acted upon by the external forces of the Ideal and in turn changes them:

> I bind the sun's throne with a burning zone,
> And the moon's with a girdle of
> pearl [...] (ll. 59–60).

It should be noticed in these lines, too, that Shelley follows the Western (not most Eastern) mythological traditions and attributes a distinctly masculine character to the sun and a feminine character to the moon. This suggests that the moon and the sun represent the essential duality of the Ideal: the sun embodies the ideal during the day and the moon, at night.

The cloud controls all the rest of earthly and heavenly phenomenon like a cosmic stage designer:

> The volcanoes are dim, and the stars reel and swim,
> When the whirlwinds my banner unfurl.
> From cape to cape, with a bridge-like shape,
> Over a torrent sea,
> Sunbeam-proof, I hang like a roof,
> The mountains its columns be (ll. 61–66).

This star-dance is an optical illusion created by broken and partly transparent cirrus clouds rushing overhead. Sea and cloud are often moved to great displays by the same whirlwinds. The clouds in stratus formations and mountains make of the world a great columnar temple hall.

In the lines that follow, a new symbol for the Ideal—which unlike the rather passive spirits that inform the moon and sun—is now an immediate participant in the workings of the cloud. This is the 'powers' which stir the cloud into destructive

35 Michael Henry Scrivener, *Radical Shelley: The Philosophical Anarchism and Utopian Thoughts of Percy Bysshe Shelley* (Princeton: Princeton UP, 1982) 238–39. I think he is mistaken when he states that: 'The activities of the cloud, its making and unmaking, are an emblem, too, of the poet who mediates between "heaven" and "earth"'. The cloud symbolizes not a poet but the process of poetry; when the cloud unmakes and makes the apostolic succession is passed on. Although David Duff is writing about *Laon and Cythna* and does not even refer to 'The Cloud', he comes closer to its larger significance than does Scrivener: 'In Shelley's poetry there are many metaphors for [...] the transmission of ideals from one mind or generation to another, this being one of his most cherished themes' (*Romance and Revolution: Shelley and the Politics of a Genre,* Cambridge Studies in Romanticism 7 [Cambridge: Cambridge UP, 1994] 207). The relation between individual poets is even better described, I think, as apostolic succession in the office of poet or within the process of poetry.

actions—hurricane, fire, and snow. The more heroic imagery of epic poetry is suggested:

> The triumphal arch through which I march,
> With hurricane, fire, and snow,
> When the powers of the air are chained to my chair,
> Is the million-coloured bow;
> The sphere-fire above its soft colours wove,
> While the moist earth was laughing below (ll. 67–72).

The seat of fertility below is blessed by the cloud; the world below is rich with its gift, as the world delights in the gifts of the poet/prophet—or more likely—the spirit of poesis: the poetic process or mind that inhabits a poet, a statesman, or a reformer. This poem works on two levels: the external world extending from the stars to the depths of the sea and the internal world of the mind extending from the level of the imagination to consciousness. This equation of the universe with the mind is found more explicitly stated by Shelley elsewhere. He wrote in his notebook that 'the mind' is as 'wide as the universe' and is 'a world within a world'.[36] Also, in the preface to the collection of poems in which 'The Cloud' is included, Shelley states:

> The imagery which I have employed will be found, in many instances, to have been drawn from the operations of the human mind, or from those external actions by which they are expressed (2: 172).

For Shelley, then, the process of the cloud signifies the process of the mind.

In the last stanza of Shelley's 'The Cloud', all the above processes are clearly placed in the wider context of the water-cycle. Here, the element of permanence in change is to be found in the cloud's process of simultaneously dying and being reborn. When the cloud has dissolved in rain, leaving a blue sky, its reconstruction through evaporation is already under way:

> I am the daughter of earth and water,
> And the nursling of the sky;
> I pass through the pores of the ocean and shores;
> I change, but I cannot die (ll. 73–75).

When the cloud expends its substance in precipitation and leaves a blue sky behind, the rebuilding process of evaporation, aided in part by wind and sunshine, starts—there is no death for the process that is the cloud:

> For after the rain when with never a stain,
> The pavilion of heaven is bare,
> And the winds and sunbeams with their convex gleams,
> Build up the blue dome of air,
> I silently laugh at my own cenotaph,
> And out of the caverns of rain,
> Like a child from the womb, like a ghost from the tomb,
> I arise and unbuild it again (ll. 76–84).

Seen thus in the context of the never-ending water-cycle, the cloud is a symbol of the collective mind of all poets. For like the cloud in the last stanza, a particular poet

36 Qtd. in Neville Rogers, *Shelley at Work: A Critical Inquiry* (Oxford: Clarendon, 1956) 15.

can find consolation for his death in the fact that he participates in the eternal processes of nature's creativity and in its recreation in art, and that he functions as the Poet for one small moment in the entire continuum of poets: he is the specific representing the general. The Poet's form, like the cloud's, may change, but his function remains permanent. Shelley may also be signifying that every great poet must be familiar with the works of his predecessors and thereby share in the collective consciousness which dates back to the beginnings of, at the very least, written literature.

This close identification of the cloud with human consciousness and, above all, this paradoxical equilibrium between permanence and change is to be found in a passage from Germaine de Staël's *Corinne*; it is true that natural phenomena is reduced to a simile—albeit, an epic one, and her emphasis is upon permanence in change in human psychology. And yet, the effect is powerful:

> what gave Lord Nelvil such a great power over his lover's [Corinne's] heart were the unexpected contrasts that lent his whole person a special charm. Everyone admired his intelligence and graceful countenance, but inevitably he appealed most to a person combining those qualities of constancy and change in uncommon harmony, a person who enjoyed impressions at once varied and stable. He never paid attention to anything but Corinne, and that very attention constantly took different forms: sometimes restraint held sway, sometimes spontaneity; sometimes perfect tenderness; sometimes a bitter gloom, proving the depth of his feeling, but unsettling the sense of security and constantly giving rise to new emotion.

> Ce qui donnait [...] un si grand ascendant à lord Nelvil sur le cœur de son amie, c'étaient les contrastes inattendus qui prêtaient à toute sa manière d'être un charme particulier. Tout le monde admirait son esprit et la grâce de sa figure; mais il devait intéresser surtout une personne qui, réunissant en elle, par un accord singulier, la constance à la mobilité, se plaisait dans les impressions tout à la fois variées et fidèles. Jamais il n'était occupé que de Corinne; et cette occupation même prenait sans cesse des caractères différents; tantôt la réserve y dominait, tantôt l'abandon; tantôt une douceur parfaite; tantôt une amertume sombre, qui prouvait la profondeur des sentiments, mais mêlait le trouble à la confiance, et faisait naître sans cesse une émotion nouvelle.

Dare one call Lord Nelvil's psyche the archetypical Romantic mind? Perhaps. Then comes the epic simile that embodies this psychology: 'A superior mind, a sensibility as fervent as it is refined, could weary of everything except a truly superior man whose constantly turbulent soul, like the sky itself, seemed now serene, now covered with clouds'.

> Une esprit supérieur, une sensibilité aussi ardente que délicate, pouvait se lasser de tout, excepté de l'homme vraiment extraordinaire, dont l'âme constamment ébranlée ressemblait au ciel même, qui se montre tantôt serein, tantôt couvert de nuages.

Tragedy, death, and chaos come to Corinne through Nelvil's rejection of her; it seems that she had been fated to fall in love with the person whom she is equally fated to lose. Corinne chooses Nelvil with the same love of the tension found in the state of equilibrium that causes Goethe's Faust to chose to create his utopia at the edge and below the level of the ocean—poised on the edge of chaos: 'Oswald—although he was always genuine, always deep and passionate—was often on the point of giving up the object of his tenderness, because long experience of sorrow gave him to believe that nothing but remorse and suffering could come of the heart's affections when they were too strong'.

Oswald, toujours vrai, toujours profond et passionné, était néanmoins souvent prêt à renoncer à l'objet de sa tendresse, parce qu'une longue habitude de la peine lui faisait croire qu'il ne pouvait y avoir que du remords et de la souffrance dans les affections trop vives du cœur (1: 729).

Oswald's fight for moderation is a struggle against the chaos of extremes—a chaos that Staël paradoxically tends both to celebrate and to see as the seeds of tragedy, which seems to be for her an excusive pattern by which to design prose fiction.

Germaine de Staël's passage just considered and Percy Shelley's presentation of the relation between water phenomena seen in the high mountains and poetic consciousness in 'The Cloud' bears comparison with two of the Mont Blanc passages already touched on in chapter one: one by Mary Shelley from her *Frankenstein*; the other found in Percy Shelley's 'Mont Blanc'.

There is an employment of the link quarter to represent human consciousness in Percy Shelley's 'Mont Blanc' (1: 229–33) where he describes how across the valley of the Arve

sail
Fast cloud shadows and sunbeams: awful scene,
Where Power in likeness of the Arve comes down
From the ice gulphs that gird his secret throne,
Bursting thro' these dark mountains like the flame
Of lightning thro' the tempest [...] (ll. 14–19).

Thus, Shelley returns by similitude to his favorite symbol for poetic/prophetic inspiration: lightning. Shelley later goes on to describe how the poet's subjective mind registers passively the external landscape and the Platonic power which it represents:

Dizzy Ravine! and when I gaze on thee
I seem as in a trance sublime and strange
To muse on my own separate phantasy,
My own, my human mind, which passively
Now renders and receives fast influencings,
Holding an unremitting interchange
With the clear universe of things around [...] (ll. 34–40).

In the remaining lines, what before were 'the ice gulps' from whence 'Power in likeness of the Arve' found its source (ll. 5–6, discussed in chapter one), is now termed 'the witch Poesy' in her 'still cave'—the subjective mind of the poet which sorts out the objective world in order to find that which it symbolizes—the Platonic eternal:

One legion of wild thoughts, whose wandering wings
Now float above thy darkness, and now rest
Where that or thou art no unbidden guest,
In the still cave of the witch Poesy,
Seeking among the shadows that pass by
Ghosts of all things that are, some shade of thee,
Some phantom, some faint image; till the breast
From which they fled recalls them, thou art there! (ll. 41–46).

In these lines, we again find the sense of poetic consciousness as a winging presence passing through everything, and the sense of cyclical return pervades the last lines.

Compare these lines from 'Mont Blanc' with the mountain passage from *Frankenstein.* Once again and more powerfully than before we are reminded that the narrator's moods are as mutable as the scene before him; it is one which contains within itself all the processes of the link quarter of the water-cycle—the evaporation, cloud condensation, and precipitation as rain or snow:

> I looked on the valley beneath; vast mists were rising from the rivers which ran through it and curling in thick wreaths around the opposite mountains, whose summits were hid in the uniform clouds, while rain poured from the dark sky, and added to the melancholy impression I received from the objects around me (94).

Mountain scenes are precisely the ones that offer such cosmic visions of nature naturing on such a comprehensive and gigantic scale. And it is precisely *natura naturans,* the processes that are Nature, that Frankenstein has intruded upon; this as much as the darkening weather may be the source of his melancholy.

Shelley's symbols for poetic consciousness taken from both cloud and French Alps appear united with particular force in Victor Hugo's 'Dictated before the Glacier of the Rhône' / 'Dicté en présence du glacier du Rhône' (2: 30–31), a lyric in *Autumn Leaves*. Although the ostensible locus of the poem is the starting point of the Rhône in the Valais region of Switzerland, Pierre Albouy states that Hugo had not seen this glacier, but that he had seen the Valley of Chamonix in 1825, five years before the composition of this poem—thereby implying that the latter was the source of inspiration for Hugo's poem.[37] I will take it as such. I do not mean to suggest, however, that the ostensible title is not without significance; indeed, this title cosmically contextualizes the glacier in the cycle—as the headwater of the great Rhône river system of the South of France; this source stream connection would probably be as evocative to the French as the spring of the Mississippi for Americans. In the opening stanzas, however, the persona places the cloud imagery in a stricter relationship with that which it symbolizes—the poetic mind—than does Percy Shelley's cloud persona. Indeed, in the imaginary landscape of Hugo's poem this relationship is carried often to the point of allegory.[38] The sun is to the processes of the cloud what divine inspiration is to the poet's creativity. Hugo's lyric has a theistic dimension certainly not found in Shelley's lyrics.

Nevertheless, the same theme of the cloud's freedom of movement that is envied in Shelley's lyrics—especially in the fragment, 'O that a Chariot of Cloud were Mine'—occurs in the opening stanza of Hugo's lyric: 'Often, when my spirit, rich in metamorphoses, floats and rolls, asleep upon the ocean of things, God—hearth of

---

37 Pierre Albouy, ed. *Oeuvres poétiques,* 3 vols. (Paris: Gallimard-Pléiade, 1964–74) 1: 1358, n.

38 With Hugo's lyric one is dealing with something other than topographical poetry; concerning this, John Porter Houston writes: 'Indeed, it is established that one of the more vivid nature pieces of *Les Feuilles d'automne,* "Dicté en présence du glacier du Rhône" [...] concerns a landscape Hugo had never seen. Hugo's imagination seems to have functioned with very little of the kind of stimulus of a particular moment and experience of nature we find in the English Romantic poets' (*Victor Hugo* [New York: Twayne, 1974] 73). I, also, think that viewing Chamonix helped prepare him for writing about the unseen Valais.

true day which is not seen by the eyes, mysterious sun from which the soul takes fire—strikes it with a ray, and, like a drop of dew, takes it up again and carries it to the skies'.

Souvent, quand mon esprit riche en métamorphoses
Flotte et roule endormi sur l'océan des choses,
Dieu, foyer du vrai jour qui ne luit point aux yeux,
Mystérieux soleil dont l'âme est embrasée,
Le frappe d'un rayon, et, comme une rosée,
Le ramasse et l'enlève aux cieux (ll. 1–6).

As in Shelley's 'The Cloud', Hugo's cloud/mind is the passive recipient of the effects of the sun/God; here, too, the sun will be shown as a more active agent in the cloud's metamorphosis than in Shelley's work, making the scientific causal relationship clearer. According to my reading, the changes of the cloud are made to correspond with the 'metamorphoses' of the poet's consciousness, which is able to circulate throughout the vast cosmic whole in great leaps and twists of imagination. Even the cosmos is signified by the metaphor of the 'ocean', as if to strengthen the hydrological connection. The process of poetic inspiration ('the soul takes fire') is signified by the sun causing a water droplet to evaporate and be carried cyclically (suggested by 'again') into the sky quarter. The lyric is marked by a sense of dynamic and powerful motion. Comparable to Shelley's lyric, the metamorphoses of phenomena and their processes portrayed are evoked through Hugo's powerful mastery of the verb.[39]

In the opening lines of the second stanza, Hugo makes the allegorical connection even clearer: the 'errant cloud' is also addressed as 'my sublime poetry': 'Then, wandering cloud, my poetic inspiration flies capriciously and without a planned path, from the west to the south, from the north to the east; and regards, from the height of those radiant vaults, the cities of the earth, and—disdainful of all—casts its shadow while departing'.

Alors, nuage errant, ma haute poésie
Vole capricieuse et sans route choisie,
De l'occident au sud, du nord à l'orient;
Et regarde, du haut des radieuses voûtes,
Les cités de la terre, et, les dédaignant toutes,
Leur jette son ombre en fuyant (ll. 7–12).

Hugo does not, however, establish the relation between the poet and his audience as does Shelley with his cloud and flower image—thereby embodying the creative and nurturing aspects of the cloud/poet's mind. Hugo's cloud does, however, offer a chariot or experiential platform from which to view the diversity of natural and human landscapes. The poet emphasizes the holistic vastness of the cloud's motion and the poet's apperceptions by means of his elaborate introduction of the four geographic directions into the verse at this point. A cloud apparently serves as a symbol of self-willed poetic whim; such a signifying use of a cloud—which in the first stanza is shown to be pulled and pushed by the light rays and sun-driven winds—runs counter to its nature.

39 Herbert Juin, *Victor Hugo*, 3 vols. (Paris: Flammarion, 1980–86) 1: 657.

This tension between signifier and signified is even more apparent in the central section consisting of an extended enumeration of the changing processes of the cloud. Especially important here, too, are the effects of light on the cloud. While the symbolic import of this effect is left open by not being overtly set forth, it seems to represent poetic inspiration which comes from external sources. Often, however, the cloud/poet's consciousness seems to articulate its metamorphoses into various shapes, including the active agency of being a thundercloud. The sense here is of a change from one cloud type to another—a cirrus to a nimbo-cumulus. The cloud is even made to will the effects the sunlight has on it. As in the Shelley poem, the diction often becomes heroic: 'Then, shining like a star in the gold of the morning, it [cloud/poet's spirit] presently detaches a fringe from its veil; after a while, like a warrior who resounds in marching, it strikes the murmuring forest with lightning; and, eventually in passing, it reddens its black armor in the furnace of the setting sun'.

Puis, dans l'or du matin luisant comme une étoile,
Tantôt elle y découpe une frange à son voile;
Tantôt, comme un guerrier qui résonne en marchant,
Elle frappe d'éclairs la forêt qui murmure;
En tantôt en passant rougit sa noire armure
Dans la fournaise du couchant (ll. 13–18).

In this stanza, then, the cloud is shown to profoundly affect other phenomena; in the lines that follow, we see how other phenomena affects it.

The poet describes how the moisture is hurled up to a higher, colder atmospheric level by an updraft over the mountains and, thereupon, is transformed into snow—thereby to become part of the great stream of ice, of the glacier: 'Finally on an old mountain—gray-haired colossus—on the Alps, a jealous wind shatters it [cloud/poet's consciousness]. That signifies nothing!'

Enfin sur un vieux mont, colosse à tête grise,
Sur des Alpes de neige un vent jaloux la brise.
Qu'importe! (ll. 19–21).

The cloud/poetry transforms all efforts to thwart or destroy it into a thing of great beauty; this is a pattern we have already seen in Hugo's 'Waterfalls'. James Hutton had begun the research that ultimately would prove that at higher altitudes air retains less moisture than it can at the lower strata of the atmosphere. Broken up, by having been forced upward into cooler atmosphere that causes condensation, precipitation, and freezing, it falls in the form of snow—crystaline flowers, as it were: 'Suspended over the gaping abyss the cloud changes itself into a sublime glacier, and the peak bristles with thousands of flower-patterns composing the crown of a giant!'

Suspendu sur l'abîme béant
Le nuage se change en un glacier sublime,
Et des mille fleurons qui hérissent sa cime,
Fait une couronne au géant! (ll. 21–24).

These lines, therefore, trace the movement of moisture from the sky to the source quarter. The mountain is also, thereby, transformed—receiving the crown of snow, which Shelley in his 'Mont Blanc' saw as a symbol of eternity.

As in Shelley's 'The Cloud', the cosmic viewpoint of the cloud allows the poet to take a synoptic view of mountain images—with their elements of the vertiginous sublime, native animals, and continued play of light effects. As with Shelley's poem (ll. 31–58), we find embedded within Hugo's vision of the water-cycle, another cycle—that of day and night. Hugo and Shelley feel compelled to have smaller cycles within the larger: 'Like the high summit of an unscalable mountain, there it raises up the formidable crest. The rainbow hesitantly plays on its steely side; and, each night, while the shadows besiege it from below, the sun—flowing over the snow like lava—lends to the glacier the appearance of a [volcanic] cauldera'.

Comme le haut cimier du mont inabordable,
Alors il dresse au loin sa crête formidable.
L'arc-en-ciel vacillant joue à son flanc d'acier;
Et, chaque soir, tandis que l'ombre en bas l'assiége,
Le soleil, ruisselant en lave sur sa neige,
Change en cratère le glacier (ll. 25–30).

The suggestions of volcano and lava are, of course, an elaborate simile to suggest the light effect of the setting of the sun in its daily cycle; however, it does suggest geology and the Huttonian cycle of the earth's crust—intentionally or not. Shelley similarly includes a reference to earthquakes.

The next stanza bears a strong resemblance to Shelley's 'Mont Blanc', in that Hugo endows the snowcap with a sense of power and eternity: 'Its face white in the night seems to be eternal dawn; the chamois—with its wing-like feet—takes fright, even the eagle—somber and silent fears it; the tempest swirls and crawls at its feet; the eye hardly dares to penetrate to its serene face—so far it reaches into the sky'.

Son front blanc dans la nuit semble une aube éternelle;
Le chamois effaré, dont le pied vaut une aile,
L'aigle même le craint, sombre et silencieux;
La tempête à ses pieds tourbillonne et se traîne;
L'œil ose à peine atteindre à sa face sereine,
Tant il est avant dans les cieux! (ll. 31–36).

Here again, we find a tension between the signifier and the signified, the symbolizer and the symbolized. Hugo, like Shelley, realizes that the eternal snow on the top Mont Blanc is not really eternal and unchanging—it only seems so under certain aspects.

Unlike Shelley's poem, the speaker of Hugo's lyric is not the cloud but instead is a human persona situated by imagination, if not by Hugo's personal experience of the specific locus—as the title has said and now the poem clarifies—before the glacier at the headwaters of the Rhône River; he, further, is imagining that his spirit or consciousness is moving throughout the whole. He freely associates his spirit with the glacial ice; they have the potential to melt and flow down through the two final quarters of the water-cycle—the river and the ocean. God is freely associated with the sun and is the theistic source of the poet's inspiration:

And alone, at these altitudes, without fear and vertigo, my spirit, oblivious to gravity, observes starry-creating day, a sky no longer blue, and it contemplates up close these starry splendors, which the night sprinkles across its somber cathedrals until a ray from God strikes it anew, it falls and changes the prisms of the glacier into streams mixed with mud; then it runs, then, awaking a thousand echoes, it falls torrentially into the world-ocean—blind and deaf chaos, immense and profound sea, when it becomes indistinguishable from the other waters.

Et seul, à ces hauteurs, sans crainte et sans vertige,
Mon esprit, de la terre oubliant le prestige,
Voit le jour étoilé, le ciel qui n'est plus bleu,
Et contemple de près ces splendeurs sidérales
Dont la nuit sème au loin ses sombres cathédrales,
Jusqu'à ce qu'un rayon de Dieu

Le frappe de nouveau, le précipite, et change
Les prismes du glacier en flots mêlés de fange;
Alors il croule, alors, éveillant mille échos,
Il retombe en torrent dans l'océan du monde,
Chaos aveugle et sourd, mer immense et profonde,
Où se ressemblent tous les flots! (ll. 37–48).

The often sordid experiences of the process and cycle of life seemingly are signified by the mud. The 'deaf and blind chaos' of death is the ocean (ll. 46–48); this, however, symbolizes the death of only one poet. The sea is seen here as the ultimate unity of earthly things, where all the waters reunite. We have found this allegorical symbol in the sea as the all-embracing Father in Goethe's 'Song in Praise of Mohammed' and as the vast receptacle in which the soul mingles with eternity in the penultimate sonnet of Wordsworth's *River Duddon* series. In Hugo's poem, the great difference lies in the fact that here the ocean (as in Shelley's poem) is not just set clearly in the context of the water-cycle, it is also set in a poem that (like Shelley's) has no other reason for being than to celebrate the cycle itself and employed it above all other symbol systems in the poem to signify poetic continuity. Set so manifestly in the cycle, the water/poetic-spirit is not fated to remain in this dark chaos of death represented by the sea. Change is optimistically mediated by being subsumed under a larger permanence.

By the last stanza, under the influence of the rays of the sun, the waters of the world's ocean evaporate, form clouds, and precipitate in rain, which represents the effect of inspiration upon the poet's consciousness: 'My thoughts are carried by a divine wind within an eternal circle, incessantly propelled. From the earthly ocean, with their waves of bitter experience, just as a ray causes the thickened clouds to raise, they eternally mount toward the sky and ceaselessly descend again from sky to sea'.

Au gré du divin souffle ainsi vont mes pensées,
Dans un cercle éternel incessamment poussées.
Du terrestre océan dont les flots sont amers,
Comme sous un rayon monte une nue épaisse,
Elles montent toujours vers le ciel, et sans cesse
Redescendent des cieux aux mers (ll. 49–54).

Hugo gives the lyric perfect structural closure by waiting for the final stanza to offer a full version of the water-cycle; this resembles the same cosmic overview with which Shelley concludes 'The Cloud'. The poet/persona of Hugo, like that of Shelley, transcends mutability in an eternally recurring, collective process of all poets symbolized by the water-cycle. Since there seems to be no direct influence by Shelley's poem upon Hugo's 'Dictated before the Glacier of the Rhône', it appears that both drew upon a common reserve of scientific knowledge to give a strikingly similar world view. The poets, thereby, found a mythic mode by which they might integrate themselves as individual, mortal poets into the continuity of poetry. This comprehensive symbol allows both writers to accept the inevitability of change and multiplicity while at the same time allowing for continuity and unity. The fact that Hugo waits for the final stanza to offer a comprehensive vision of what he has just suggested lends to the expansive poem a satisfying sense of closure and of culmination. It also makes it clear that Hugo had a full, if perhaps a popular-science, conception of the hydrological-cycle.

While the cloud and water-cycle symbol appears not to have the hold on Hugo that it does on Shelley, 'That Heard on the Mountaintop' / 'Ce qu'on entend sur la montagne' (2: 24–26) contains a passage that is oracular even for Victor Hugo. Read in the light of the English lyrics we have just examined and of the locus of the speaker on a mountaintop overlooking a sea- and skyscape, this poem seems to use the sky image in the water-cycle context to embody the music of eternal creation: 'It was a music ineffable and profound, which, like a fluid, ceaselessly flowed around the earth, and in the vast skies—renewed by its waves—revolved in ever-increasing cosmic cycles, up to the place where its billows lose themselves in the shadows with time, space, form, and number'.

C'était une musique ineffable et profonde,
Qui, fluide, oscillait sans cesse autour du monde,
Et dans les vastes cieux, par ses flots rajeunis,
Roulait élargissant ses orbes infinis
Jusqu'au fond où son flux s'allait perdre dans l'ombre
Avec le temps, l'espace et la forme et le nombre (ll. 19–24).

In these lines, then, it is as if Hugo were offering the Romantic alternative to the Pythagorean conception of astronomy and harmony; this time it is meteorology/hydrology and the musical harmony: 'Like another atmosphere dispersed and boundless, the eternal hymn covered all the submerged globe'.

Comme une autre atmosphère épars et débordé,
L'hymne éternel couvrait tout le globe inondé (ll. 25–26).

Stars, planets, and crystal spheres seem to have been replaced by the ocean and the winds of a terrestrial envelope of air: 'The earth, enveloped in this symphony, as it sails in the air—adrift in harmony—and in meditation, I heard the harps of the atmosphere, lost in this roar as in a sea'.

Le monde, enveloppé dans cette symphonie,
Comme il vogue dans l'air, voguait dans l'harmonie,
Et pensif, j'écoutais ces harpes de l'éther,
Perdu dans cette voix comme dans une mer (ll. 27–30).

This poem about the symphony of nature became the program of possibly the most revolutionary of the High Romantic tone symphonies.

We have already seen that Franz Liszt—programmatically basing his musical vision on Hugo's verbal one—was able to suggest the sea in his *That Heard on the Mountaintop / Ce qu'on entend sur la montagne*. The question is: can music really suggest a cycle? A composer can suggest cycles by having arpeggios or runs that move several tones up within an octave, followed by several running back down to the home note. Other times, the two series of notes can run up past the home note and back down again past it. It seems no accident that this is precisely the way that Liszt has constructed the dominant motifs of almost every one of the sections. Thematics are not the only means: harmonics are also important. It is, of course, quite normal for nineteenth-century music to begin and end in the same key, as this first of Liszt's tone poems does, and yet it might be noted that not all his tone poems return to the home key. While many of his later tone poems move freely through unrelated and unusual keys, this first one seems more consciously organized around the circle of fifths. The harmony moves circularly from the flat keys to the sharp keys; it also moves around from the heavily flatted keys to those with few or with no flats at all (C major and a minor). Since this piece appeared, it has been a critical commonplace to react negatively to the fact that it is excessively divided into short episodes in differing tonal regions; but it seems to me that only such short tonal sections could allow Liszt to suggest the harmonic cycle—and even with this, he has had perforce to leave gaps by not passing through every major/minor step in the cycle; rather, Liszt contents himself with dividing the cycle into quarters and moving from one of these quarters to the other.

Liszt's cycling harmonics are most particularly interesting for our purposes. In the following analysis, I have only noted the passing harmony when it points out the general movement: the piece begins in $E^{\flat}$ major with oscillations to its relative minor c; it then moves clockwise around the cycle of fifths to D major (with two sharps); it then moves much further in the same direction to B major (with its six sharps); to $F^{\sharp}$ major (with six sharps) and $C^{\sharp}$ major (with seven), oscillating then between the two; then after passing for a considerable period through the key of $e^{\flat}$ minor (marked with six flats, with one as it were cancelled out), it moves clockwise back to a heavily chromatic $E^{\flat}$ major; then clockwise to $B^{\flat}$; the same direction to a heavily chromatic G; the same direction again to B (with five sharps); after passing through a chromatic $A^{\flat}$, it moves clockwise to $B^{\flat}$; to passing $E^{\flat}$; to C; and so on it goes, throughout the work, with only one slight retrograde, perhaps added for variety. This depiction will make it clear (keys employed by Liszt, even in passing, are in boldface type):

$F^{\sharp}$ ($d^{\sharp}$) / $G^{\flat}$ ($e^{\flat}$)

B ($g^{\sharp}$)/$D^{\flat}$ ($b^{\flat}$) **$C^{\sharp}$** ($a^{\sharp}$)/$D^{\flat}$ ($b^{\flat}$)

E ($c^{\sharp}$) **$A^{\flat}$** (f)

A ($f^{\sharp}$) **$E^{\flat}$ (c)**

D (b) **$B_{\flat}$** (g)

G (e) F(d)

C (a)

There are many other keys throughout the cycle that are suggested by Liszt in passing harmonic sections; however, such momentary harmony is hard to distinguish from a general chromatic palette. This was Liszt's first program work for orchestra; it is extraordinary that he would have decided on such unprecedented harmonic cyclings; after all, Classical composers rarely used the upper (heavily sharped or flatted) area of the cycle of fifths. It is true that in his other twelve tone poems he moved freely around the cycle; however, I can find none of these later works that have the same very distinctive cyclical pattern, moving in essentially the same direction. Liszt's historically unprecedented harmonic structuring seems to fit the program poem by Hugo: one can imagine few better ways of suggesting 'a music ineffable and profound, which, like a fluid, ceaselessly flowed around the earth, and in the vast skies—renewed by its waves—revolved in ever-increasing cosmic cycles' / 'une musique ineffable et profonde, / Qui, fluide, oscillait sans cesse autour du monde, / Et dans les vastes cieux, par ses flots rajeunis, / Roulait élargissant ses orbes infinis [...]'. Music and poetry, however elusive their correlations, offer few closer parallels of structures.

I have shown the use of the cloud quarter of the water-cycle in a few very extraordinary works in the Romantic arts. It remains to be asked about the Romantic employment of this symbol—how widespread was it in the Western European world and in the Americas?

Friedrich Rückert's 'Die Wolke' (45)[40] contains some interesting parallels and contrasts with Goethe's, Hugo's, and Shelley's cloud poems. The persona describes how as he was leaning on the birch trunk, he sees a white and extensive cloud undulate overhead; he says, 'I imagined I was able to fly there with it [the cloud]':

Hin mit ihr zu fliegen
Hab ich mich empor gesehnt (ll. 5–6).

As with Goethe's messenger cloud in the Howard poems, and, with that in Shelley's poem, the cloud here is a bringer of philosophy and poetry. Rückert's cloud, too, is used to symbolize a messenger bearing to the persona-as-poet his 'orders' (i.e., inspiration) from the gods: 'Delightfully you steer your boat, Cloud, messenger of the gods, blown from rosy dawn to the sunsets glow; if I had the command of your command of magic power!'

Lieblich steuerst du dein Boot,
Wolke, Götterbote,
Angehaucht von Morgenrot,
Und vom Abendrote;
Stände zu Gebote
Mir dein Zaubermachtgebot! (ll. 7–12).

What is, above, an open symbolic relationship between poetic inspiration and the cloud becomes somewhat clearer in the next stanza: here the cloud is shown passing through various metamorphoses which resemble a dream—as in Coleridge's 'Fancy in Nubibus, or The Poet in the Clouds'—and is shown creating forms to fill the void. Here, too, the cloud/poet is shown in its/his cyclic process of eventually becoming part of the ocean—as Shelley's cloud-as-poet passes 'through the pores of the ocean and shores' and recreates himself in his cenotaph, 'the blue dome of air', and as Hugo's cloud-as-poetic-thoughts 'eternally mounts toward the sky, and ceaselessly descends again from sky to sea': 'Transforming yourself like a dream you fill the void with form. One moment, you fill the celestial space with battle and troops, the next you raise cliffs in a blue sea and stir up foam'.

40 Friedrich Rückert, *Gedichte*, ed. Johannes Pfeiffer (Hamburg: Schröder, 1953).

Dich verwandelnd wie ein Traum,
Füllest du die Leere
Mit Gestalt, den Himmelsraum
Bald mit Schlacht und Heere,
Bald im blauen Meere
Ragst du Fels, und stiebst du Schaum (ll. 13–18).

Finally, Rückert's last stanza bears comparison with the cloud's proclaiming in Shelley's poem: 'I bear light shade for the leaves when laid / In their noon-day dreams. / From my wings are shaken the dews that waken / The sweet buds every one': 'Whatever the soul may desire, you show in a picture, before the sun on a hot day you serve as a shield and from your gentleness the spring grove begs for dew'.

Was die Seele wünschen mag,
Zeigest du im Bilde,
Vor der Sonn am heißen Tag
Dienest du zum Schilde,
Und von deiner Milde
Bettelt Tau der Frühlingshag (ll. 19–24).

Above all, like Shelley, Rückert does not use the cloud as an allegorical symbol but rather as an open or multivalent symbol. The cloud does not have to represent the poet or his inspiration, but the relationship between cloud and poet is only implicit in the similar nature of their creative tasks. On the other hand, Rückert's 'Die Wolke' more closely resembles Hugo's poem in that the cloud is portrayed in general terms and not with the strong element of scientific description found in Shelley's 'The Cloud', as well as in Goethe's Howard poems.

In *The Book of Thel* (127–30),[41] Blake uses the cloud-in-the-cycle symbol for a much broader purpose than that of the poets just discussed. Whereas Goethe's symbol represents the poet/scientist in relation to the scientific process and those of Rückert, Hugo, and Shelley represent the poet/prophet in relation to the poetic process, Blake uses this symbol for the souls of all men in relation to the larger patterns of the life and death processes. Although Frye warns against an 'exclusive' allegorical interpretation of this work and writes that 'the analogies suggested' by the poems 'are more tenuous', he offers one possible interpretation: the poem is about the cycle of life, a cycle in which the titular being, Thel—perhaps symbolizing 'any form of embryonic life'—refuses to take part.[42] Although only the cloud passages are sufficiently developed to concern us here, it might be mentioned in passing that the work contains other water images and symbols (e.g., 'the river of

41 *The Complete Writings of William Blake*, ed. Geoffrey Keynes (London: Oxford UP, 1966). My attention was called to Blake's use of the cloud symbol in *Thel* by Bachelard, *L'Air et les songes* (220, where it is mentioned in passing). Harold Bloom finds the cloud symbol to be important to the poem as a whole: 'The intensification of mutability in the cloud makes it a more eloquent apologist for the process of natural decay' (*Blake's Apocalypse: A Study in Poetic Argument* [Garden City: Doubleday, 1963] 56). I also found useful the discussion of the poem within the larger context of Blake's works in Laurence S. Lockridge, *The Ethics of Romanticism* (Cambridge: Cambridge UP, 1989) 166–78.

42 Northrop Frye, *Fearful Symmetry: A Study of William Blake* (Princeton: Princeton UP, 1947) 232–33.

Adona', l.8; 'the lotus of the water', l.10; 'shadows in the water', l.13; 'a wat'ry weed', l.20). The cloud symbol is referred to several times in the opening lines of the poem: first, where Thel compares herself to the cloud by the emphatic means of a simile:

> 'Ah! Thel is like a wat'ry bow, and like a parting cloud [...]' (l. 12).

Second, again in the voice of Thel, where this simile is repeated; however, this time, the process of evaporation is indicated, as well as that of precipitation suggested:

> 'But Thel is like a faint cloud kindled at the rising sun:
> 'I vanish from my pearly throne, and who shall find my place?' (ll. 40–41).

She soon addresses a cloud, again drawing a simile between her existence and that of the cloud:

> 'O little Cloud', the virgin said, 'I charge thee tell to me
> 'Why thou complainest not when in one hour thou fade away:
> 'Then we shall seek thee, but not find. Ah! Thel is like to thee:
> I pass away: yet I complain, and no one hears my voice' (ll. 48–51).

The cloud is integrated into and thus reconciled to the processes of nature, of which neither is true of Thel the virgin. Here, Blake is, of course, being atavistic about women.

Blake describes this cloud in terms that are mythopoetic and are far from the scientific descriptiveness of Goethe or even Shelley:

> The Cloud then shew'd his golden head & his bright form Emerg'd,
> Hovering and glittering on the air before the face of Thel (ll. 52–53).

The renewing process of evaporation is suggested by the clouds being horses of one of Blake's own mythological figures. We have seen the cloud/horse analogy in 'Die Heideschenke' and in the cloud passage from *Endymion*; but now something far more is meant, for it is not the cloud per se that is the symbol, but the cloud is placed clearly in context with the process that in turn connects it with the rest of the hydrological-cycle:

> 'O virgin, know'st thou not our steeds drink of the golden
> springs
> 'Where Luvah doth renew his horses?' (ll. 54–55).

The cloud poses another question:

> 'Look'st thou on my youth,
> 'And fearest thou, because I vanish and am seen no more,
> 'Nothing remains?' (ll. 55–57).

Just as Shelley's cloud/poet brings 'fresh showers for the thirsting flowers' and 'the dews that waken / The sweet buds every one'—just as it changes, but 'cannot die' and rises and dissolves again, thus too does Blake's cloud/soul find its place in the cycle of water / of life (with one exception: dew in Blake's poem is shown evaporating; in the Shelley, precipitating):

> 'O maid, I tell thee, when I pass away
> 'It is to tenfold life, to love, to peace and raptures holy:

'Unseen descending, weigh my light wings upon balmy flowers,
'And court the fair-eyed dew to take me to her shining tent:
'The weeping virgin, trembling kneels before the risen sun,
'Till we arise link'd in a golden band and never part,
'But walk united, bearing food to all our tender flowers' (ll. 57–63).

Being left outside the cyclic continuity of things, Thel feels useless; she only can 'smell the sweetest flowers' but cannot 'feed' them like the cloud (ll. 64–66).

The cloud ironically suggests that if she were mortal and not a heavenly being, then she might be able to achieve her unfulfilled promise of creating life:

'Then if thou art the food of worms, O virgin of the skies,
'How great thy use, how great thy blessing!' (ll. 72–73).

These lines establish the relationship between the symbols of this poem and the thing symbolized: the life cycle of the cloud symbolizes the life cycle of humans. These lines also make it clear that Blake's central purpose in using the cloud symbol is in essence the same as that of Shelley in 'The Cloud' and that of Hugo in 'Dictated in the Presence of the Rhone Glacier'. When placed within the greater context of the water-cycle, however, the mutability of the cloud seems to reconcile the poets to the stages of birth, death, and decay by placing them in the context of a cyclic continuity. Although all *are* subject to change, all *is* one greater permanence.

Much more than Blake's poem, the American Philip Freneau's 'The Brook of the Valley' (3: 376–77) offers an allegorically direct and concise vision of the sky link in the evaporation/precipitation cycle. The brook represents the 'Emblem, thou, of restless man' (l. 48). The water droplets

may return once more
To visit scenes they knew before;—
Yonder sun, to cheer the vale
From the ocean can exhale
  Vapors, that your waste supply,
Turn'd to rain from yonder sky;
Moisture, vapors, to revive
And keep your margin all alive (ll. 25–32).

This most concise of all presentations of the water-cycle in poetry prepares the way for an equally direct use by two painters—one an American and the other a German.

Like Constable and Turner, Friedrich observed clouds and rendered them with considerable accuracy. William Vaughan asserts that pure, scientific objectivity in rendering the external world has nothing to do with Friedrich's treatment of clouds:

> Only in 1816, when Friedrich refused to assist in Goethe's scheme for studying clouds—saying that this would be the 'end of landscape painting' did the poet come to realize that Friedrich's vision, with its emphasis on a psychic response before nature, was essentially different from his desire to achieve an objective understanding of the organic forces that determine natural phenomena.[43]

---

43 William Vaughan, *Friedrich,* Tate Gallery 26; see also Kurt Badt, *John Constable's Clouds* (London: Routledge, 1950) esp. 47.

This is not to say that Friedrich did not observe clouds and reproduce them with some accuracy, for he did; nor is it to say Goethe never used natural phenomena for symbolic purposes, which I hope by now is clear that he did again and again. In all cases, it must be admitted, the symbolic purposes to which the water phenomena could be employed are more important than the accuracy with which they are reproduced.

In Friedrich's *Drifting Clouds / Ziehende Wolken,* the natural and the symbolic aspects are combined. The top three-fourths of the picture is covered with a sky of rising stratus clouds—the tops of which are rendered by a smooth, grey and white facture. As is appropriate for the waif-like stratus clouds, the paint has been lightly applied to give the feeling of transparency. The clouds are painted over blue; distant mountains are also rendered in extremely vibrant blues, which offers a strong contrast to the very dark green of the foreground. The hazy sky beyond is painted with a grey tinted with purple and pink—a color combination usually associated with dawn. The stratus clouds are analytically rendered with a transparent application to suggest rapidly changing clouds. But scientific analysis is somewhat outweighed by the symbolic purposes of the painting. Concerning the contrast between the darkened and narrow foreground and the extremely bright and wide background, Börsch-Supan interprets the landscape as an emblem of Christian Paradise.[44] In the heavenly ascent of the hero's soul (composed in 1830) that concludes *Faust, Part Two* (5), Pater Seraphicus sings of a group of infants that have died at birth: 'What is this morning wisp of cloud floating through the spruces' waving tresses; do I divine what its interior contains? It is a throng of young spirits'.

> Welch ein Morgenwölkchen schwebet
> Durch der Tannen schwankend Haar?
> Ahn ich, was im Innern lebet?
> Es ist junge Geisterschar (11890–93).

Like Goethe, Friedrich knows his science of clouds, but they both can employ them mythopoetically; Friedrich here employs them as symbols of the destiny of the human soul. The painter loaded his works with elaborate allegory—even when it is pure landscape, with human figures.

Although Friedrich's works are in a Christian allegorical tradition, they do contain a vision of change that takes them beyond the more limited confines of that tradition. In reaction to another of Friedrich's paintings, Friedrich Wilhelm Basilius von Ramdohr, Chancellor of Saxony, found in it the 'calamitous spirit of the present times'.[45] There is in *Drifting Clouds* an overwhelming sense of a weather change—change is ideologically, even symbolically, to be feared by most conservatives.

Similar to *Drifting Clouds* in coloration (but without blue), in its composition (but now with human figure), and perhaps in symbolic meaning is the *Wanderer over the Sea of Mist / Wanderer über Nebelmeer.* The mist is made up of the same stratus clouds. The landscapes of both paintings and the added figure in *Wanderer over the*

---

44 Helmut Börsch-Supan, *Caspar David Friedrich,* 2nd ed. (Munich: Prestel, 1975) 124.

45 Qtd. by Hugh Honour, *Romanticism* (New York: Icon-Harper, 1976) 28.

Sea of Mist find a striking parallel in Wordsworth's Mt. Snowdon passage in 1805 *The Prelude*:

> on the shore
> I found myself of a huge sea of mist,
> Which, meek and silent, rested at my feet (13: 43–44).

In these respects both Wordsworth and Friedrich come as close as poetry can to realizing the cosmic spirit of Diderot's advice to the young artist, which is significant enough here to requote in my translation: 'Turn your observations toward the mountaintops. Those that begin to pierce the ocean of clouds. Run; quickly climb upon some high hill; and from there contemplate the surface of this ocean which gently undulates over the earth'. In Diderot's advice, as in the work of the English poet and that of the German painter, the clouds are seen as a great ocean; but more importantly, the three share a common desire to see the great unity behind all these phenomena.

The unity of the processes of nature stated in Wordsworth's 1805 Mt. Snowdon passage is implied particularly in Friedrich's *Drifting Clouds* where the clouds are rising from land to sky to form one link in the water-cycle:

> A meditation rose in me that night
> Upon the lonely Mountain when the scene
> Had pass'd away [...].

In these lines and those that follow, Richard Bourke finds that 'Wordsworth's project resembles most an exemplification of the putative principles of the French Revolution—liberty housed as an abstraction in the privacy of the soul [...]'.[46] As in Friedrich's painting where the viewer of nature has his face turned away from the actual viewer of the painting, Wordsworth is offering to his readers the chance to stand in his place: in short, both poet and painter have left open a vantage point for Everyman/woman to stand and contemplate the scene. This is not the hegemonic, Baroque vision of the God-chosen saint of Bernini's *St. Theresa of Avila*; this is a democratic epiphany.

It is my reading of the Mount Snowdon passage that political temporality is left behind as this mountain vision continues. Peering from the mountain top and from out of his consciousness, the poet reaches that perfect Spinozistic epiphany, what Karl Kroeger calls 'proto-evironmentalism'. The poet has a vision where all of Spinoza's infinite mind of God—whose infinitude exists as such because it contains all there is—does not allow room for creature or creation: they are one with the creator:

> and it appear'd to me
> The perfect image of a mighty Mind,
> Of one that feeds upon infinity,
> That is exalted by an under-presence,
> The sense of God, or whatsoe'er is dim
> Or vast in its own being, above all

---

46 Richard Bourke, *Romantic Discourse and Political Modernity: Wordsworth, the Intellectual and Cultural Critique* (New York: St. Martin's, 1993) 254.

> One function of such mind had Nature there
> Exhibited by putting forth, and that
> With circumstance most awful and sublime,
> That domination which she oftentimes
> Exerts upon the outward face of things,
> So moulds them, and endues, abstracts, combines [...] (ll. 66–79).

This one passage does in microcosm what the entire *Prelude* does: it mixes political hopes with ecological ones. Indeed, it affirms my belief that Jonathan Bate's Red and Green concerns, if viewed *sub species eternitatis,* are one. There can be no separation between man and nature and Wordsworth's ecological deity—for which the water-cycle is an emblem.

It is true, of course, that Wordsworth's poetic passage, like the painting of Friedrich with which we compared it, is limited to one portion of the water-cycle; in order to find the cloud symbols in the water-cycle context, one must turn to the *Cycle of Life's Ages / Lebelsalter-Zyklus.* In 'Skeleton in the Stalactite Cave' / 'Skelette in der Tropfsteinhöhle', the waters of the earth and sky are connected: formations hanging from the roof of the cave symbolize the water locked in the earth and symbolize, juxtaposed as they are with the skeletons, the human souls locked in earth awaiting resurrection. The clouds seen in the background are the unlocked waters of the sky and symbolize the resurrected soul. In 'Engel in Anbetung', the juxtaposition of angels and clouds makes clear that the clouds are drawn upwards by the sun in the same way that souls are thought to be drawn up by God.

Clouds and angels similarly dominate Thomas Cole's *Voyage of Life: Old Age* and the parts of the allegorical 'program' that we have not already quoted and considered:

> Portentous clouds are brooding over a vast and midnight ocean. [...] Directed by the Guardian Spirit, who thus far has accompanied him unseen, the Voyager, now an old man, looks upward to an opening in the clouds, from whence a glorious light bursts forth; and angels are seen descending the cloudy steps, as if to welcome him to the Haven of Immortal Life. [...] The chains of corporeal existence are falling away; and already the mind has glimpses of Immortal Life. The angelic Being, of whose presence, until now, the Voyager has been unconscious, is revealed to him; and, with a countenance beaming with joy, shows to his wondering gaze scenes such as the eye of mortal man has never beheld (29).

A voyage of life is different from a cycle of life in that the one implies a beginning and an end, the other does not. I have quoted his program to admit that Cole makes no direct reference verbally to the evaporation process of the water-cycle. Nevertheless, it is the fourth panel of a work that has treated all the other aspects of the hydrological-cycle. We must remember also that the eventual assent of the soul comes at the end of a story told by the four panels. In the final version, even more than in the oil sketch, the clouds not only extend down to the surface of the ocean but they seem to merge with the water itself. This contiguity, indeed merger, of the sea and clouds is not without significance—it does suggest a link, a unity in process. All this taken together suggests that evaporation is the analogue here to the ascension of the soul.

As already discussed in the introduction and chapter two, *The Voyage of Life* and the *Cycle of Life's Ages* contain a paradoxical use of the water-cycle—one that creates a tension with what is being represented. As indicated in the introduction, I made use of a study by Stephen Jay Gould that finds an interesting tension in Thomas Burnet's *Sacred Theory* use of the cycle metaphor. The Judeo-Christian history, moving as it does from Eden to Apocalypse, and its vision of the soul, moving from birth to resurrection, does not fit very well with the constantly recurring implications of a true cycle. Unlike the cycle, the former are one-time events. Gould views 'the *Sacred Theory* as a playground for Burnet's struggle to combine the metaphor into a unified view of history that would capture the salient features of each—the narrative power of the arrow and the immanent regularity of the cycle'.[47] The way of combining the 'time's arrow' metaphor with that of 'time's cycle' is that of 'arguing for repetition *with a difference*'.[48] What Gould is describing contains the same paradox of Vico's and Hegel's upwardly spiraling cycle. Friedrich's and Cole's Christian employment of the cycle, like that of Wordsworth, of Liszt, and of Hugo, contains this paradoxical clash of eternal recurrence with progressive movement from point A to point B. Only a Hindu conception of reincarnation would fit conveniently within a naturally recurring cycle metaphor in its pristine simplicity. I think that this complexity adds, rather than detracts, from the richness of the works of these Romantic creators.

The Voyage of Life and the *Cycle of Life's Ages,* along with the two great skyscapes of Shelley and Hugo, all perfectly crystallize the Romantic desire to be integrated into the greater unity of things. For this reason, I feel justified in calling the water-cycle a Romantic symbol. It was this concrete cyclic symbol that—in its parts or in its totality—lent itself so well to an expression and, indeed, a resolution of the Romantic yearning for permanence in a universe of seething change. We have even seen that Liszt, in his tonal version of a Hugo poem, can suggest a musical version of the cycle. No other painters of the period seem to have used the spring in the water-cycle context quite as extensively as do Cole in his *Voyage of Life* and Friedrich in his *Cycle of Life's Ages*. In the two versions of his *Schmadri Brook / Schmadribach,* however, Joseph Anton Koch presents a large waterfall being fed by melting glaciers and smaller rock-born waterfalls; all are placed in a larger context. The canvas includes various cloud levels and types from stratus to cirrus clouds. In the foreground, the resulting mountain torrent is shown being joined by another. Thus, in this painting three of the four quarters of the cycle are represented. William Vaughan writes that Koch's ''universal picture', with its mighty sweep through the strata of nature from the foreground stream to the snow-capped mountaintops, is above all a feat of control. Each form is embedded in its plane, there is no mystery to the lighting or atmospherics; everything is organized and classified'.[49] The phrase which Vaughan refers to in the above, 'universal picture' or 'Weltraumbilder', is Koch's own.[50] Although this painting is a detailed delineation of a specific

---

47 Stephen Jay Gould, *Time's Arrow/Time's Cycle: Myth and Metaphor* (Cambridge: Harvard, 1987) 41.

48 Gould 56 (Gould's italics).

49 Vaughan, *Romantic Art* 180.

50 William Vaughan, 'Caspar David Friedrich', in *Friedrich,* Tate Gallery, qtd. from 15.

locus,[51] the picture's holistic grasp of the natural phenomena—especially water—far transcends topography in scope.

Joseph Turner's *Morning Amongst the Coniston Fells, Cumberland* is almost equally comprehensive in its suggestion of mountain sources in the water-cycle context. A mountain torrent, which runs down the whole center of the canvas beginning around the center to the left, cuts the painting in half. It creates a nervous zigzag pattern down the painting before it disappears into divided waterfalls at the bottom middle slightly to the left. In the darkened foreground, there is another waterfall; it is placed at the lower left middle; it is counterbalanced by a cloud formation (cumulus) at the upper right middle. This cumulus formation is rendered by a golden-white impasto application of the paint. The painting also contains stratus clouds forming on the side of the mountains at the upper center and the upper left. In the sky in the upper left, there are cirrus clouds. The waterfall in the distance is rendered by a stipple application of white on a bright blue field. In the foreground the waterfall is rendered by a striated, dry-brush application over a dark surface. As with the paintings of Koch, there is in *Coniston Fells* a comprehensive picture of the water-cycle—lacking only the sea quarter. The stratus clouds, in nature, would rise from the mountain; either they would continue to rise to become cirrus clouds, or they would form cumulus clouds and eventually precipitate as rain or snow. Then, this water either would melt from ice fields and glaciers and would flow as run-off into the streams, or it would cycle through the earth to burst forth as springs.

Both Koch and Turner show this process holistically, in so far as they are able by means of a single static picture. In their attempts to achieve cosmic paintings, *Weltraumbilder,* within the limits of their art, they evidence the same yearning for a comprehensive symbol which I will reveal in Goethe's 'Commemoration of Howard', in Shelley's 'The Cloud', and Hugo's 'Dictated in the Presence of the Rhone Glacier'. By having a unified vision of the workings of things, all these artists could feel that things human were part of what Hugo terms 'the ocean of things' / 'l'océan des choses'. It is, however, in a painting by Turner that one gets the most profound sense of how the cycle of nature and the things of man unite into one possibility.

Romantic poems, compositions, and even paintings emphasizing the sky quarter are insignificant in quantity when compared to works treating the river—especially the waterfall—and the sea; nevertheless, the artifacts treating the sky are surpassing in their spectacular beauty and profound originality. Chapter four demonstrated that, to a greater extent than with the other quarters, the sky artifacts of Romantic culture mark the most radical departure from the past through artistic exploitations of actual scientific theory—either known firsthand or known via popularized versions. Chapter four completely contextualized chapter one—since the cloud is the ultimate source behind the earthly source. Chapters two and three treated phenomena that were immediately comprehensible to audiences, without any full-scale scientific framework built upon systematic observation and theory-building. Even prior to the beginnings of the science of hydrology in the sixteenth and seventeenth century, prior, also, to the general dissemination of this new knowledge in the eighteenth,

---

51 Hans Engelhard, 'Kunstgeschichte und Geologie: Der Wasserfall in Gemälden des 17, bis 19. Jahrhunderts', Diss. Cologne, 1975, 135–42.

everyone between 1760 and 1870 would almost certainly be aware of the connection between springs and rivers—between rivers and seas. The same was, however, not true of the link between the ocean and the spring—between the sea and the sky. Indeed, even among scientists, an accurate knowledge of the way precipitation generates springs and otherwise feeds streams was not known until the decades around 1700, and a full understanding of the cloud and its processes of evaporation and precipitation was not developed until the decades around 1800. Such a full and accurate comprehension of these phenomena demanded a context of scientific observation and method. Perhaps the word 'matrix' is closer than 'context' to what I am attempting to delineate for the simple reason that most Romantics did not (although a few actually did) participate in the actual scientific method and process, or even have direct knowledge of it; rather, strands of hydrological theory were woven into the intellectual fabric around them. It was just such a comprehensive theory encompassing these same processes that the quasi-cycle of Vaughan's poem lacked. The poems analyzed in this chapter were created during a time when the science of the sky quarter of the hydrological-cycle, meteorology, was flourishing at the hands of Howard and others. Both literally and figuratively, then, the sky quarter has formed the culmination of this study.

The Romantic culture, as has already been discussed throughout the foregoing pages, was the birth age of meteorology. This is no accident. It was only due to the emphasis that Romanticism placed upon *dynamic becoming* that provided the needed philosophical ground from which to study a complex system such as weather. James Hutton, who contributed most to this new meteorology, viewed weather as a great machine. The empowering engine that drove the great cosmic engine of the water-cycle was the heat of the sun; this fact constitutes a nexus that relates other sciences and technologies to the water-cycle studied by hydrologists and meteorologists. Hutton's macro-cycle finds its correspondence in the micro-cycle of his contemporary and fellow Scot—James Watt. The whole process of the steam engine might be seen as a water-cycle in miniature. Sections of Watt's single-acting steam engine of 1788 very roughly approximate the quarters of the water-cycle: the fire under the boiler might be the sun; the boiler, the sea; the pipe that carries the steam to the cylinder, the sky/link; the condenser, the rain; the pipe carrying water back to the boiler, the stream. The study concludes on a work where the micro-cycle and macro-cycle are united: Turner's *Rain, Steam and Speed.*[52] A locomotive, surrounded by a heavy rain, is shown speeding toward the spectator. Just as the rain is part of a titanic natural moisture machine driven by the heat of the sun, so too steam is the product of the fiery boiler that Turner suggest with streaks of bright red—both cycle through their dynamic process. Rightly or wrongly, Romantic man felt that the early changes he was making or at least envisioning were part of a larger vision of the cosmic changes of nature. His microcosms were moving synchronously with the macrocosms of Nature to such a degree and on such a scale that a sixteenth-century alchemist would have been astonished.

---

52 Although none treat the water-cycle and the steam connection, the following studies are useful: John Gage, *Turner: Rain, Steam, and Speed* (New York: Viking, 1972); William Rodner. *J.M.W. Turner: Romantic Painter of the Industrial Revolution* (Berkeley: U of California P, 1997) 140–60; James Hamilton, *Turner and the Scientists* (London: Tate Gallery, 1998).

CONCLUSION

The Water-Cycle in Romantic Adaptive Systems and (Proto)ecology

Chapter four, treating as it has done the link that gave closure to the cycle, also gave closure to most of the ideas treated in this study, making this conclusion less necessary than usual. I, however, think it would be useful to treat a few of these ideas outside of the water-cycle and place them in the larger context of complexity theory, of ecology, and of Romantic culture in its full scope.

M.H. Abrams, in *Natural-Supernaturalism* has addressed this desire of the individual thinker or artist to rejoin the whole. In speaking of the appearance of this yearning in Marx's early 1844 Manuscripts, Abrams, among others, finds that the economist saw the real negative force in industrialization as 'essentially separation, the division of the integral man and society into self-centered, isolated, and hostile parts'.[1] As I have tried to show throughout this study, many Romantic artists found in the hydrological-cycle the model of science—combined in varying degrees of complexity with other models from art, rhetoric, and politics—a way of negotiating the chaos of despair due to fragmentation. Three types with varying properties have emerged.

Having undertaken my research with a literary bias, it is ironic that the three great Romantic painters make up the perfect exemplars of the three ways that creative intellects of Romantic culture employ the hydrological-cycle as a coping mechanism or as a complex adaptive system. While each approaches the problem differently, Turner, Constable, and Friedrich all give expression to the paradox of permanence in change, which can be compared to similar expressions of this paradox in the poetic and musical works examined. The world views of the three painters often differ considerably, but these all become resolved by means of this paradox. For Friedrich, the change of this world is embraced by the permanence of God; for Turner, the only permanence is the continuity of change itself, and he found affirmation in celebrating this change at its most cataclysmic. For Constable, shifting, substantial, and changing natural phenomena are made permanent both in art and in the larger cycle context.

Tradition Order: The Water-Cycle as a Judeo-Christian Adaptive System

Friedrich's world view is neither pessimistic nor optimistic, but Christian. The culminating link of his *Cycle of Life's Ages* comes in the clouds with angels. This series uses the water-cycle as an allegorical narrative of the life of Man set in relation to God. Thomas Cole similarly employs the water-cycle with the voyage theme to embody human existence. His cycle is sprinkled with rather feeble nineteenth-century ornamentations of Christianity—the Guardian Angel and an

1 M.H. Abrams, *Natural Supernaturalism: Tradition and Revolution in Romantic Literature* (New York: Norton, 1971) 324.

operatic finale with the opening of clouds to receive a soul lost in dark despondency. Wordsworth at the conclusion of a similar allegorical series employing the water-cycle (*Duddon River*), offers the following conclusion which gives expression to a Christian world view paralleling that of Friedrich: '*Through love, through hope, and faith's transcendent dower, / We feel that we are greater than we know*'. These lines crystallize both the yearning of Wordsworth, Friedrich, Cole, Freneau, and their coevals to reconcile themselves to the life and death processes by placing them in a larger, eternal process and to view individual changes as parts of the larger permanence—a Christian permanence. They—as I have sought to demonstrate—achieve a certain richness in their superimposition of the eternal return of the cycle with the one-time-only trip of a Christian Soul on earth or with the linear construction of Christian movement from Creation to Apocalypse. Composers are not so easily typed. Wagner and Liszt both, just before their deaths, buried their mythoi in a kind of Christianity; yet, their move to a shifting and unstable harmony under an endlessly meandering melody line makes me want to place them in the next category.

Lovers of Chaos: The Water-Cycle as a Change-Based Adaptive System

Turner and Byron loved chaos; they viewed it as the epicenter of creativity. This acceptance of the fact that the only thing certain and continuous on earth is change was at the center of Turner's world view, which is partly conveyed in a poignant comment on his feelings toward time and change made in old age shortly before his death: 'Old Time has made sad work with me [...] I always dreaded it with horror, now I feel it acutely'.[2] In addition to this almost pathological fascination with time and change, Turner had what Graham Reynolds refers to as his 'lifelong preoccupation with the frustration of human plans'.[3] Above all, there are the fragments of Turner's unfinished epic, *The Fallacies of Hope,* of which the following might be taken as an example:

Man the easy bark which sailing
On the unblest treacherous sea
Where Cares like waves in fell succession
Frown destruction oer [sic] his days
Oerwhelming crews in traitorous ways

Thus thro life we circling tread
Recr[e]ant poor or vainly wise
Unheed[ing] grasp the bubble Pleasure
Which bursts his grasp or flies.[4]

2 Qtd. in Graham Reynolds, *Turner* (London: Thames and Hudson, 1969) 206.

3 Reynolds 205.

4 Qtd. in Jack Lindsay, *Turner: His Life and Work* (New York: Harper, 1966) 73.

Little wonder that Turner's works are dominated by, or deal almost entirely with, water phenomena in a state of seething turmoil. I think it is appropriate that Lockspeiser places Wagner (I would also say Liszt) with the nihilism of Turner.[5]

In poetry, close parallels to Turner's world view are found in the works of Byron, Hölderlin, Staël, and Lenau. All four poets use symbols of water in the violent flow of falls or oceans. With such symbols one can give expression to the most pessimistic conception of things: 'Yet it is our fate never to rest in any place; miserable human beings dwindle and fall headlong from one hour to the next', writes Hölderlin, before crystallizing this *Weltanschauung* by means of the waterfall symbol, 'hurled like water from precipice to precipice through the years deeper into uncertainty'. Like Turner, all three felt the fascination with the tremendous and cataclysmic flux which they saw as the dominant principle of *rerum natura*. In the face of this almighty flux, the only thing to do was to celebrate it in art by bidding it, like Byron: 'Roll on, thou deep and dark blue Ocean, roll!' Byron, unlike Shelley, never felt the need to contextualize specific water phenomenon in the hydrological-cycle. And for Byron, the more turbulent the change in the water phenomena the better: the sea, with its titanic and often chaotic flux and reflux, was his favorite.

As I made clear throughout this study, I take a very rigorous view of historical judgments in the realm of political-economics. I feel that expecting from Romantics or from anyone a clear perception of political progress, before 1871, is dubious and, before 1848, absurd. Liszt's aspirations for Hungarian freedom from Hapsburg imperialism, Turner's visions of change in the face of Regency and Early Victorian complacency, or Byron's hope for Greek freedom from Ottoman domination are limited goals—all these, when seen in their historical context, made their contributions to the larger march of historical progress. The fact that these particular Romantics have placed an inordinate emphasis on change and, indeed, chaos, does not invalidate their world view. It is preferable to Edmund Burke's mediation between chaos and change, and it is just as progressive as that of Lamartine or Hugo. The modern scholar Robert Clark has suggested a danger in 'essentialising the struggle of chaos against order, and of not recognizing that chaos is the name that political order always gives to otherness'.[6] Radical Otherness must surely always have a place in all progressive thinking and creating. The celebration of change in Heine's *The North Sea* might seem pessimistic when taken by itself; however, when seen as the transition to a new and higher level of revolutionary political awareness, the joy in change, even chaos, takes on new significance.

Balancers on the Edge of Chaos: The Water-Cycle as an Aesthetic Adaptive System

Constable did not share Byron's or Turner's interest in cataclysmic changes, but concentrated his energies on capturing the nuances of more subtle changes. But for

---

5 Edward Lockspeiser, *Music and Painting: A Study in Comparative Ideas from Turner to Schoenberg* (New York: Harper, 1973) 54, 63.

6 Robert Clark, 'Introduction: Closing (with) Jane Austen', *'Sense and Sensibility' and 'Pride and Prejudice'* (New York: St. Martin's, 1994) 8.

him, a universe subject to constant and subtle changes was no less fleeting, insubstantial, and, therefore, terrifying than it was for Turner. But Constable saw the possibility for achieving permanence in art. It is extremely significant that the effect, which he created by his technique of juxtaposing sharply contrasting shades, and his world view, which he expressed by means of the former, were so bound together that he employed the same term for both—chiaroscuro. In his short treatise on landscape, he writes that he aspires

> to arrest the more abrupt and transient appearances of the CHIAR'OSCURO IN NATURE [...] to render permanent many of those splendid but evanescent Exhibitions, which are ever occurring in the endless variety of Nature, in her eternal changes.[7]

In this passage, the artist gives clear expression to the paradox central to this study: art gives permanence to change in nature, which is itself permanently changing. His humanized landscape, like that found in Austen and Goethe, is a celebration of the human artifice of hydrology. Beethoven's *Pastoral Symphony* is just such an affirmation of art in the face of the cyclical changes of nature: the brook movement celebrates nature as the muse of music; the shepherd's hymn of the fifth movement resolves the tension of the thunderstorm movement that precedes it.

Constable's system and Beethoven's are paralleled in Goethe's 'Permanence in Change' and 'Song and Form'. In the latter, Goethe used the symbol of the most mutable of substances as being crystallized or given form by art: thus, in this symbol there is a tremendous tension between the formless, lively element before being brought under control *and* the same element after it has been forced into absolute stasis. Similarly, Goethe uses the stream image in 'Permanence in Change' to give expression to the titular concept of the poem: the poet/persona makes permanence out of the change around him. His mind may cease functioning and his body may decay, but the forms they create do not. Above all, when an older poet dies, another poet is ready to take up the place left vacant—in what might be termed a bardic apostolic-succession. Constable takes the most mutable phenomena of water and clouds—the chiaroscuro in nature—and freezes them for all times in his art. In the permanence of art, Constable and Goethe find optimism.

We must also recall those four great cosmic visions that celebrate the processes of science, prophecy and poetry: Goethe's 'Song in Praise of Mohammed', his later *Commemoration of Howard,* Shelley's 'The Cloud', and Hugo's 'Dictation in the Presence of the Rhone Glacier'. The process of creation is shown to be passed on, in apostolic succession, by each generation of thinkers/artists. The symbol joined generations of artists and thinkers to one another, and, most vitally, it saved the Romantics from the isolation and fragmentation they feared. Only such a cosmic and ever-recurring vision as that offered by the hydrological-cycle could integrate their sense of the important process of the human sciences and arts. Liszt gives this powerful vision of the water-cycle permanent expression by means of the harmonics based on the cycle of fifths—his harmony cycled around through various unprecedented key changes. Politically, Liszt, along with Shelley, Lamartine, Freneau, Jefferson, and Hugo, fit awkwardly with Austen, Goethe, and Constable. They

7 Qtd. in William Vaughan, *Romantic Art* (New York: Oxford UP, 1978) 197.

welcomed change in politics: yet, compared with Byron or Turner they all convey certain qualms in the face of change—above all the threat of chaos behind. Optimistic, unlike Byron and Turner, they wanted change to be taken with a stiff dose of permanence. The permanence of the hydrological-cycle offered them a way of representing the mediation in behalf of orderly change between two forms of chaos on either side, which seemed to be threatened by too much or too little change. None of them feared revolution, but all preferred gradualism within a larger pattern. The water-cycle in its great cosmic motion seems to have been for them an emanation from a higher, transcendent Power.

Other Emergent Properties of Romantic Culture

The three basic, therefore perforce reductive, types of adaptive systems in a culture of unprecedented change constitute a simplicity; however, I would not be true to my guiding theory if I failed to discuss the corollary issues which constitute the complexity. These complicating factors do not, I think, undercut the simplified categories, which when identified as such have their uses. My study has brought more clearly to the surface than before the essential dialectics between those Romantics who saw the cosmos as a great unity, which they often did holistically, and on the other hand, those Romantics who were dualist in that they believed in a split between what is of nature and what is of humanity—which might be termed *dualist* in approach. The Romantics' attitude toward landscape and their attitude to contemporaneous technological changes both depend upon this holistic/dualist dialectic.

The dualist approach is essentially that of Wordsworth, Hugo, Turner, and Liszt. Things belonging to the human and the natural are carefully distinguished. Nature is something apart from man; it is located in wild places—the mountain and the seas. Human presence is an intrusion. Only the chosen few—the shepherd, the poet, and the madman—are allowed in the holy of holies of the wilderness. When there is too much of the human taint, then, the wilds become denaturalized—violated. There is a strong feeling of elitism. The middle class and the proletariat of the cities must be kept from violating the sanctity of Wordsworth's Lake Country, a country that is blessed by its infertility—its exclusion from the extremes of agricultural development. Wordsworth offers a diluted, idiomatic version of Pantheism; however, it includes only himself and nature within the Godhood, and he excludes places where humans make their mark—like the Stour Valley, let alone London. For Hugo and it seems for Liszt, Nature and Humans are split—incompatible. Although not as extreme as these others in matters of politics and technology, Turner portrays a nature that is hostile to man. The Alpine storm that wreaks havoc upon Hannibal's army or on a little chalet preserves an exclusive realm beyond human control. This celebration of the wilderness is found in the dogmatically dominant approach to the ecology of today; it is an incomplete view—it must be supplemented by yet another; this other is also contained in Romanticism.

As a sidelight to his great field application of complexity theory to the irrigated rice fields of Bali and its hydrologic system, Lansing finds in Karl Marx's 1844 papers the concept of 'humanized landscape'. What Marx seems to have meant by 'humanized nature' Lansing finds in the irrigated fields of Bali, and I find in Constable's renderings of Hampstead Heath or—better yet—of the canalized River Stour flowing through Suffolk's grain-rich Dedham Vale. This vision, which is holistic, offers, I think, the other half of the vision: one that potentially can include all humans; it is not exclusive. The purely human is seen as part of nature and, even, by some, as part of nature's God—in a pantheist, or more precisely, as a Spinozist vision. It approaches the central tenet of the ethics of the philosopher Baruch Spinoza: in order to reconcile oneself to the hard ways of the world, one must look at it under the aspect of the eternal (*sub specie eternitatis*) rather than dwelling on the specific pains of temporality. Most importantly, humans must realize that their destiny lies in the great whole of God(dess)/Nature (*Deus/a sive Natura*). Several Romantic (proto)ecologists adapted Spinoza's philosophy into their own brand of Pantheism for demonstrating humanity's place within the whole. Goethe often articulates just such a vision.

The Marx-Lansing conception of 'humanized landscape', on the other hand, offers us a less metaphysical mode of viewing this inclusive ecological vision than that shared by Goethe and Spinoza. Constable's Stour River paintings embody this interworking of human industry and nature. The way he places the farmers, lockmen, and bargemen within the landscape is firstly a proto-ecological statement and only secondarily a socioeconomic one—if this latter exists at all. Jane Austen offers us verbal constructions of just such a human/nature unity, in her very selective, rare, and always strategically significant descriptions of landscape. Hydrology, and often hydraulics, are the central, controlling point in all these landscape visions of Austen, Constable, and Goethe. But the waters of the earth are not the only key. Constable, Goethe, Shelley, and Beethoven also use the meteorological aspects to cinch the wholeness of this (proto)ecological revisioning of the man-Nature-God unity. They demonstrate nevertheless that this relationship is a precarious one. Goethe shows how human intervention can throw the whole balance into chaos. Human effort is shown perched on the edge of elemental forces—which also threaten chaos of their own. All these alternative visions of unity can be brought to the surface by a complexity-directed examination of the technological, economic, aesthetic, and environmental implications involved in the hydrological workings of the flow of the empowering waters and in the meteorological study of the heat/water/steam machine we call *weather*. Out of this multiplicity clustered around the water-cycle, I seek to create one vision of a change walking the fine line between order and chaos.

PRIMARY BIBLIOGRAPHY

The following list seeks to distinguish between the individuals who form the focus of this study and those subsidiary ones who are introduced to show that the concerns of the primary thinkers, artists, and statesmen were not just confined to them. In the generalized nature of a synthesis, however, even these focal individuals will not be treated in any exhaustive degree. I will further use this section as an opportunity to fix these individuals within the very complex and varied phenomenon reduced into the rubric 'Romantic Culture' in the title of the study; this is shot through with a multiplicity of movements and styles—including centrally the (Neo)Classicism so often dialectically opposed to Romanticism. The following also constitutes a list of editions or provenance of their works. References will be made hereafter in the text to these primary sources; secondary sources will be cited in endnotes. Only paintings discussed in the text that have dates between 1760 and 1870 are listed below; all others will be treated in endnotes. In keeping with the musicological convention, I will cite no specific edition for the musical compositions (usually, I have used the very accessible Dover editions, which are facsimiles of standard editions of the works). Concerning those individuals and works lying chronologically outside my period of study who and which have been introduced for the purposes of pointing out contrasts or establishing continuity, I will provide information either in the text or the notes.

Focal Thinkers, Artists, and Statesmen, 1760–1870

Jane AUSTEN (1775–1817) seems on the surface only to fit into the English Romantic period by an accident of chronology; she appears to partake of the Neoclassicism of the eighteenth century; however, if one looks deeper one sees that her narratives are rich with Romantic elements, which are further happily blended with Neoclassical ones. Austen, also, would not immediately come to mind for her use of natural landscapes—but, I feel, when she did use them, they are strategically placed to give them a power beyond what is expected.

The Works of Jane Austen. Ed. R.W. Chapman. Rev. by Mary Lascelles. 3rd ed. 6 vols. Oxford: Oxford UP, 1969.

Ludwig van BEETHOVEN (1770–1827) first created in the Classical style of Haydn and Mozart; however, he early began to develop the vocabulary of Romantic music.

George Gordon Noel, Lord BYRON (1792–1824), was the most widely celebrated beyond England of the second generation of High Romantics. His poetic ideal, however, is represented by the Neoclassical qualities of Alexander Pope; therefore, especially in his *Don Juan,* Byron often writes in a Romantic-Neoclassical style.

The Complete Poetical Works. Ed. Jerome J. McGann. 7 vols. Oxford: Clarendon, 1980–93.

Thomas COLE (1801–48) was an American Romantic painter, who was fascinated by water phenomena and by cycles. Indeed, it is his extraordinary use of the water-cycle in his great painting series, *The Cycle of Life* that, in part, caused me to broaden this study from Western Europe to include the Americas.

The Voyage of Life. [Complete with preparatory sketches.] Munson-Williams-Proctor Institute. Oil on canvas. 52 × 78 (132 × 198).
'Youth' from *The Voyage of Life*. 1840. Oil on canvas. 52 × 78 (132 × 198). National Galley of Art, Washington.
Painter's Exhibition Program Notes for *Cycle of Life* in: *Thomas Cole, 1801–1848: One Hundred Years Later*. Hartford: Wadsworth Atheneum; New York City: Whitney Museum of American Art, 1949. 27–30.
The Notes (as well as the paintings and the studies) are reproduced in Earl A. Powell. *Thomas Cole*. New York: Abrams, 1990. 84–92.

Samuel Taylor COLERIDGE (1772–1834) was an important link between the Storm and Stress generation of German poets and the first generation of England's High Romantic poets.

The Complete Poetical Works of Samuel Taylor Coleridge. Ed. Ernest Hartley Coleridge. Oxford: Clarendon, 1912.

John CONSTABLE (1776–1837) was a Romantic painter of landscape, who linked this tradition in England with that in France—where he had his first great critical success. In his later life, the landscapes are clearly Romantic in their stormy scenery; however, his earlier and more placid landscapes can best be termed Neoclassical-Romantic.

Flatford Mill from a Lock on the Stour. C. 1810–11. Oil on canvas. 15.2 × 21 (6 × 8¼). Victoria and Albert Museum, London.
The Mill Stream. C. 1814. Oil on canvas. 71.1 × 91.5 (28 × 36). Ipswich Museum.
Rough Sea, Weymouth. C. 1816. Oil on canvas. 22.5 × 55.5 (8⅞ × 21⅞). Millon Family. Upperville, Virginia.
Scene on a navigable river (Flatford Mill). 1817. Oil on canvas. 101.7 × 127 (40 × 50). Tate Gallery, London.
Landscape: Noon (The Hay-Wain). 1821. Oil on canvas. 130.5 × 185.5 (51¼ × 73). National Gallery, London.
Cloud Study: Cirrus. C. 1822. Oil on paper. 48 × 59 (18⅞ × 23¼). Victoria and Albert.
A Boat Passing a Lock. 1824. Oil on canvas. 142.2 × 120.7 (56 × 47½). Walter Morrison Collection, Sudeley Castle, Winchcombe, Gloucestershire.
Brighton Beach 'Squaly (sic) Day'). 1824. Oil on paper. 12 × 29.7 (4¾ × 11⅝). Victoria and Albert.
Chain Pier, Brighton. 1824–27. Oil on canvas. Tate. 127 × 183 (50 × 72).
Stonehenge. 1835. Watercolor. Victoria and Albert. 38.7 × 59.1 (15¼ × 23¼).

J. Hector St. John de CRÈVECOEUR (1735–1813). A French-American writer who is not usually identified with Romanticism; however, his astute feeling for the environment warrants his inclusion in this study. There is as yet no definite modern edition of the original, the English version of his *Letters from an American Farmer*; however, the following edition is based on the 1783 version and the *Sketches* is the Yale UP text edited by H.L. Bourdin, R.H. Gabriel, and S.T. Williams:

Letters From an American Farmer and Sketches of Eighteenth-Century America. New York: Signet, 1963.

Denis DIDEROT (1713–84) is of two ages. His published works fit somewhat conveniently into the Enlightenment; his posthumous works contain many Romantic elements. His privately circulated art criticism collectively entitled *Salons* contain forward-looking writings on the elemental in painting—particularly of mists, clouds, and storms.

Salons. 2nd. ed. 4 vols. Ed. Jean Seznec. Oxford: Clarendon, 1979.

Philip FRENEAU (1752–1832) is interesting both as a political poet and a journalist. He, like Jefferson, was fearless in the face of change; but, unlike Byron, he was optimistic.

A Collection of Poems on American Affairs and a Variety of Other Subjects Chiefly Moral and Political. Ed. Lewis Leary. Facsimile. Delmar: Scholars Facsimiles, 1976.

Caspar David FRIEDRICH (1774–1840) is perhaps the arch-Romantic painter of Germany; he was a keen observer of natural phenomena, but painted it for symbolic, indeed full-blown allegorical purposes.

Spring and fountain series: 'Emilias Kilde'. 1797. Ink and watercolors. 21.8 × 16.7 (8⅝ × 5⅝). Kunsthalle, Hamburg.
Der Mönch am Meer. 1809. Oil on canvas. 108 × 170 (42½ × 67) Schloss Charlottenburg, Berlin.
Kreuz im Gebirge. C. 1812. Oil on canvas. Kunstmuseum, Düsseldorf.
Kreuz im Walde. C. 1813 or c. 1835. Oil on canvas. 42 × 32 (16½ × 12⅝) Staatsgalerie, Stuttgart.
Der Wanderer über dem Nebelmeer. C. 1818. Oil on canvas. Kunsthalle. 94.8 × 74.8 (37¼ × 29½)
Ziehende Wolken. 1821. Oil on canvas. Kunsthalle. 18.3 × 24.5 (7¼ × 9⅝)
Das Eismeer, once wrongly called *Die gescheiterte 'Hoffnung'.* 1824. Oil on canvas. 96.7 × 126.9 (38 × 48⅞) Kunsthalle.
Lebensalter-Zyklus, oder Die Jahreszeiten. Brush-applied sepia over pencil. Kunsthalle. The extant portions of the 1826 version:
'Frühling'. 19.1 × 27.3 (7½ × 10½).
'Sommer'. 19 × 27.1 (7½ × 10⅝).
'Meer mit aufgehender Sonne'. 18.7 × 26.5 (7⅜ × 10½).
The extant portions of the 1834 version:
'Herbst'. 19.1 × 27.5 9 (7½ × 10¼).
'Winter'. 19.2 × 27.5 (7½ × 10⅞).

'Skelette in der Tropfsteinhöhle'. 18.8 × 27.5 (7⅜ × 10⅞).
'Engel in Anbetung'. 18.5 × 26.7 (7½ × 10½).

Johann Wolfgang von GOETHE (1749–1832) created great works in a succession of styles from Rococo, through the Early Romanticism of the Storm and Stress, into his Weimar (Neo)Classical period; however, even in the latter he was evolving or formulating much of the Romantic expressive syntax of poetry and prose. He also qualifies as a scientist. As an administrator and to some extent a statesman, he made important contributions to the Duchy of Weimar and, perhaps, to the liberality of thought that 'Weimar' connoted in the German political tradition.

Gedenkausgabe der Werke, Briefe und Gespräche. Ed. Ernst Beutler et al. 24 vols. Zürich: Artemis, 1949.

Heinrich HEINE (1797–1856) was a later Romantic German poet, who was keenly interested in politics.

Heinrich Heines Sämtliche Werke. Eds. Oskar Walzel, James Fränkel, Ludwig Kräke, and Albert Leitzmann. 10 vols. Leipzig: Insel, 1911–15.

Luke HOWARD (1772–1864) made his crucial contribution to hydrology and to the understanding of the hydrological-cycle with his catalogue of clouds in their relation to the other aspects of weather.

On the Modifications of Clouds, and on the Principles of Their Production, Suspension, and Destruction. Facsimile printed in *Neudrucke von Schriften und Karten über Meteorologie und Erdmagnetismus*. Ed. G. Hellmann. 1893; Nendeln: Klaus-Thomson, 1969.

Victor HUGO (1802–85) the supreme representative of Romantic literature in its purest extreme form and in all the major genres. I will limit my consideration to selections from his *Les Feuilles d'automne* (1831) and from *La Légende des siècles* (1859–83).

Oeuvres complètes de Victor Hugo. Eds. Paul Meurice, Gustave Simon, and Cécile Daubray. 45 vols. Paris: Imprimerie Nationale, 1904–52.

James HUTTON (1726–1797), most know for his work in geology, set forth a theory of rain, thereby making a crucial contribution to meteorology and to a complete understanding of the hydrological-cycle. He presented this in two papers: the first read on 2 February 1784 and the other read 12 April 1784 before the Royal Society of Edinburgh.

The Theory of Rain, Transactions of the Royal Society of Edinburgh for 1784.

Thomas JEFFERSON (1743–1826) is important as a political statesman—both on his own account and for his effect on the career and writings of Philip Freneau. Of all political thinkers, Jefferson and Freneau were not in the least threatened by change, rather they welcomed it. They both make Hegel, Marx, and Engels seem anti-change, anti-revolutionary because the two Americans saw change without any final utopian closure. They both, however, do not partake of the pessimistic and cynical attitudes of Byron and Turner on the subject of change. His aesthetic and scientific writings, also, have their place in this study.

The Portable Thomas Jefferson. Ed. Merrill D. Peterson. New York: Viking, 1975.

Alphonse de LAMARTINE (1790–1869): his *Poetical Meditations / Méditations poétiques* (1820) were an early contribution to French Romanticism; he was an outstanding and effective French statesman in national politics—especially during the start of the 1848 Revolution.

Oeuvres poétiques complètes de Lamartine. Ed. Marius-François Guyard. Paris: Gallimard, 'Bibliothèque de la Pléiade', 1963.

Nikolaus (Franz Niembsch) LENAU (1802–50) was a German poet born and raised in Hungary. He was very conservative in his politics. In 1832, he made voyages to America, which resulted in a set of poems on Niagara Falls.

Sämtliche Werke und Briefe. Ed. by Walter Dietze after Eduard Castle. 2 vols. Leipzig: Insel, 1970.

Franz LISZT (1811–1889) was an international composer with German and Hungarian roots. His tone poems have in recent years come to be recognized as among the great monuments of Romantic music.

Juan MELÉNDEZ VALDES (1754–1817) is a Spanish poet who fused Neoclassical elements with Romanticism; he is perhaps the most important Spanish poet of his day.

Poésias. Ed. Pedro Salinas. Clásicos Castellanos, number 64. Madrid: Espasa-Calpe, 1941.

Jean-Jacques ROUSSEAU (1712–78), technically speaking, may not be a Romantic himself; however, it is certain that he contributed immensely to the syntax of the Romantic and to their attitude toward natural phenomena.

Les Rêveries du promeneur solitaire. Ed. Henri Roddier. Paris: Garnier, 1960.

MARY SHELLEY (1797–1851) is included for her elemental landscapes in *Frankenstein*; however, I make no claim of saying anything comprehensive about her complete oeuvre—let alone making exhaustive use of the avalanche of secondary material that has appeared on her in recent decades; I can only say that I have taken care to seek out and examine those Mary Shelley scholars who treat subjects tangential to my concerns.

Frankenstein, or the Modern Prometheus. Rev. ed. Ed. Maurice Hindle. Harmondsworth: Penguin, 1992.

PERCY Byssche SHELLEY (1792–1822) was a leading member of the Second Generation of English High Romantics. His interest as an amateur scientist led him into a detailed study of the meteorological theories of his time. With the exception of Goethe, no one made greater use of water phenomena.

The Complete Works of Percy Bysshe Shelley. Eds. Roger Ingpen and Walter E. Peck. The Julian Edition. 10 vols. London: Benn, 1927.

Germaine Necker de STAËL (1766–1817) made occasional, if significant, use of water symbolism, but her great importance to my study is her integrated methodology for treating the interworkings of the components of culture, as contained in the first part of her *On Literature as Regards its Interrelation with Social Structures / De la Littérature considérée dans ses rapports avec les institutions sociales*.

Oeuvres complètes et posthumes de Madame la Baronne de Staël-Holstein. 3 vols. (1861; Geneva: Slatkine, 1967).

Friedrich Leopold Graf zu STOLBERG (1750–1819), a Storm and Stress poet, who wrote interesting water poetry. He is included, in large part, because Van Tieghem and others place such an emphasis on him as an innovator in elemental nature poetry. My data indicate that he and other members of Storm and Stress are an early and important manifestation of Romanticism; I fear that my findings cause me to go against the tradition of German literary scholars who place Early Romanticism later in the time of Eichendorff and Novalis.

Gesammelte Werke der Brüder Christian und Friedrich Leopold Grafen zu Stolberg. Authorized edition. Hamburg: Perthes, 1827.

Joseph Mallord William TURNER (1775–1851) is the arch-Romantic painter of England, with a definite penchant for the elemental. His rendering of sunrises and sunsets over water, as well as his great snow and rain storms, make him very crucial to this study.

From *The South Wales Sketch Book*. 1795. 'Melincourt Fall, Vale of Neath'. Pencil and watercolor. 26.4 × 20.3 (10⅜ × 8). Turner Museum, Tate Gallery, London.

Morning amongst the Coniston Fells, Cumberland (1798), oil on canvas, Tate Gallery, London.

Snowstorm: Hannibal and His Army Crossing the Alps. 1812. Oil on canvas. 144.8 × 236.2 (57 × 93). Tate Gallery, London.

From *Sky Studies* (c. 1820) Series CLVIII: 53 and 54 of sunset sequence, both: 12.4 × 24.8 (4⅞ × 9¾). Turner Museum, Tate Gallery, London.

The Bay of Baiae, with Apollo and the Sybll. 1823. Oil on canvas. 146 × 237.5 (57½ × 93½). Tate Gallery, London.

From *Colour Beginnings*, also called *Colour Structures* (c. 1820 to c. 1830): 'Sunrise' (CCXXII: 64, watermark 1823) 33 × 47 (13 × 18½); 'Sunrise at Sea' (68, watermark 1823), watercolors, 29.2 × 39.4 (11½ × 15½). Turner Museum, Tate Gallery, London.

Sketch for *Ulysses Deriding Polyphemus*. 1829. Oil on canvas. 60 × 89 (23⅝ × 35⅛). Tate Gallery, London.

Ulysses Deriding Polyphemus. 1829. Oil on canvas. 32.7 × 204.5 (52¼ × 80½). Turner Museum, London.
Snowstorm: Steam-boat off a Harbour's Mouth Making Signals in Shallow Water, and Going by the Lead. 1842. Oil on canvas. 91.4 × 122 (36 × 48). Turner Gallery, London.
Rain, Steam and Speed—The Great Western Railway. 1844. 90.8 × 122 (35¾ × 48). Turner Gallery, London.

William WORDSWORTH (1770–1850), an English poet of the Early-Romantic generation, known by its contemporaries as the 'Lake School'. He openly acknowledged his break with Neoclassicism, especially its treatment of nature.

The Prelude or Growth of a Poet's Mind. Ed. Ernest de Selincourt. 2nd ed. Rev. by H. Darbishire. Oxford: Clarendon, 1959.
Wordsworth's Poetical Works. Eds. Ernest de Selincourt and Helen Darbishire. 5 vols. Oxford: Clarendon, 1940–49.
The Prose Works of William Wordsworth. Eds. W.J.B. Owen and Jane Worthington Symyser. 3 vols. Oxford: Clarendon, 1974.

Other Thinkers, Artists, and Statesmen, 1760–1870

Gustavo Adolfo BÉCQUER (1836–70) was the great poet of the late flowering of Spanish High Romanticism; he was included to show the breadth of the Romantic use of water symbols, specifically those of the Byronic sort.

Libros de los gorriones. Eds. Guillermo G. Gallent, Rafael de Balbin, and Antonio Roldán. 1868; Madrid: Ministerio de Education y Ciencia, 1971.

William BLAKE (1757–1827) was an English poet, engraver, and watercolorist. While his visual works have a Neoclassical concentration on line and on balanced composition, he drew his main inspiration from High Gothic forms. His poetry rejects the Enlightenment values of rationalism and materialism, and it is conveniently typed as Romantic; however, its style—determined by the poet well back into the eighteenth century—is in the spirit of the Pre-Romantic Miltonists. His lyrics draw heavily upon Elizabethan popular forms.

The Complete Writings of William Blake. Ed. Geoffrey Keynes. London: Oxford UP, 1966.

William Cullen BRYANT (1794–1878) is included not for his own poetry—which does employ water symbols, albeit in a rather pedestrian way—but for his fine translation of Heredia's *Niagara*.

'Ode to Niagara'. *Hispanic Anthology*. Ed. Thomas Walsh. New York: Putnam, 1920.

Edmund BURKE (1729–1797) saw politics as a dialectical complexity of balancing permanence and change.

The Works of Edmund Burke. 8 vols. London: Bell, 1906–10.

Robert BURNS (1759–96) made important Scottish contributions to Early Romanticism; I have included him for one of his remarkable, but rarely remarked on landscape 'pencil sketches' in words.

Burns: Poems and Songs. Ed. James Kinsley. London: Oxford UP, 1969.

Salvatore CAMMARANO (1801–52) was a frequent adapter of Romantic writers. He loosely adapted Scott's *Bride of Lammermoor* into *Lucia di Lammermoor* for Donizetti.

Françoise-René de CHATEAUBRIAND (1768–1848) promoted Catholicism in prose that is taken as one of the early monuments of French Romanticism; he, like Staël, found a key source of his style in Rousseau.

Atala: Edition Critique. Ed. J.M. Gautier. Textes Littéraires Française. Geneva: Droz, 1973.

Frederic Edwin CHURCH (1826–1900) was an American painter who is included for his *Niagara.*

Niagara (1857). Oil on canvas. 108 × 229.9. (42½ × 90½). The Corcoran Gallery of Art, Washington D.C.

Eugéne DELACROIX (1798–1863) stands as the arch-Romantic of French Painting who specialized in figure painting; however, he painted few but excellent landscapes with significant water imagery.

The Sea from the Heights of Dieppe. C. 1851–54. Oil on canvas. 34.9 × 51.1 (13¾ × 20⅛). Private collection, Paris.

Gaetano DONIZETTI (1797–1848) was one of the Romantic operatic composers in Italy and in France. He is included for the fountain music in his *Lucia di Lammermoor,* inspired by Scott's novel.

Joseph Freiherr von EICHENDORFF (1788–1857) was one of the key High Romantic poets in Germany; this group (often referred to as Early German Romantics) employed source symbols with supernatural trappings that set them off from the more natural use of Goethe and certain of the English Romantics.

Eichendorffs Werke. Ed. Richard Dietze. Leipzig and Vienna: Bibliographisches Institut, n.d.

Francisco de GOYA y Lucientes (1746–1828) painted in a whole range of styles from his Rococo cartoons for the Royal Tapestry Factory, through his Neoclassical portraits, to the Romanticism of his later works. His landscapes are few, but are inevitably centered on water.

Landscape with Waterfall. Etching with aquatint. 16.6 × 28.5. (6½ × 11¼). Art Institute, Chicago.

Johann Gottfried HERDER (1744–1803), a German thinker, used water symbols to describe his organic cycle theory of human culture.

Werke in Zwei Bänden. Ed. Karl-Gustav Gerold. 2 vols. Munich: Hanser, 1972.

José Maria de HEREDIA (1803–39), a Cuban Romantic poet, wrote 'Niágara' (1824), which is a remarkable lyric in its own right, but also for its place in Romantic water symbolism. Of course, this is not the Parnassian poet of the same name (1842–1905) who wrote in French.

The Literature of Spanish America. Ed. Angel Flores. New York: Las Americas, 1966.

Johann Christian Friedrich HÖLDERLIN (1770–1843) is a German poet whose output bears an uneasy tension between Neoclassical and Romantic elements.

Sämtliche Werke: Stuttgarter Hölderlin Ausgabe. Ed. Friedrich Beissner. Stuttgart: Kohlhammer-Cotta, 1943–53.

John KEATS (1795–1821), an English poet, who is conveniently placed with the second generation of Romantics centered around Shelley. While his early works derive from the Protoromantic Spenserians and while he certainly wrote works filled with Neo-Gothic elements ('The Eve of St. Agnes' and 'La Belle Dame sans Merci'), he also wrote some of the most Neoclassically perfect—in the deepest sense of capturing the concrete essence of Greek Classicism—poems in English ('Ode on a Grecian Urn' and, above all, 'To Autumn'); in order to find an equivalent use of this tranquil mode of Neoclassicism among his contemporaries, one must look to Goethe to find comparable poems.

The Poetical Works of John Keats. Ed. H.W. Garrod. 2nd ed. Oxford: Clarendon, 1958.

Joseph Anton KOCH (1768–1839) is not usually taken as a Romantic painter; however, for me, his cosmic conception of the water-cycle places him with Cole, Liszt, Shelley, Hugo, Freneau, Turner, Friederich, and other Romantics.

Wasserfall. 1796. Oil on canvas. Kunsthalle, Hamburg.
Schmadribach. 1821 version. 132 × 110 (52 × 43). Bayerische Staatsgemäldesammlungen, Munich.

Giacomo LEOPARDI (1798–1837) is the supreme Italian Romantic poet.

I Canti di Giacomo Leopardi. Ed. Alfredo Straccali. Florence: Sansoni, 1912.

Jacques Philippe de LOUTHERBOURG (1740–1812) is a topographical artist whose elements of the sublime and painterly facture of water indicate the beginnings of Romanticism; Diderot admired his early paintings and seems to have used them in formulating his Early Romantic theories on the visual rendering of the elemental.

The Falls of the Rhine at Schaffhausen. C. 1775. 135 × 198 (53 × 78). Oil on canvas. Victoria and Albert Museum, London.

Herman MELVILLE (1819–91) is included for comparative purposes. Although I could not do justice to *Moby Dick,* I have included an evocative and significant ocean passage from that work.

The Writings of Herman Melville. Eds. Harrison Hayford et al. 15 vols. Evanston: Northwestern UP; Chicago: Newberry Library, 1968–89.

Felix MENDELSSOHN-Bartholdy (1809–1847) was the composer of several concert overtures that contributed to the development of what would later be called the tone poem.

NAPOLEON I Bonaparte (1769–1815) was an obvious agent of change on a grand scale; much recent monographic research concentrates on his influence on literature and the other arts.

NOVALIS, Friedrich Leopold von Hardenberg (1772–1801), created a highly supernatural source—representative of High German Romanticism.

Novalis Werke. Ed. Gerhard Schulz. Munich: Beck, 1969.

Francesco Maria PIAVE (1810–76) was Verdi's most frequent librettist, usually adapting the works of Byron, Hugo, or Spanish Romantics into lyrical form.

Rafael POMBO (1833–1912), the Columbian Poet, is one of the key, but late figures of Latin American Romanticism. 'En el Niágara: Contemplación' is an important contribution to the Romantic tributes to that great falls.

Poesías de Rafael Pombo. Ed. Antonio Gomez Restrepo. 2 vols. Bogota: Imprenta Nacional, 1916.

Friedrich RÜCKERT (1788–1866), a minor but very popular German Romantic, is included to demonstrate just how widespread the poetic usage of the water-cycle was.

Gedichte. Ed. Johannes Pfeiffer. Hamburg: Schröder, 1953.

Johann Christoph Friedrich von SCHILLER (1750–1805) was an Early Romantic German (Storm and Stress) playwright and poet, who in his last years came to some extent under the influence of Goethe's Weimar Neoclassicism; his water symbolism is a rarity and is included for contrast.

Schillers Werke: Nationalausgabe. Ed. Julius Petersen, Friedrich Beissner et al. Weimar: Böhlau, 1943–[not yet completed].

Franz Peter SCHUBERT (1797–1828) made some powerful settings of several of Goethe's elemental lyrics.

Sir Walter SCOTT (1772–1832), a Scottish historical novelist and poet; he further influenced Romantic sensibilities internationally by having so many of his novels adapted for opera librettos.

The Bride of Lammermoor. Ed. Fiona Robertson. Oxford: Oxford UP, 1991.

Giuseppe VERDI (1813–1901) is generally deemed to be the arch-Romantic composer of Italy. He had a keen understanding of literature and drew heavily on Romantic writers as the literary source for the libretti that were crafted by others, but under his careful direction.

Richard WAGNER (1813–1883) shows a definite cleverness in creating through lyrics, music, and drama a sense of the elemental in his mythological multi-arts *Ring* cycle. He both wrote his literary texts and set them.

Richard WILSON (1714–82) was a Neoclassical landscape painter, a follower of Poussin, who as time went by made topographical forays into the sublime.

Niagara Falls. Early 1770s. Oil on canvas. 152.4 × 182.9 (60 × 72). Woverhampton Art Gallery and Museum.

Holt Bridge on the River Dee. 1762. Oil on canvas. 148.5 × 193 (58 × 76). National Gallery, London.

José ZORÌLLA y Moral (1817–93) was a Spanish Romantic poet and playwright.

Obras completas. Ed. Narciso Alonso Cortes. Valladolid: Santaren, 1943.

Translations

I examined many translations, but I closely consulted only the following three; they alone seemed true to the letter and the spirit of the original; nevertheless, my own translations are significantly different from these:

Godman, Stanley, trans. Johann Wolfgang von Goethe. The Luke Howard Poem Series. Qtd. in Kurt Badt. *John Constable's Clouds*. London: Routlege, 1950. 12–15.

Hulse, Michael, trans. Johann Wolfgang von Goethe. *The Sorrows of Young Werther*. Harmondsworth: Penguin, 1989.

Trask, Willard R., trans. Johann Christian Friedrich Hölderlin. *Hyperion, or the Hermit in Greece*. 1959; New York: Ungar, 1984.

BIBLIOGRAPHICAL ESSAY ON SECONDARY WORKS TREATING THE WATER-CYCLE IN THE HUMANITIES (1760–1870)

Some of the reasons for the inclusion of the poets listed in this study will emerge by the examination of the comparative studies which treat the usage of water phenomena in the poetry of the later eighteenth and early nineteenth century. There is a dearth of such studies, and even those available usually form only a part of more general ones. None offers a comprehensive view of water images and symbols in the context of the entire water-cycle. Except for a flurry of interest in Germany at the turn of the century and isolated studies since, especially in France and America, studies of water images and symbols by and large have not been comparative but have been limited to monographs of national literatures and individual poets. All images and symbols, including those of water, have been neglected by comparative methodology. According to Henry H.H. Remak, 'Recent scholarship on West European Romanticism has been strong on the investigation of specific Romantic elements, particularly images and metaphors, in one or more writers, but usually, alas, of the same nationality'.[1] One great exception to this otherwise bleak scene is the work of the great pioneer of comparative method, at least in its traditional guise of motif typing: Paul Van Tieghem. Since he dominates the field and since his works have not been translated into English, I will quote rather freely from him in my translations (with the original in the notes).

A comparative study concerning springs and fountains has yet to appear, and rivers and waterfalls have fared only slightly better. Rivers and their usage as allegorical symbols are treated only briefly in a single comparative essay.[2] Paul Van Tieghem is the only comparatist to have treated the waterfall in Western European literatures. He does so in several pages of the chapter concerning mountains and their various corollary images in his *Le Préromantisme*.[3] Van Tieghem limits his study to waterfalls in alpine landscapes and fails to point out how, in such a landscape, they often precipitate directly from their source—a rock-born spring or

1 Henry H.H. Remak, 'Trends of Recent Research on West European Romanticism', in *'Romantic'—A Word and its Cognates*, Hans Eichner, ed. (Toronto: U of Toronto P, 1972) 475–89. Qtd. from 486.

2 H.M. Wilson, '"Rivers and Waters" in Literature', *Contemporary Review*, 170 (1946): 235–38. This is a pithy, thoughtful, and thought-provoking—if short, undeveloped, and often incoherent—essay. Of particular interest to my study is Wilson's discussion, based on passages from Dante and Coleridge, of the river symbol: 'To the thoughtful, the journey of a river from its tiny source to full-flowing strength and the final swallowing-up of its lesser waters in the greatness of the sea serves as a parable of the life of man, as the final merging of personality in the vaster sea of Being whence it arose, has always been the supreme truth to mystics' (235).

3 This treatment of mountains is found in the last volume, entitled *Le Sentiment de la nature dans le préromantisme européen* of Paul Van Tieghem's *Le Préromantisme* (Paris: Nizet, 1960) 4: 174–86; waterfalls, 183–85. Should anyone be surprised at my translating and quoting so amply from Van Tieghem, they are truly comparative studies of water in literature by one of the great pioneers of modern comparative methods and are works not translated elsewhere into English.

glacier. Indeed, in another section of the same chapter, he treats the latter of these, but not as a source.[4] In the eighteenth century, these waterfalls, he writes, 'were novelties that for the first time made for themselves a place in literature, where they had not up to then penetrated, but for the partly artificial cascades'. Continuing in the same vein, he states that Thomson was the first to use the falls and that when he did so, 'he reveals the sudden fall, the clouds of foam with their "silver mist"' and conveyed the motion as well as the sound of the falls with an expressive harmony.[5] Moreover, Van Tieghem finds considerable historical significance in Stolberg's 'Der Felsenstrom'.[6]

A more comprehensive treatment of the waterfalls is offered by Dimitrij Cizevskij in his comparative study of Slavic literatures, 'A New Attitude toward Nature—Waterfalls',[7] which does not limit itself to alpine waterfalls, but also treats falls which occur in the midcourse of streams. This work is significant in that Cizevskij suggests some important keys in distinguishing between a descriptive usage of the waterfalls in late eighteenth-century poetry and the symbolic usage of them in early nineteenth-century poetry. Cizevskij does this by first examining the eighteenth-century Russian poem *The Waterfall,* in order to show how the 'pictorial and colorist poet Derzavin made use of the picture of the waterfall as a landscape'. His observations on Derzavin have useful side applications, if one uses the term *image* instead of *picture,* since the former conveys the sense of the latter—a 'mind-picture' as it were—without the misapplication of terms best reserved for the visual arts. According to Cizevskij, even highly descriptive poetry is not merely that, because poets can draw upon the tradition of older symbolic usages of the falls—here, the Baroque use of water as a symbol for change. He does this by showing how 'Derzavin connects his favorite Baroque theme, the "perishability and transitory character of all earthly existence", with the waterfall'.[8] He then compares Derzavin's 'pictorial style' to 'the "psychologization" of a natural phenomenon' in Bestuzev-Marlinskij. Unfortunately, Cizevskij fails to offer a full explanation of what he means by 'psychologization'; he merely speaks of Marlinskij's 'subjective reminiscence'. He seems to suggest pantheistic imagery when he writes that 'the waterfall's waves are already living beings' but does not attempt to compare or distinguish this use from that of the water as a symbol or from its use as a 'subjective impression'.[9] He mentions

4 Van Tieghem, *Préromantisme* 4: 185–86.

5 Van Tieghem, *Préromantisme* 4: 183. Originals read: 'étaient des nouveautés qui pour la première fois se faisaient une place dans la littérature où n'avaient pénétré jusqu'alors que les cascades en partie artificielles'; 'il montre la chute brusque, les nuages d'écume avec leur "brouillard argente"'.

6 Van Tieghem, *Préromantisme* 4: 184. Alfred Biese also places emphasis on this ode and others by Stolberg in his study, *The Development of the Feeling for Nature in the Middle Ages and Modern Times* (1905; rpt. New York: Franklin 1964) 258–59.

7 In Dmitrij Cizevskij, *On Romanticism in Slavic Literature* (The Hague: Mouton, 1957) 11–28. This study of waterfalls is limited in its comparative scope in that Cizevskij concentrates on Russian literature and then, at the end, makes a passing reference to lines from the Czech poet K.H. Mácha and to a two-line fragment by Mickiewicz (27–28).

8 Cizevskij 12–13.

9 Cizevskij 16–17.

Kjuchel'beker's use of the waterfall as a symbol for the poet in his ode on the death of Byron, and he also mentions this Russian poet's use of the rainbow image in connection with the waterfall. While English literature lies outside the scope of his study, Cizevskij might have taken this opportunity to note in passing that these symbols/images may well be direct influences derived from Byron's own poetry.[10]

Flowing water in falls or streams, while it certainly appears in the works of Classical-Romantic composers has received little treatment in a larger, comparative context. One exception to this occurs in H.G. Schenk's comprehensively interart study—*The Mind of the European Romantics: An Essay in Cultural History.* He includes a brilliant sketch of a comparison between Wordsworth and Beethoven: 'It is perhaps worth noting [...] that Ludwig van Beethoven's *Pastoral Symphony* (1808) was almost contemporaneous with the original version of *The Prelude,* which was written in 1805. The analogies between those two works are astounding'. It is at this point that Schenk seizes on 'By the Brook' or second movement of this Sixth Symphony, which is useful to my study in chapter two; then, he goes on: 'What Easdale, Gasmere or Rydal Water meant to Wordsworth, the secluded Wiesental valley, now named after the composer, close to Heiligenstadt in the wooded surroundings of Vienna, was to Beethoven'.[11] I must look elsewhere for the treatment of another Romantic metaphor to find my model for incorporating music, as with science, with the other arts—namely, David L. Montgomery's *Musical Quarterly* article, 'The Myth of Organicism: From Bad Science to Great Art'.[12] The organic metaphor, which I said above followed Herder's water-cycle metaphor, was a dominant theme among the Romantics.

As for art history, the waterfall has received extensive and international treatment in a recent German dissertation by Hans Engelhard.[13] Although it is perforce of a general nature, Kenneth Clark's 'The Worship of Nature', a segment of his television script *Civilization: A Personal View,* notes some valuable and suggestive parallels between Coleridge and Friedrich, and among Goethe, Constable, and Turner.[14] Clark also points out how the 'emotional response to nature' on the part of poets and artists of the time had its origin in the works of German landscape painters and English landscape poets, such as James Thomson, 'who was, in his day, the most famous poet in Europe', and reached its culmination with Rousseau in the Lac de Bienne passage from his *Les Rêveries du promeneur solitaire*: 'In listening to the flux and reflux of the waves, he [Rousseau] tells us, he became completely at one with nature, lost all consciousness of an independent self, all painful memories of

---

10 Cizevskij 18–19.

11 H.G. Schenk, *The Mind of the European Romantics: An Essay in Cultural History* (Garden City: Anchor-Doubleday, 1969) 167.

12 David L. Montgomery's, 'The Myth of Organicism: From Bad Science to Great Art', *Musical Quarterly* 76 (1992): 17–66.

13 Hans Engelhard, 'Kunstgeschichte und Geologie: Der Wasserfall in Gemälden des 17. bis 19. Jahrhunderts', diss. Cologne, 1975. See also John Wilmerding et al., *The Waters of America: 19th-Century American Paintings of Rivers, Streams, Lakes, and Waterfalls* (New Orleans: Historic New Orleans and New Orleans Museum of Art, 1984).

14 Kenneth Clark, *Civilization: A Personal View* (New York: Harper, 1969) 269–91, see esp. 279, 285.

the past or anxieties about the future'.[15] Reductive though these observations are of a complex, multifaceted use of the lake symbol, they still point up a passage from a work and a theme central to my study. The vision of flux and reflux found in this work offers an important example of permanence and change.

In the context of comparative studies, the sea—offering as it does the grandest vision of flux and reflux—has fared better than the source, the stream, the falls, and the lake. Scholarly work on this motif is extensive enough to guarantee me almost something like a literary canon of pre-Romantic and Romantic sea poetry. In Alfred Biese's article, 'Die Poesie des Meeres und das Meer in der Poesie', which surveys the use of the sea in poetry from Homer to the nineteenth century, Stolberg's 'An das Meer' is shown to make innovative use of the sea in eighteenth-century poetry.[16] Biese also assigns important places to Wordsworth, Byron, Lamartine, and Heine, as well as offering thoughtful insights on them.[17] Even more helpful than Biese's article to my literary selections, indeed, than all comparative studies concerning the sea is Van Tieghem's 'The Beginnings of Sea Poetry in the 18th Century'.[18] The author finds the full poetic awareness of the sea to have come late relative to that of other natural imagery: 'It is known, by in large, that the poetic sentiment about the sea found its literary expression later than the landscape, and even later than that of mountains'.[19] In the article, Van Tieghem examines selected eighteenth-century poetry to find the turning point from the descriptive usage of the sea to the expressive, if not fully symbolic, usage. He locates this turning point in the 1770s and attributes it, in great part, to the German poets: 'it is during this period that there seems to be, among the German writers, an unprecedented, awakening sensitivity towards aspects of the sea, of rendering to this perception and the new feelings to which it had given rise'. Van Tieghem goes on to illuminate the central paradox of this poetry: the fragmentary nature of these pre-Romantics' poetic products combined with their powerful suggestions of some large cosmic whole, which they can suggest but not fully embody: 'These notations, generally brief, come to suggest a profound emotion before the grandeur of the vision that we might call a cosmic apperception and to suggest an artist's capability to seize and render the picturesque beauty of the view'.[20] Van Tieghem, like Cizevskij, is inclined to

---

15 Clark, *Civilization* 270–72.

16 Alfred Biese, 'Die Poesie des Meeres und das Meer in der Poesie', *Preußische Jahrbücher* 88 (1897): 279–301.

17 Biese, esp. 294–99.

18 Paul Van Tieghem, 'Les Débuts de la poésie de la mer au XVIIIe siècle', *Neophilologus* 23 (1937–38): 437–43; hereafter, cited as 'Les Débuts'. This article is reprinted with minor alterations in *Le Préromantisme* 4: 174–86. With several exceptions, I have quoted from the article because the ideas are stated there more concisely, although I have carefully checked each quotation with the corresponding passages in the book for any significant alterations.

19 'On sait, à peu près et en gros, que le sentiment poétique de la mer a trouvé son expression en littérature plus tard que celui de la campagne, plus tard même que celui de la montagne' (Van Tieghem, 'Les Débuts' 437; also in *Le Préromantisme* 4: 199).

20 'c'est pendant cette période que paraissent s'être éveillés spécialement chez des écrivains allemands le goût et l'art de peindre les aspects de la mer et d'exprimer les impressions et les sentiments qu'il font naître. Ces notations, brèves en général, attestent tantôt une émotion

describe the visually evocative nature of late eighteenth-century water imagery in terms borrowed from the graphic arts.

Just as he did in his study of falls, Van Tieghem places considerable emphasis on the poetry of Friedrich von Stolberg, where, he says, one finds 'the first texts where one can truly breathe the poetry of the sea'.[21] By the real poetry of the sea, he means a phenomenon which is one important focal point of my study: the Romantic poets' usage of water as a catalyst for the state of 'rêverie': 'The poet turns the sea into a confidant; he listens to it with drunken delight. His soul is elevated by it; with it he plunges back into the primal depths of Nature'. Van Tieghem, then, proceeds to catalogue some of the imagery that will pervade the sea poetry of our time-frame and will culminate in Heine's *North Sea* series: 'At the hour when the reddening sun plunges into its breast, the lapping of the waves creates a cradle song for the exhausted world'.[22] What was begun in the late eighteenth century did not culminate until early in the nineteenth century: 'It was reserved for the Romantics to possess the temperament, as well as the theory—namely, for Byron, Lamartine, Heine, Geijer, [Hugo]—that would allow them to give the sea the prominent place, which (in spite of the efforts [of the pre-Romantics] which we have just summarized) it had not possessed before them'.[23] The importance which Van Tieghem attaches to the sea poetry of Byron, Lamartine, Heine, and Hugo was taken into consideration in my selection of poets to be treated in chapter three. Of a somewhat lesser value is Van Tieghem's short survey of lakes in the mountain-landscape chapter of his *Préromantisme*. Here, he discusses what he terms the discovery of the lake by literature during the second half of the eighteenth century. He assigns considerable credit for this discovery to Goethe's 'An dem See' and Rousseau's Lac de Bienne passage; these, thus, constitute my opening to the lake section.[24]

In their treatments of sea images, Van Tieghem and Biese have only sought to isolate and analyze a certain sensibility, which implies a passive appreciation rather than an active symbolization of natural phenomena. Felix Melchior[25] and W.H.

---

profonde devant la grandeur du spectacle et pour ainsi dire un sentiment cosmique; tantôt une aptitude artistique à saisir et à rendre le pittoresque du tableau'. This summary is not found in the article but only in Van Tieghem, *Le Préromantisme* 4: 211.

21 The poetry of Friedrich von Stolberg is where one finds 'les premiers textes où l'on respire vraiment la poésie de la mer' (Van Tieghem, 'Les Débuts' 441–42; also in *Le Préromantisme* 4: 206–07).

22 'Le poète se fait le confident de la mer; il l'écoute avec ivresse. Son âme s'élève avec elle, avec elle se replonge dans l'abîme mystérieux de la Nature'. 'A l'heure où le soleil rougissant se plonge dans son sein, le clapotis de ses vagues fait entendre un chant berceur à la terre fatiguée' (Van Tieghem, 'Les Débuts' 441–42; also in *Le Préromantisme* 4: 207).

23 'Il était réservé aux romantiques de tempérament, sinon de doctrine, à Byron, à Lamartine, à Heine, à Geijer, [à Hugo,] de lui faire dans leurs vers la place éminente que, malgré les efforts que nous avons résumés, elle n'avait pas acquise avant eux' ('Les Débuts' 443); Hugo is added to this list in Van Tieghem, *Le Préromantisme* 4: 221.

24 Van Tieghem, *Le Préromantisme* 4: 180–83.

25 Felix Melchior, *Heinrich Heines Verhältnis zu Lord Byron* (Berlin: Felber, 1903) 106–17. In a short section of this study, Melchior shows how these poets share a common love for employing the sea as a symbol for personal as well as political revolt. Otherwise, the central core of this study is not relevant to my own since his comparative analysis is devoted almost

Auden[26] have each written comparative studies treating this symbolization of the sea. Unfortunately, neither study attempts any real breadth in its European scope.

Although clouds have not fared well in receiving the attention of comparatists, they have fared better than the other phases of the water-cycle. Van Tieghem establishes the following interesting connection between the eighteenth-century poets' 'discovery' of the mountains and, on the other hand, their being struck by sky images—among these are the clouds: 'The sky, the cloud, the stars endowed with a new aspect to the eyes of the plain-dwelling poets by the clear mountain air, like the contrast with the white magnificence of the field of snow, were all frequent objects of surprise and admiration'.[27] He discusses the late eighteenth-century discovery of one image which is important for the analysis of a poem by Hugo and that of passages from one by Wordsworth—the image (and, potentially, the symbol) of 'the sea of clouds': 'The "sea of clouds" is the other phenomenon discovered by this generation: we experience this phenomenon when we arrive on a towering mountain peak; we are surrounded on all quarters by clouds that break like sea-foam at our feet, drown all the valleys, and only allow the peaks to emerge'.[28]

As in the case of his other studies, here Van Tieghem is concerned only with the cloud as image. The only extended comparative study seeking to examine the cloud as symbol is a chapter in Gaston Bachelard's *Air and Dreams: Essay on the Imagination of Movement,*[29] the usefulness of which is severely limited by the vagueness with which he describes the state of 'réverie' in which poets approach clouds.

Finally, although Werner Ross discusses weather in his 'Literarische Wetterlagen', this is still a topic which implies the cloud and rain processes of the cycle.[30] Ross detects an unprecedented interest in the thunderstorm around 1770. Basing his

---

exclusively to the voyage and the sea as a milieu for a hero's actions.

26 W.H. Auden, *Enchafed Flood, or the Romantic Iconography of the Sea* (New York: Vintage-Random, 1950). The analyses section of this study deals with symbols (e.g., voyages, ships, 'the Stone', and 'the Shell' of Wordsworth's *The Prelude*) and works (e.g., poems of Baudelaire, *Don Quixote,* and *Moby Dick*). Auden offers some relevant generalizations on the sea symbol: 'The sea or the great waters, that is, are the symbol for the primordial undifferentiated flux' (7). Auden sets forth the general definition of what he calls the 'Romantic attitude' toward the sea as a simple fascination for its danger and challenge (13–14).

27 'Le ciel, les nuages, les astres, avec l'aspect, si nouveau, aux yeux des habitants des plaines, que leur donnent la pureté de l'air et souvent le contraste avec le blanc éblouissant des champs de neige, sont pour ces pionniers un objet fréquent d'étonnement et d'admiration' (Van Tieghem, *Le Préromantisme* 4: 179).

28 'L'autre phénomène que cette génération découvre avec étonnement est la 'mer de nuages' qu'arrivés sur un sommet nous dominons, qui nous entourant de toutes parts, déferlant à nos pieds, noyant toutes les vallées, ne laissant émerger que les cimes' (Van Tieghem, *Le Préromantisme* 4: 180).

29 Gaston Bachelard, *L'Air et les songes: Essai sur l'imagination du mouvement* (Paris: Corti, 1943) 212–24.

30 Werner Ross, *Teilnahme und Spiegelung: Festschrift für Horst Rüdiger* (Berlin: de Gruyter, 1975) 1–13. Hereafter cited as 'Wetterlagen'. Also there is a related article by Ross: 'Vom Schwimmen in Seen und Flüssen: Lebensgefühl und Literatur zwischen Rousseau und Brecht', *Arcadia* 3 (1968) 263–91.

argument in great part on passages in Goethe's *Werther*, he writes, 'The thunder-storm was to become one of the great themes of the approaching Storm-and-Stress literature'.[31] In Goethe's works, taken as a whole, he finds the beginning of the metaphorical usage of weather: 'From *Werther* up to *Faust* is the beginning: weather as an aspect of immediate experience was always present as a topic, as metaphor, always [...] a source of inspiration'.[32]

Strangely enough, in the two comparative studies which treat water images and symbols from a general vantage point, the authors—Gaston Bachelard[33] and Bernhard Blume[34]—fail to treat their subject matter in the water-cycle context. Frye's brief but suggestive remarks in the passage quoted at the outset of this study offer the best general treatment of water symbols in the context of their cycle. Frye's remarks are supplemented by observations in a monograph on Shelley's 'The Cloud' by Horst Meller, who points out two German lyrics which, like Shelley's, use the water-cycle.[35]

The only thoroughly comparative study of water images and symbols in literature and painting is Kurt Badt's *Wolkenbilder und Wolkengedichte der Romantik*, translated into English with the reductive title of *John Constable's Clouds*.[36] Kurt Badt set forth the thesis of his book as follows: 'To perceive and to describe the influence which science had on imagination and the creative arts of painting and poetry at the beginning of the nineteenth century is the real purpose of this essay'. He goes on to

31 'Das Gewitter wird eines der großen Themen der kommenden Sturm-und Drang-Literatur' (Ross, 'Wetterlagen' 8).

32 'Vom *Werther* bis zum *Faust*, das war der Anfang: Wetter als ein Stück unmittelbarer Erfahrung immer präsent, als Gegenstand, als Metapher gar [...] als Inspirationsquelle' (Ross, 'Wetterlagen' 11).

33 Gaston Bachelard, *L'Eau et le rêve: Essai sur l'imagination de la matière* (1942; rpt. Paris: Corti, 1970). Bachelard concentrates this study on such phenomena as water drops, mud (i.e., water, of course, as a linking agent of dirt) and baptism—all such *foci* are in keeping with the alchemic, occult, or religious concerns of his study but are beyond the pale of mine.

34 Bernhard Blume, 'Lebendiger Quell und Flut des Todes: ein Beitrag zu einer Literaturgeschichte des Wassers', *Arcadia* 1 (1966): 18–30. In this study, Blume is concerned almost exclusively with voyages, shipwrecks, baptism, and similar images. The following studies by Blume concentrate on voyage shipwreck motifs: 'Das Bild des Schiffbruchs in der Romantik', *Jahrbuch der deutschen Schiller-Gesellschaft* 2 (1958): 145–61; 'Die Insel als Symbol in der deutschen Literatur', *Monatshefte* 41 (1949): 239–47; 'Die Kahnfahrt: Ein Beitrag zur Motivgeschichte des 18. Jahrhunderts', *Euphorion* 15 (1957): 355–84; 'Sein und Scheitern: Zur Geschichte einer Metapher', *Germanisch-Romanische Monatsschrift*, Neue Folge 9 (1959): 277–87.

35 Horst Meller, 'The Cloud', in Teut Andreas Riese and Dieter Riesner, *Versdichtung der Englischen Romantik: Interpretationen* (Berlin: Schmidt, 1968) 269–93. According to Meller, 'Gesang der Geister über den Wassern' of Goethe shares with 'The Cloud' an 'Ideenkomplex von den unablässig kreisenden Wassern' (280). Meller, however, fails to pursue this parallel, to analyze either poem in light of the water-cycle, or to show how the cycles in the two poems differ totally in kind from one another.

36 Kurt Badt, *John Constable's Clouds* [1949; rpt. Berlin: de Gruyter, 1966] trans. Stanley Goodman (London: Routledge, 1950); qtd. from the latter. I have employed this translation because it is an excellent one and is highly regarded by English and American comparatists and art historians.

say that he chose a specific phenomenon to better demonstrate the connection: 'That it became a *Study of Clouds* was because the historical evidence concerning the influence of meteorology is clear and positive. We must therefore consider the meteorologists themselves'.[37] Although he does it in the vaguest terms, Badt establishes a valuable dichotomy between those artists who treat nature scientifically—in his words, those who 'regarded nature without any symbolic intent, with a kind of submissiveness', on the one hand, and, on the other 'those painters of the time who were definitely Romantic, who were painters of mood, opposed to all scientific encroachment on the territory of their artistic feeling'. Into the former group, he puts Wordsworth and Constable; in the latter, Friedrich and Koch. Only Goethe completely escapes this reductive dichotomy. Badt does allow some complexity, or rather, inconsistency, to Wordsworth, in whose later works, 'the power of thought won the upper hand over the power of observation, and human destiny became more significant than the beauty and expressive power of nature'.[38] Yet he fails to realize that the only poetry which does not fit into Wordsworth's 'later works' as he defines them are the poet's earliest works, his topographical poetry of the earlier 1790s and before. Badt also fails to take into account the later works of Constable, when he writes: 'The "hour of man", which finally decided the direction of Wordsworth's later writings, never came for Constable. Nature remained his exclusive interest and without ever arousing a mysterious spiritual exaltation in him'.[39]

37 Badt 95, 62, respectively.

38 Badt 22.

39 Badt 98–99.

www.ingramcontent.com/pod-product-compliance
Ingram Content Group UK Ltd.
Pitfield, Milton Keynes, MK11 3LW, UK
UKHW041143300926
15471UKWH00011B/112

9 789042 014275